High Performance
Self-Consolidating
Cementitious Composites

High Performance Self-Consolidating Cementitious Composites

Prof. Ganesh Babu Kodeboyina
Former Director, Central Building Research Institute
(CSIR-CBRI), Roorkee
and
Professor, Indian Institute of Technology Madras, Chennai.

CRC Press
Taylor & Francis Group
Boca Raton London New York

CRC Press is an imprint of the
Taylor & Francis Group, an **Informa** business

CRC Press
Taylor & Francis Group
6000 Broken Sound Parkway NW, Suite 300
Boca Raton, FL 33487-2742

First issued in paperback 2020

ISBN-13: 978-0-367-57211-2 (pbk)
ISBN-13: 978-1-138-06304-4 (hbk)

Library of Congress Cataloging-in-Publication Data

Names: Kodeboyina, Ganesh Babu, author.
Title: High performance self-consolidating cementitious composites / Ganesh Babu Kodeboyina.
Description: Boca Raton : Taylor & Francis, a CRC title, part of the Taylor & Francis imprint, a member of the Taylor & Francis Group, the academic division of T&F Informa, plc, 2018. | Includes bibliographical references and index.
Identifiers: LCCN 2017042109 | ISBN 9781138063044 (hardback : acid-free paper) | ISBN 9781315161310 (ebook)
Subjects: LCSH: Cement composites.
Classification: LCC TA438 .K63 2018 | DDC 620.1/35--dc23
LC record available at https://lccn.loc.gov/2017042109

Visit the Taylor & Francis Web site at
http://www.taylorandfrancis.com

and the CRC Press Web site at
http://www.crcpress.com

Contents

Preface

The improving urbanization standards forced the construction industry to look for not only the most economical of the construction materials but also avenues to ensure high performance, necessitating a significant transformation in its outlook. Self-consolidating concretes brought about a lasting change in the construction setup, even if it was only by tweaking some of the existing materials, design, and production methodologies already well known to the industry.

High Performance Self-Consolidating Cementitious Composites attempts to bring together some of the basic intricacies in the production of the complete range of self-consolidating cementitious composites, with a proper understanding of the contributions of different materials and their combinations. To the discerning reader the two fundamentals aspects - the modulation of rheology of cementitious materials paste through superplasticizers and viscosity-modifying agents, and the interaction of this paste phase with filler components like coarse and fine aggregates - to ensure a self-compacting mass are explained. Also, it was presumed a priori that given the higher fines content and the self-compactability for ensuring a defect-free structure and recognizing the fact that additional fines are generally supplemented through materials like fly ash, the resulting concrete will automatically be of high performance. However, there is a need to understand the pore filling and pozzolanic effects of the powders and the different supplementary cementitious materials in vogue appropriately, to ensure that both the strength and the performance could be adequately predicted, which is also explained in detail herein.

Another important factor that is probably not fairly conspicuous and is often overlooked is that with the limited budgetary provisions available for most research efforts, even in prestigious research laboratories and educational institutions, coupled with the limited time and manpower available, studies in general are limited to specific targets in each case. An overarching program of research, like the "Concrete in the Oceans" program of the United Kingdom, will need support for over a decade with several specific aspects looked at by different institutes to ensure a realistic and comprehensive understanding. However, there are a lot of efforts directed toward different goals by several members of the research community. Now, the other way to find a solution for this problem was to comprehensively put together these diverse research efforts of the different investigators, identify the broad fundamentals underlying the different aspects of the problem, and systematically analyze the information to arrive at a more holistic understanding. This approach, more easily said than done, needs a compilation of the available results into a data platform; it also needs categorization of the results under acceptable scientific criteria through earlier experience, and finally we can arrive at broad guidelines for an appropriate understanding of the problem. In line with this, an effort was made in this book to compile a large database on self-compacting concretes from the literature, categorize them to represent specific groups, and suggest methods for arriving at self-compacting concretes with a wide range of materials and over a wide range of consistencies. Apart from this, *High Performance Self-Consolidating Cementitious Composites* tries to look at the performance and limitations of this regime of concretes and suggests methods to address some of these appropriately. Naturally, in some cases a few of the aspects discussed were based only on the limited information available from the earlier investigations and may have to be verified before extending them to reflect the behavior of the entire spectrum of self-compacting concretes.

The fact that it is not possible to summarize even partly the volume of information available from various conferences, seminars, and publications available on a topic that has generated an enormous amount of interest in the research and development community as well as the construction industry was well recognized. Naturally, the basic idea was to only highlight the critical parameters and interactions between these parameters to ensure the production of high-performance, self-consolidating cementitious composites with greater confidence. The second aspect was to collate a relevant group of concretes in each particular case and put together a series of ready reckoner charts to have a guideline for arriving at an appropriate self-consolidating cementitious composite of choice. Finally, it was attempted to present briefly the relationships between various green and hardened state characteristics. *High Performance Self-Consolidating Cementitious Composites* also presents a broad outline of the present state of the art, in terms of the different national recommendations and codal provisions, the views and perceptions of the industry, along with an insight into the basic materials technology. Specifically, some of the graphs (nomograms) could help the practicing engineers on-site to understand and modulate the concrete composites being produced directly. Thus, instead of summarizing the information available in literature, the book envisages to be a tool for explaining the basic relationships and the interactions between the constituents to assist students, researchers, and industry professionals.

Acknowledgments

Self-consolidating concrete has attracted the attention of both the industry and the research community because of the ease of placement and time saved in concreting operations. Naturally, this prompted several efforts toward the design, production, and placement of the material. A critical and explicit understanding of the insights into any such concrete requires understanding of the contributions from several earlier researchers on the various facets that originally resulted in the development of the high-performance cementitious composites of today. I heavily relied on the wealth of knowledge that originated from North America and, more importantly, Europe in the attempts to present the topic in my own perspective. Many of the concepts that are discussed and elaborated in these pages have in the background the several discussions I had with these researchers, teachers, and students, even if some of them were not directly related to the making of this book.

At the very outset, I should thank profusely all my teachers starting from school to the point of obtaining my doctorate as well as several other professional colleagues who always encouraged me in my efforts to just learn. It is only appropriate to acknowledge the fact that among several of these tall figures is the name of one distinguished personality that stands out, Prof. Dr.-Ing. P. Srinivasa Rao, Department of Civil Engineering, Indian Institute of Technology, Madras, who inculcated in me an intensely critical thinking, strict attention to detail, and persistence, being a guide and mentor during my time as his research student for my master's and doctoral programs and also later as his associate at the institute.

I also gratefully acknowledge that it was indeed an honor to be a fellow of the Alexander von Humboldt Foundation, which helped me to not only interact with several eminent professors and researchers from German universities but also helped me in learning the German language so that I can have a direct understanding of the literature published, apart from being a member of their research and social community. Some of the specific intricacies in the modifications of concrete are hidden in the earlier literature that was largely published in German, and for an individual educated in the English medium, there is limited possibility of getting to know them. During a later visit as an Indo-German fellow, I had the opportunity to be associated with Prof. Dr.-Ing. Peter Schiessl, then heading the Institut für Bauforschung, RWTH Aachen University, Aachen, working on fly ash and corrosion of steel in concrete. It was indeed this association that paved the way for the development of the concept of dual factors in cementitious efficiency of most pozzolanic admixtures, which forms the central theme of this book. I thank him with all my heart for all the long and open discussions on the several aspects of fly ash in concrete and also corrosion of steel in concrete.

xiii

Also, over the years, it was indeed a pleasure to be associated with a very large number of students, who during their research efforts inadvertently posed several questions that needed to be answered. They also participated in many ways in refining and reorienting probably some of the known facts to be put in their proper perspective. Even among this very large group of well over a hundred, I should specifically appreciate and thank the efforts of Bhojaraju ESTR Chandrasekhar, a PhD scholar, particularly for his contribution in shaping this book in the present form through his several unending objective discussions and also, more importantly, through his untiring efforts to help in processing and producing most of the graphs and nomograms into the present form that really make them stand out as the hallmark of this book.

Finally and most importantly, I express my deep sense of gratitude to my family, my wife Devi, daughter Deepti, and son Srikanth Aditya, for their love and understanding of my absence during the several academic and professional commitments over the years.

Author

Professor Ganesh Babu Kodeboyina obtained his bachelor's degree in civil engineering with distinction from the Andhra University and joined the Indian Institute of Technology (IIT) Madras for a master's degree in structural engineering. He continued at IIT Madras and obtained a doctoral degree, working in the area of behavior of partially prestressed concrete structural members. He then joined Structural Engineering Research Centre as a scientist and was involved in several projects including large diameter prestressed concrete pipes, ferro cement and fiber reinforced concrete, apart from being the principal investigator on the United Nations Development Programme (UNDP)-sponsored project on Polymer Concrete Composites. He was awarded the Alexander von Humboldt (AvH) Foundation Fellowship during this period to undertake postdoctoral research in Germany.

However, during the same period, he joined IIT Madras as a faculty member of the Department of Ocean Engineering, working in the area of ocean structures and materials in marine environment, and undertook his AvH fellowship program. After his return to Germany even while he was engaged in active research on offshore structures, with special interest, he singlehandedly developed the most successful "Structures, Materials, Applications and Rehabilitation Technologies" laboratory of the department, which he headed until retirement. Apart from being an active research worker producing more than 20 doctoral theses, and several other master's theses on the various topics related to ocean structures, and in particular on materials in marine environment, he was also a consultant to the industry on several different aspects. To mention a few, as a structural engineer, he proof-checked designs of more than 50 prestressed concrete bridges and suggested the repair strategies for another 100 bridges of the National Highway Authority of India, designed and rehabilitated several port and harbor facilities apart from damage assessment, and was involved in the rehabilitation of a large number of operating industrial structures. Even so, he was better known for his contributions to high-performance cementitious composites and was a retainer consultant to nearly all the major cement producers as well as construction chemical manufacturers in the country. In fact, he investigated the performance characteristics of almost all of the cements that were produced by these industries. Apart from this, he was a part of the UN common fund–sponsored program, and was entrusted with the evaluation of the corrosion performance of galvanized reinforcement both in the laboratory and at the Madras port field site organized by him exclusively for this purpose.

These achievements resulted in him being appointed as the director of the Central Building Research Institute (CBRI), Roorkee, a prestigious constituent

National Laboratory of the Council of Scientific and Industrial Research (CSIR) of the Government of India. It was here that, apart from other research activities as director, he undertook several rehabilitation projects, notably the rehabilitation of the only existing double shell built by the famous Prof. Kurt Billig, the very first director of the CSIR-CBRI. These efforts culminated in his being involved in several activities related to the preservation and rehabilitation of many of the centuries-old temples and also other world heritage sites in the country like the Taj Mahal, Konark, and many others, an activity close to his heart, or his premier passion.

He later returned to IIT Madras to complete his tenure at the institute and was once again involved in research and consultancy in areas of ternary cements, Ultra-High-Performance Concretes (UHPCs), and polymer composites. He also undertook a study on the residual service life and service-life extension measures of the Fast Breeder Test Reactor (FBTR). Incidentally, he was responsible for the design and testing of thermal effects on heavyweight concretes that had to be almost self-consolidating in the enclosed top shield of the Prototype Fast Breeder Reactor (PFBR). He was rated a very good teacher by several groups of students and is also passionate about fine arts, ancient texts on yoga, and temple vastu, to name a few.

1

Introduction

1.1 Concept

Concrete has been the preferred material for construction because it can be molded into any shape dictated by the structural configuration requirements while being the most economical. Cementitious composites have undergone several changes over the years, due to the advancements in cement production processes resulting in higher-grade cements and the advent of modern pozzolanic and chemical admixtures, leading to strengths reaching the levels of steel in compression, while having significantly improved performance characteristics even in the most aggressive of environments. However, being a highly complex composite with so many materials of varying sizes, shapes, and textures, the prediction of the strength and performance of the resultant material has always been a matter of serious concern, without stringent control on the constitutes. The large and ever expanding database of research findings with several different local constituents, be it aggregates or the cements produced or the admixtures utilized in these concrete composites, also causes a certain amount of confusion in understanding the material.

In the context of the normally vibrated concretes, be it the general-purpose concretes, high-strength concretes, no slump concretes, or even lightweight concretes, recommendations are available for arriving at a concrete of a specific strength and consistency required for a particular application in the construction activity. In brief, the methodology adopted was to define the water–cement ratio in terms of the strength required for the structural member. Depending on parameters such as the construction practice, member size, and dimensions, the consistency or the workability of concrete was fixed. The workability ranges from no slump to a collapse slump based on the compaction procedures adopted. Depending on the constituent materials, particularly the maximum size of the aggregate, the water content is defined for that workability. After this the coarse aggregate and fine aggregate proportions are fixed or a continuously graded aggregate is recommended for each maximum size of the aggregate. All these were possible through an understanding of the gradation of aggregates and the amount of paste and mortar contents including water, which were all chosen from a very large database of the experimental evaluations and field experiences that are available on these types of normally vibrated concrete systems. One important factor that was also learned during these investigations was that an inadequate compaction of the resulting mass can significantly affect the strength of the composite ultimately. Several compaction procedures such as vibro-compaction, high-frequency vibration, vacuum dewatering, spinning, and pressure application have all been used effectively, depending upon the member details and construction practices available. Naturally, the emphasis is more on appropriate concrete compaction methodologies to alleviate the problems associated with the lowering of strengths during these processes. Though the most desired parameter in defining the characteristics of concrete is strength, which is related to the water to cement ratio, the fact that it can be produced with various workabilities at different water contents was also well established. Slump is a measure of the workability, and structural concretes with no slump to collapse slump characteristics can be produced in practice with water contents ranging from 150 to 215 kg/m^3, as recommended in the ACI guidelines (ACI 211.1, 1997).

In this complex scenario wherein the concrete has undergone several significant modifications to cater to the varying needs of the construction industry, the modification that has attracted the attention of the industry in recent times is the concept of self-compacting or self-consolidating concrete (SCC), a concrete that consolidates under its own weight. This concept of SCC, originally articulated by Okamura (1995), came into existence in an effort to design a concrete mix that could consolidate and fill without any difficulty the massive end anchorages of the Akashi-Kaikyo bridge with its highly impenetrable surface reinforcement cage and anchorage details. These SCCs have also been referred to in the literature as SCCs, self-placing concretes, and self-levelling concretes by a few, depending upon the requirements and perception. Semioli (2001) opines that a less dramatic terminology such as self-accommodating concrete might be more appropriate to describe the self-levelling and self-compacting properties that permit easier installation and working at sites for Lafarge's "Agilia," which was termed a self-placing concrete by them. He also discussed its use for both horizontal and vertical applications as well as in the repair and rehabilitation of an old courthouse structure. Szecsy (2002) presented a broad discussion on the nomenclature as well as the technology. He opines that each of these different names defines a specific aspect of the type of concrete, namely, SCC—a concrete that consolidates through gravity to achieve maximum density without the need for vibration, self-levelling concrete (SLC)—a concrete that can seek a level grade within the formwork, and self-placing concrete (SPC)—a concrete that has ease of placement and has the ability to be both self-compacting and self-levelling. The various admixtures that help in achieving self-compactability (mostly polycarboxylate-based superplasticizers in recent years) and a few tests that establish the different workability characteristics have also been presented. Also, others followed a similar approach in defining what is expected of SCC over the years. To be more specific the name carries the entire meaning of the expectation from the cementitious composites developed.

1.2 Historical Development

One of the significant developments in the area of concrete technology could be traced to the advent of superplasticizers, though certain classes of plasticizers have always been available from times immemorial, even from the early lime concrete technology periods. This change is essentially because concretes with even a limited slump of just 25–50 mm could be readily reduced to a fluid consistency with collapse slump, if only it is ensured that there is enough sand and fines (even cement as in the case of high-strength concretes of lower water–cement ratio) in the mix. The simple yet often forgotten rule in the development of these concretes was to increase the sand content by about 10% and also to have a good gradation in the coarse aggregate fractions. These are then termed as superplasticized concretes; while the same superplasticizers could be utilized in concretes where the slump was nearly retained at the original level of 25–50 mm by appropriately reducing the water content (almost by about 20% if possible) and be termed as high-range water reducing admixtures, resulting in significantly higher strengths in these concretes. This in a way promoted the initial development of high-strength concretes, which were presumed, even if wrongly, to be an approach for realizing high-performance concretes due to their lower water–cement ratio and the resultant discontinuous capillary pore structure that could inhibit the permeation of moisture and environment.

The second significant development in concrete technology is the introduction of silica fume (an industrial waste from the ferrosilicon industry) as a pozzolan and the consequent development of very high-strength high-performance concretes. Simultaneously a lot of interest was generated in the use of secondary cementitious materials or mineral admixtures, which not only ensure a saving of cement and economy but also impart a higher performance. The need for an effective utilization of the abundantly available industrial wastes like fly ash from thermal power stations as an industrial waste and slag from steel plants (in the form of ground granulated blast furnace slag, GGBS) also contributed to the formulation of binary cement composites. These pozzolans in conjunction with the available superplasticizers paved the way for an entirely new regime of concrete and concrete composites that proved to be of a significantly higher performance with the possibility of also achieving very high strengths. Many other materials such as metakaolin, zeolite, and rice husk ash (RHA) have also come to be used as pozzolanic materials.

SCCs have essentially evolved in the background of these significant changes that took place in the past few decades. Okamura published a few studies prior to his proposed mix design methodology for SCCs leading to its use in the end anchorages of the Akashi-Kaikyo bridge (Okamura, 1994, 1995, 2003). The method, simply stated, attempted to achieve a highly workable and thixotropic mix through augmented fines and superplasticizers while reducing the coarse aggregate content. After this initial semi-empirical approach of modifying the conventional concrete design, researchers appear to have looked for avenues to achieve self-compactability though several different approaches, which are later, delineated. It looks obvious that the primary objective of any design methodology proposed to date is to achieve a cohesive and stable SCC without much thought given to its strength or performance characteristics.

A closer look at the broad spectrum of concrete composites will clearly show that SCCs, considered to be new in the present day context, are not in any way totally unknown to the industry. Several references could be found to concretes of collapse slump, sometimes termed as flowing concretes, self-levelling concretes, and grouts that exhibit slumps of 200–250 mm and even beyond, which were formulated to have no bleeding and segregation either through manipulations in cement content or by using thixotropy-imparting agents like vinyl- or cellulose-based materials. These were far more common as concrete composites suitable for underwater application with the necessary anti-washout characteristics even in the continuous or oscillatory flow regimes occurring in river, estuarine, or ocean environments (Gerwick, 2015). Some such precursors to SCCs have been described by Collepardi (2001, 2005, 2007). Apart from these facts, spatiality grouts of self-levelling nature have been in the market for quite some time now, and in applications requiring higher thickness of concrete for repair and rehabilitation these have been used incorporating 50%–60% well-graded coarse aggregate of a maximum size up to 10–12 mm. Such concretes were also developed in the laboratory for applications such as jacketing repairs of corroded columns and beams in both industrial and marine structures.

Some other aspects that are often quoted as factors that distinguish an SCC from a normally vibrated concrete (NVC) are the use of fly ash (maybe up to about 30%–40% of the total cementitious material or sometimes other fine materials as powder extenders) and a viscosity modifying agent (VMA). However, concretes with supplementary cementitious materials like fly ash, particularly to ensure reasonable economy in the normal strength concrete production, are well known. It is also known that to control the increased wetting water requirements of the increased surface area due to the presence of these powder extenders superplasticizers are often the only solution unless the total cementitious materials are kept well within the limits. This aspect of the need for superplasticizers becomes obvious in the case of superfine or nano pozzolans like silica fume, nanosilica, and calcined clays.

Several research reports are already available in the literature presenting concrete composites of very high workability, even with flowing characteristics, with many of these superfine high-end pozzolans. It is also to be noted that the design of these concretes was based on the presently available mix design methodologies of normally vibrated concretes, with only minor adjustments, essentially to the water and superplasticizer requirements wherever necessary. It is probably appropriate to point out at this particular stage that several publications have also taken into account the cementitious strength efficiency of these pozzolans, to design with confidence concretes of the desired strength and performance.

Having looked at all these perspectives one wonders if there is any special concrete composite that needs to be presented under the name of SCC. Extending a little further into what is known, it is obvious that even if we were to term a group of concretes in a special class called SCCs, one cannot be far wrong in saying that these can be designed with the available knowledge and the proposed methodologies for normally vibrated concretes. The later part of this book deals with some of these aspects and also delineates a methodology based on the existing concrete mixed design procedures to ensure a perfectly satisfactory SCC.

1.3 Definition

Self-compacting concrete, as the name indicates, is one that has the property of self-consolidating, self-placing, or self-levelling as indicated by several of the alternative names it has in the literature. However, some of the national bodies felt it is prudent to define it as a concrete that confirms to certain qualities, essentially to set it apart from the traditionally vibrated concretes, which in themselves will never come under one single group as can be seen by referring to them as normal concrete, heavyweight concrete, lightweight concrete, high-strength concrete, and even no slump concrete just to name a few. Some of these definitions are presented here to give a clear picture of the perceptions of each of these national bodies.

The Japanese recommendations (1999) define SCCs as concretes with self-compactability, which is defined in the Japanese recommendations as "the capability of concrete related to the placeability of concrete, with which it can be uniformly filled and compacted in every corner of the formwork by its own weight without vibration during placing." This definition essentially talks about its filling ability, though the terms deformability, passability, placeability, and segregation are also defined in the text of the report.

The European guidelines for SCCs (2005) recognize this as "the concrete that is able to flow and consolidate under its own weight, completely fill the formwork even in the presence of dense reinforcement, whilst maintaining homogeneity and without the need for any additional compaction." It obviously recognizes filling and passing ability, apart from segregation resistance, as the essential characteristics.

Probably a fairly inclusive definition of SCC is presented by the ACI (2007), which states that "Self-consolidating concrete (SCC) is highly flowable, non-segregating concrete that can spread into place, fill the formwork, and encapsulate the reinforcement without any mechanical consolidation." It also states that, in general, SCC is concrete made with conventional concrete materials and, in some cases, with a VMA. Apart from all other factors recognized in the earlier definitions, this recognizes the importance of the encapsulation of reinforcement, which is an essential prerequisite for the structural performance of reinforced concrete constructions.

As already stated, the primary aim in each of these definitions, as well as every effort of the research and developmental activities related to SCCs, appears to center around the aspect of self-compactability without giving credence to the fact that this could only be a further extension to the already existing concrete composites at the various consistency levels as defined in the various codes. The fact that it occupies the entire void space in a formwork appears to be reason enough for it to be referred to as a concrete of high performance though this alone is not adequate in practice. Apart from this, if it contains a pozzolanic material like fly ash even just at about 30% as recommended in most research efforts, one can comfortably consider this to be a concrete of higher performance compared to the concretes without such a pozzolanic replacement using other simple forms of powder extenders that may not help in ensuring such a high performance.

It may not be out of place to have a look at the characteristics that are essential to term a particular concrete as a self-compacting or a SCC—a concrete with the high workability required to flow through densely reinforced structural elements under its own weight and adequately fill voids without segregation or excessive bleeding and without the need for vibration to consolidate it. These characteristics are basically recognized by the Precast/Prestressed Concrete Institute (2003):

Filling ability (flowability)—ability to flow and fill under its own weight all spaces within an intricate formwork, containing obstacles, such as reinforcement.

Passing ability—ability to flow through openings between reinforcement approaching the maximum size of the coarse aggregate without segregation or blocking, and

Stability (segregation resistance)—ability to remain homogeneous during transport, placing, finishing and also through the initial phase of hardening.

1.4 Formulations and Classifications of SCCs

The available procedures for the proportioning of self-consolidating or SCCs could be categorized broadly into three main types. The first is the powder type—essentially increasing the powder content while reducing the water to power ratio to ensure that the aggregates are supported in such a viscous medium of paste and mortar. Alternatively, it was felt that it was possible to generate a similar effect by adding a VMA to provide the same segregation resistance to a concrete of high fluidity. Lastly, a combination of both these methodologies could be used to ensure an optimal utilization of the effects of both.

In every one of these formulations the use of additional powder material content is almost always taken for granted. In the first place the increased powder content can be achieved in two ways either by increasing the cement content, which is highly uneconomical and has its own limits even if one were to consider the increase in strength appropriately, or this can be achieved by using waste powders like stone dust to do the needful. A simple alternative that is presently available is to realize this increase through low-cost, ubiquitously available industrial waste by-products like fly ash and GGBS, which will also perform as low-end secondary cementitious materials. Some researchers also suggest the use of limestone powder readily available in the cement plants as inert filler, though the limestone powder was also observed to be exhibiting some chemical reactivity. Apart from these materials there are several other high-end pozzolanic materials like silica fume, metakaolin, rice husk ash, calcined clays, and so on, which can also be used for this purpose more effectively producing high-strength concretes.

Another important aspect is the fact that while most research reports in general try to classify the concrete as self-compacting once the slump flow is above 500 mm, there are three different slump flow regimes or ranks defined in the various national recommendations. These were also classified into subgroups of SCC based on other characteristics like viscosity or passing ability. SCCs are classified into three different categories or class ranges by JSCE (1999), essentially based on their slump flow characteristics (650–750; 600–700; 500–650 mm as Ranks R1, R2, and R3), basically to suit the needs of the construction conditions, like the minimum gap between reinforcement and the amount of reinforcement to cater to the filling ability needs of the concrete required. Similar classifications based on the parameter of interest are also available in other national specifications like the European and the American. It is in fact said that the choice of the specific characteristic and the subclass or regime of SCC that needs to be adopted essentially stems from the requirements in the field, like the dimensions and intricacy of the structural member, thickness, density of reinforcement, and so on. This in fact is very similar to that for traditionally vibrated concretes, which can be produced with various workabilities or slump, a measure of the workability. Most of the design procedures reported for producing SCCs have concentrated on proportioning of the mix constituents to achieve the primary requirement of self-compactability, but have not focused on a procedure to design concrete of a desired strength and workability that characterizes the design of conventional concretes with or without the chemical and mineral admixtures, not to mention the durability aspects.

Having considered all these perspectives, one wonders if there are any special concrete composites that need to be presented under the name of SCCs. Extending a little further into what is known, it is obvious that even if we were to term a group of concretes a special class called self-compacting or SCCs, one cannot be wrong in saying that these can be designed with the available knowledge and the proposed methodologies for normally vibrated concretes. The later part of this book deals with some of these aspects and also delineates a methodology based on existing concrete mixed design procedures to ensure a perfectly satisfactory SCC.

One important aspect that is worth mentioning at this stage is that this study was realized in such a way to ensure that the information collated and the databanks established are comprehensive and are representative of the overall picture and yet they are separated to ensure an appropriate recognition of the intrinsic chemical dynamics of the cementitious systems analyzed. This is to ensure true and representative relationships between the appropriate parameters. The parameters have all been chosen in such a way that they are the most acceptable to date and were seen to have an unequivocal validity in the other systems of cementitious materials that were investigated over the years. Needless to say that the approach is not purely mathematical or empirical as could happen in a case where it is purely governed by the numerical system being analyzed through totally fuzzy or neural logic methods without connecting them to the physical and chemical aspects of the system in question, which could lead to serious misinterpretations. In fact both the spectral width of the data and also the parameters of investigation have always been tempered by an overall understanding of not just the material science and technology of concrete, but also an understanding of its requirement as structural material and its performance in the environment in which these structures are expected to perform. In addition, the possibility of representing not just the concepts but also the outcomes in them in a ready to use format for the practitioner was also attempted wherever possible.

1.5 Potential and Limitations

In the present scenario, SCCs can be treated as essentially a newer group of concretes that do not require any compaction energy through vibration or other means. However, as the emphasis is on self-compactability, the more important and defining structural requirements of strength and performance have not attracted adequate attention. Naturally, the resulting SCCs, unless appropriately addressed otherwise, may not comply with any specific strength or performance class as required in the structural practice. The discussions in the preceding part of this chapter clearly bring out the fact that SCCs of various strength grades can be appropriately designed if one has better understanding of the specific relations as in the case of traditionally vibrated concretes.

At this stage probably the most important requirement, to fully understand and realize the complete potential of SCCs, is to look at the total picture through a proper appreciation of the concretes reported so far in the literature and arrive at acceptable guidelines and recommendations, not just to arrive at an SCC with the required self-compactability but to be able to design it with an appropriate confidence having the required flowability and also resulting in a concrete of a specific strength and performance grade for a specific structural application. In fact many design efforts essentially started with a very similar objective of obtaining a concrete of a particular strength grade, but the experimental investigations by their own admission have resulted in concretes of different strength grades. Also, the performance characteristics reported, if at all, have not been related to the constituent material characteristics and quantities to ensure an appropriate understanding and adoption. In a few cases, the mechanical characteristics like the tensile strength and modulus are related to the strength, which were observed to be nearly following the accepted relationships used for normally vibrated concretes. This in a way was the justification for an approval of the use of SCCs in structural concrete applications. Even so there are a few who consciously advocated caution in adopting SCCs without an appropriate verification. Some specific aspects like early age characteristics, strength development, shrinkage, bond with reinforcement, and creep, relaxation, and anchorage stress distributions particularly for prestressing applications need a better understanding. There is an urgent need to address some of these aspects at least on a case-to-case basis to ensure safety of the structural systems or alternatively suggest ameliorative measures that could address these concerns based on an appropriate background in such cases from previous experience. It is these important yet fundamental concerns that instigated and initiated the present study to look at the entire spectrum from a structural engineering application standpoint. At the same time appreciating the material science characteristics that are required to ensure an appropriate system of SCCs for practical applications in all the relevant flow regimes is also needed.

1.6 Future Prospects

This concept of SCC, originally articulated by Okamura (1994), was seen to be mainly to harness its advantages such as the elimination of vibration and decreasing time and labor costs along with a reduction in noise pollution during compaction. It is presumed that it could ensure better durability due to compaction effectiveness and the additional fines that generally are of a pozzolanic nature. The basic difference of SCCs from conventional concretes is in ensuring the three main attributes of filling, passing, and segregation of these concretes through modifications in the proportioning of the constituents, like increasing the fines or by the addition of a VMA that imparts the required segregation resistance.

Most design procedures for SCCs have concentrated on proportioning of the mix constituents to achieve the primary requirement of self-compactability, but have not focused on a procedure to design the concrete of a desired strength and workability that characterizes the design of conventional concretes with or without chemical and mineral admixtures, not to mention aspects of durability. This study is an attempt to discuss the various parameters that influence the making of SCCs and to propose a few simple and effective guidelines to arrive at the same with specific strength and durability characteristics.

References

ACI 211.1-91, Standard practice for selecting proportions for normal, heavy-weight, and mass concrete, ACI Manual of Concrete Practice, American Concrete Institute, Farmington Hills, MI, 1997, 38pp.

ACI 237R-07, Self-consolidating concrete, ACI Manual of Concrete Practice, American Concrete Institute, Farmington Hills, MI, 2007, 34pp.

Collepardi, M., A very close precursor of self-compacting concrete (SCC), in *Supplementary Volume of the Proceedings of Three-Day CANMET/ACI International Symposium on Sustainable Development and Concrete Technology*, San Francisco, CA, September 16–19, 2001.

Collepardi, M., Self-consolidating concrete in the presence of fly-ash for massive structures, in *Proceedings of Second International Symposium on Concrete Technology for, Sustainable February—Development with Emphasis on Infrastructure*, Hyderabad, India, February 27–March 3, 2005, pp. 597–604.

Collepardi, M., Collepardi, S., and Troli, R., Properties of SCC and flowing concrete, in *Proceedings of Special Session in Honor of Prof. Giacomo Moriconi, Sustainable Construction Materials and Technologies*, Coventry, U.K., June 11–13, 2007, pp. 25–31.

EFNARC, *The European Guidelines for Self-Compacting Concrete*, Farnham, Surrey, U.K., 2005, p. 68.

Gerwick, B.C., Marine Foundations—Underwater concrete—Mix design and construction practices, 2015.

JSCE, Recommendation for self-compacting concrete, in T. Uomoto and K. Ozawa (Eds.), Concrete Engineering Series 31, 1999, p. 77.

Okamura, H. and Ouchi, M., Self-compacting concrete, Japan Concrete Institute, *Journal of Advanced Concrete Technology*, 1(1), 2003, 5–15.

Okamura, H. and Ozawa, K., Self-compactable concrete for bridge construction, in *International Workshop on Civil Infrastructural Systems*, Taipei, Taiwan, 1994.

Okamura, H. and Ozawa, K., Mix-design for self-compacting concrete, *Concrete Library, JSCE*, 25, 1995, 107–120.

Precast/Prestressed Concrete Institute, *Interim Guidelines for the Use of Self-Consolidating Concrete in Precast/Prestressed Concrete Institute Member Plants*, Chicago, IL, TR-6-03, 2003, 165pp.

Semioli, W.J., Self-placing concrete, *Concrete International*, 23, December 2001, 69–72.

Szecsy, R.S., What's in a name? *The Concrete Producer*, Washington, January 2002, 51–58.

2

Constituent Materials

2.1 Constituent Materials and Availability

After the initial euphoria of having thought to have entered into a new era of self-compacting concrete (SCC) technology that will immediately redress at least many of the problems associated with the performance of structures, the industry realized that SCC is in no way a panacea for all the inappropriate practices, but in fact is only another of the several varieties of concrete composites. It was also understood that, if appropriately utilized, it could certainly result in quite a few benefits depending upon the application. It is indeed a fact that the constituent components of SCC are all the same as was used in the making of normally vibrated concretes. This in a way makes the job of having to compose information about these constituents very simple at the outset. However, this study is not an attempt to review what is already available in the different texts on concrete technology presently but to consider them with the viewpoint of making the resulting SCC into a high-performance cementitious composite, the primary aim of putting together this book in the present form.

Actually this commonality of materials endears and gravitates everybody to the use of SCC, as they feel that it is in no way different from normal concrete and that with minor modifications one can easily achieve an SCC of acceptable quality. However, even a few initial efforts in this direction will prove that it is not possible to achieve a perfectly self-compacting composition that has both the required thixotropy and consistency to flow without segregation and bleeding so required. Naturally, without being highly conscious of the properties of the constituents and their different combinations it will be difficult to produce SCCs effectively.

Moreover, a concrete composite is made up of essentially two components: The first is understandably the more expensive and limited volume cementitious binder part, which coats and binds the less expensive and larger volume filler part. An effective balance between these two components is essential as the porosity of the resulting mass will enforce the strength as well as the performance characteristics of the material composite. The interplay between these two and their effectiveness and participation can be moderated through a few materials called admixtures, be it chemical or mineral. In the light of this the following discussions will broadly present the entire system in three parts, the cementitious binders including the supplementary cementitious materials, the fillers in terms of the aggregates along with the mineral powders that assist in packing, and finally the chemical admixture components that ensure the metamorphic transformation of the cementitious binder component to facilitate in its proper coating of the filler grains in the system, helping an effective packing density of the resulting composite while ensuring its consistency and workability in the green state.

Apart from this, it is known that the concrete construction is obviously related to the lead distance of the materials available for construction. The important reason that structural concrete composites attract so much attention is that they are made mostly with locally available materials (if only as nothing but fillers), while the cement binder is factory produced under strict quality control regimes that are dictated by the various national codes and standards. It is thus obvious that one should also have a broad perspective and understanding of the type of materials and their qualifications required to ensure that they are effective in SCCs.

It is also apt at this stage to be aware that the following paragraphs regarding the various constituents in SCC present only a broad picture of the materials and their characteristics. However, one can see that the implications of some of these characteristics on the behavior of the composite can only be explained at the appropriate point during the discussions on the mixtures, the design, the green and hardened state characteristics, and the service life performance.

2.2 Cements and Characteristics

The commercially available Portland cements constitute this component. The discussions in principle do not look at highly specialized and special purpose cements that are produced to be components in custom-made products and are excluded from discussions mostly. Portland cements of ASTM Type I and II that are commonly used in conventional structural concrete are also the cements for SCC. They are produced in significantly large quantities in most countries and are easily available. Typically their fineness is around 300–400 m^2/kg and are produced by using 95%–100% cement clinker, with additives other than the cement manufacturing materials restricted to a maximum of 5%. This is an important fact that one should keep in mind to ensure that in the later discussions the other fillers and pozzolans can be accounted for appropriately evaluating the possible compressive strength of the SCC produced. It is also to be noted that cement is produced in several strength grades, 32.5, 42.5, and 52.5, as prescribed by the Euro norms. The values suggest the strength of these cements at 28 days, tested in mortar with the specified standard sand at the specific water–cement ratio. Even the maximum and minimum strengths to which the cement should confirm are specified. Such a strength classification is nonexistent in the ASTM standards but one can safely assume that for all practical purposes at least for SCCs of the normal strength range of around 30–60 MPa, the 42.5 strength grade is mostly adequate and preferable as it will not have the higher fineness that may require a little more water for wetting as will be discussed later. It may also be essential to be informed that the cement strength test methodologies in each of the different national standards vary significantly and the strength classification and the nomenclatures used may be at variance. This aspect was discussed earlier in detail by Ganesh Babu et al. (1992). There are several other cements available in the market today—such as the rapid-hardening cement, low-heat cement, sulfate-resisting cement—which are all produced with modifications in the components of cement, which is not being discussed in the present context. Some of them, particularly the rapid-hardening cement, have been used for the making of SCCs, and the reasons and advantages or requirements will be discussed later.

Apart from this, blast furnace slag cement (BFSC) and pozzolanic cements based on fly ash are also quite common. At this stage it is appropriate to go a little further and inform that if in case there is a need to use the above pozzolanic cements, it is absolutely essential to understand the type of pozzolan and its percentage in cement to ensure that these are appropriately factored in to arrive at the strength of the resulting SCC. This aspect of accounting for these pozzolanic materials is discussed in detail later. It is also appropriate to look at the recent Euro standard classification of cements as proposed in EN 197 (2000) to ensure that the reader is aware of the current status. This code delineates a total of 27 product compositions from what it terms as the family of common cements and groups them into five groups—Portland cement, Portland-composite cement, blast furnace cement, pozzolanic cement, and composite cement (CEM I–V). Table 2.1 presents a broad summary of this complete range of the 27 common cement compositions according to this standard. It clearly delineates the percentage of Portland cement clinker in the cement and also the pozzolanic materials that constitute the remaining portion of the cement produced. The additional 5% additive that was discussed earlier could still be a part of the same cement, and the manufacturer is at liberty not to disclose this particular information to the public. This can also be factored in carefully into the evaluations of the strength of these cement composites used for the manufacture of SCC.

Similar to the Euro norms, ASTM C 150 (2009) also classifies Portland cement into five basic types: General-purpose cement (Type I), Moderate sulfate resisting cement (Type II), High early strength cement (Type III), Low-heat cement (Type IV), and Sulfate resistant cement (Type V). Apart from this ASTM also presents four basic classes of blended cements and a

TABLE 2.1

Classification of Euro Cements as Proposed in EN 197-1 (2000)

Type	Portland Cement	Cement Clinker (K)	GGBS (S)	Silica Fume (D)	Pozzolan (P/Q)	Fly Ash (V/W)	B. shale/L. stone (T)/(L/LL)
				Main Constituents Composition (% Mass)			
CEM I		95–100					
CEM II/A-S	Portland slag cement	80–94	6–20				
CEM II/B-S		65–79	21–35				
CEM II/A-D	Silica fume	90–94		6–10			
CEM II/A-P/Q	Pozzolana	80–94			6–20		
CEM II/B-P/Q		65–79			21–35		
CEM II/A-V/W	Fly ash	80–94				6–20	
CEM II/B-V/W		65–79				21–35	
CEM II/A-T/L/LL	Burnt shale/limestone	80–94					6–20
CEM II/B-T/L/LL		65–79					21–35
CEM II/A-M	Composite	80–94			6–20		
CEM II/B-M		65–79			21–35		
CEM III/A	Blast furnace	35–64	36–65				
CEM III/B		20–34	66–80				
CEM III/C		05–19	81–95				
CEM IV/A	Pozzolanic	65–89			11–35		
CEM IV/B		45–64			36–55		
CEM V/A	Composite (P/Q/V)	45–64	18–30		18–30		
CEM V/B		20–38	31–50		31–50		

Notes: Undeclared minor additional constituents are also allowed up to 5% in all types of cements.
Pozzolan—P is natural and Q is natural calcined; and Fly ash—V is siliceous and Q is calcareous.
Additive—T is Burnt shale; L and LL are limestone with 0.2% and 0.5% total organic carbon.

number of special cements like White cement in its fold. It is not envisaged at this stage to discuss the implications of the use of the cements in SCCs, apart from saying that for most practical purposes Type I cement is most commonly used. The effects of the other types of cements on the behavior of SCC composites can be reasonably understood if one goes into the composition and characteristics presented in these codes and by appropriately accounting for the cementitious efficiency of the additional components.

Another important aspect that should be kept in mind is that fly ash referred herein is only fly ash available from the electrostatic precipitators of a major thermal power plant and in no case should be confused with pond ash or ash from any other source. It is also to be noted that fly ash may sometimes be introduced at the grinding stage and because of its fineness it may not really get affected in the grinding process. However, use of pond ash even with intergrinding of any form is not acceptable. It is also to be noted that in recent years with the significant advancements in grinding technology fueled by the interest to achieve rapid strength gain most cements are supplied with a fineness of 300–400 m^2/kg (Blaine), and the addition of even finer supplementary cementitious materials like fly ash (about 420 m^2/kg) and silica fume (about 20,000 m^2/kg, BET) will make the surface area that needs to be wetted much higher necessitating higher water content, even though the finer materials do assist in the flexibility in the concrete and also make it cohesive and thixotropic.

2.3 Simple Powder Extenders

At the very inception of the concept of SCC, Okamura proposed to have a higher volume of powder (cementitious materials) while reducing the coarse aggregate content appropriately. It is also a fact that the powder content consisted of cement and fly ash resulting in a cementitious compound of lower heat of hydration, while ensuring economy. While it is most appropriate to use low-end pozzolanic materials like fly ash and ground granulated blast furnace slag (GGBS) to supplement the powder requirements in the paste (discussed later), a few researchers have used chemically inert powders in limited quantities, which help in mostly retaining the original strength of the composite. A few others have used calcareous fillers like limestone powder or certain uncalcined clays, which could exhibit slight latent hydraulic activity. This in fact has never been really quantified. These mineral powder extenders help in ensuring a more cohesive and segregation-resistant matrix that will not also bleed excessively. However, one should be cautious as these effects are only possible within a limited range wherein the additional water required for wetting such fine powders can be adequately countered with a small quantity of the superplasticizer required. Later investigations on such concretes with materials like limestone powder showed that even such inert fillers when they are in a microfine powder form could exhibit a small quantity of reactivity like pozzolans and these aspects are specifically analyzed and discussed later. Calcium carbonate-based mineral additives used as fillers essentially in cementitious emulsions and paints can also be used for this purpose. However, as already discussed, the effects of water demand for a similar fluidity that may still be a matter for concern in obtaining an effective SCC. It is also to be recognized that many a time some of the dry powders will absorb a substantial amount of water and in other cases the porosity of the larger particles in the system (as in courser rice husk ash [RHA]) will also influence the water requirements, which will affect the final strength of the concrete.

2.4 Supplementary Cementitious Materials

The most appropriate and probably the most cost-effective solution to achieve increased powder content will be to use industrial by-products like fly ash and GGBS as supplementary cementitious materials. These pozzolanic or sometimes latent hydraulic additives (if they contain significant quantities of CaO) added appropriately in limited quantities will result in excellent cementitious compositions that not only help in imparting better green state characteristics like cohesion and segregation resistance while being self-compacting, they also do not change the strength characteristics of the matrix significantly. They will result in concretes of the normal strength grades for which they were designed but will have significantly improved performance characteristics.

In general, these mineral admixtures are called as pozzolans. According to ASTM C 618 (1994), a pozzolan is defined as "a siliceous or siliceous and aluminous material which, in itself, possesses little or no cementitious value but which will, in its finely divided form and in the presence of moisture, react chemically with calcium hydroxide at ordinary temperature to form compounds possessing cementitious properties." Pozzolans can be broadly divided into two major types, namely natural pozzolans and artificial pozzolans. Natural pozzolans include volcanic ash (glass), volcanic tuft, calcined clay or shale, and raw or calcined opaline silica. Calcined as well as uncalcined clays like kaolin, (metakaolin), zeolite, and so on are also a part of these. The artificial pozzolans include fly ash, silica fume, GGBS, and RHA. The use of natural pozzolans as binders in combination with lime and other additives was known for such a long time that in fact we do not have any information on the excellent cementitious recipes that were used in several of the monuments of yesteryears that have defied the effects of the environment for centuries if not millennia. In recent years, the use of artificial pozzolans as supplementary or complimentary cementitious materials in concrete is increasing because of the fact that their presence in concrete improves the properties both in the green and hardened states. Also the use of these materials represents eco-friendliness while being economical.

2.4.1 Pulverized Fuel Ash or Fly Ash

Pulverized fuel ash or fly ash as a mineral admixture in concrete is of great utility to the present day construction industry due to its availability in abundance. The characteristics of coal ash primarily depend on the geological and geographic factors related to the coal deposit, the combustion condition, and the collection devices. According to the ACI Committee 226 (1987), the following are the primary factors that influence the effectiveness of the use of fly ash in concrete. They are the chemical and phase composition of fly ash and of the Portland cement, alkali-hydroxide concentrations of the reacting system, the morphology of the fly ash particles, the fineness of the fly ash and of the Portland cement, the development of heat during the early phases of the hydration process, and the changes in mixing water requirement. For an effective utilization, it is necessary to know about the characteristics of fly ash and their effects on the properties of concrete.

The physical characteristics of fly ash like particle shape and size, density and color mainly depend on the type of collection system and the combustion temperature of the pulverized coal. The fineness of fly ash will have an influence on the pozzolanic reactivity, workability, bleeding, and segregation of concrete. It also affects the water content and air entraining admixture demand in concrete. Normally the use of fly ash decreases the water demand but fly ashes of higher fineness may require an increase in water demand at the higher percentage levels. It was observed that the fineness of fly ash as measured by the 45 micron sieve retention was the most significant parameter influencing the suitability of fly ash for applications in concrete (Dhir, 1986). Most of the standards require that the residue on 45 micron sieve should not be greater than 34%–50%. It was also reported that a maximum limit of 20% residue is reasonable because modern furnaces can easily comply with such a requirement (Mehta, 1986). Fly ash particles are typically spherical, ranging in diameter from 1 to 150 micron. The particle shape and size mainly depend upon the mineralogical phases and collection system. The particle shape affects the water demand of a standard paste and as the number of spherical particles increases, the water demand decreases. It was also reported that as the size of fly ash decreases the requirement of water for mortars containing fly ashes increases (Wesche, 1990). The density of fly ash depends on the constituents (iron, silicon, aluminum, and silica) and higher carbon contents tend to lower the density. It generally varies between 1.97 and 2.89 gm/cc. The color of fly ash may vary from light tan or grey to almost black depending on the type and quality of the coal used and the combustion process.

The main chemical constituents of fly ash are SiO_2, Al_2O_3, Fe_2O_3, and CaO, which are responsible for its pozzolanic activity. It also consists of MgO, TiO_2, Mn_2O_3, P_2O_5, SO_3, Na_2O, and unburnt carbon. There is a possibility for variation in composition from plant to plant and even in one plant from time to time. The general variation in three principal constituents will be as follows: SiO_2 (25%–60%), Al_2O_3 (10%–30%), and Fe_2O_3 (5%–25%). There are some differences in these standard requirements in the case of SO_3 and loss on ignition. The structure of fly ash is mostly noncrystalline (glossy about 60%–90%) due to the rapid cooling of the ash from the molten state. According to ASTM, C 618 fly ashes are classified as class C and class F, mainly based on the CaO content. The fly ashes produced from lignite or sub-bituminous coals having more CaO are classified as class C (generally CaO > 10%), whereas the ashes from bituminous or anthracite coals having less CaO are classified as class F (generally CaO < 10%). Regarding the requirements of class C and F fly ashes, SiO_2 + Al_2O_3 + Fe_2O_3 should be 50% in case of class C whereas 70% in case of class F. ASTM C 618 specifications also suggest the same limitation for class F but for class C it was only 50%. The intention to specify this limit is to ensure that sufficient reactive glossy constituents are present. A lower requirement is necessary for class C because the calcium oxide content may be substantial. However, many researchers found that there was little effect of these three compounds on the performance of concrete (Dhir, 1986; Mehta, 1986; Wesche, 1990). It is well known that in Portland cement pastes the hydration of crystalline MgO under autoclave conditions leads to expansion and cracking. According to the phenomenon mentioned earlier, the maximum limit on the MgO content was specified to Portland cement. In fact it is not valid in the case of fly ash because the MgO in fly ash is either in noncrystalline form (glass) or in the form of nonexpansive mullite phase (Mehta, 1986). However, some organizations have specified the maximum limit in the range of 4%–5% to prevent expansion. The limit on the SO_3 content is imposed on the basis of its expansive nature causing deterioration of concrete due to formation of ettringite. It was stated that the SO_3 content resulted in high early strength (Mehta, 1986). The CaO content in fly ash mainly depends on the type of coal used for combustion. The CaO content of fly ashes from bituminous or anthracite coal rarely exceed 7% whereas for sub-bituminous or lignite coal it is more than 10% and as high as 30% (Mehta, 1986). It was stated that the presence of alkalies (Na_2O, K_2O) in fly ash causes efflorescence and alkali–aggregate reaction, which would cause expansion. But a study by Mehta (1986) revealed that the alkali aggregate reaction is affected mainly by the amount of alkalies present in the cement than the amount of alkalies in the fly ash. The effectiveness of fly ash, in reducing alkali–aggregate reaction, is dependent on the available alkali content of fly ash, cement, and cement replacement. The low-temperature burning (1000°C and below) causes the presence of some volatile residue, which is termed as loss on ignition (LOI). The carbon content is the most important component of LOI and it affects the water requirement for mortar and concrete application. In general, the fly ash with low carbon content will produce more workable cement composites. The carbon content of fly ash, which has high porosity and a very large specific surface, causes the absorption of water and organic admixtures in concrete such as water reducing agents, air entraining agents, set retarders, and so on (Wesche, 1990). Different standards have given different ranges of limitations on LOI. These limitations range from 5% to 12%. ASTM C 618 has allowed only a maximum of 6% in the case of class C and 12% in the case of class F, if the test results of the laboratory are satisfactory. Overall, Figure 2.1 gives the different characteristics of fly ash, their influence, and the maxima of specifications from the different national and international standards. In general, there is no specific agreement that any specific property limits the performance of fly ash in concrete.

2.4.2 Ground Granulated Blast Furnace Slag

Ground granulated blast furnace slag (GGBS) as cement replacement material has assumed prominence not only due to the improved performance characteristics of concrete composites but also due to the significant economy achieved in the production of cement or concrete composites. GGBS is used in three ways: intergrinding, grinding, and mixing or even site mixing. The first two are used in the production of PBFS. The major disadvantage of these two methods is that the proportions of slag and cement have been essentially fixed at the time of production, thus limiting its use in various environmental conditions. This actually gave rise to the different types of cement produced with GGBS. Site mixing involves the mixing of GGBS in a desired proportion during the production of concrete providing greater flexibility with which the users can tailor-make concrete depending on application. By proper proportioning and a judicial choice of the replacement percentage, one can produce GGBS concretes of excellent durability. Moreover, the high alumina content in this material admixture improves the chloride binding capacity of concrete, thus making this material best suited for marine environment.

Among the physical characteristics of GGBS, fineness is a very important parameter, which influences the reactivity, released in the development, and water requirement. In general it varies from 375 to 425 m^2/kg as the grinding process involves a substantial amount of energy, much higher than that for cement, and the limitations are based on economic considerations. The specific gravity is around 2.90 with the bulk density varying from 1200 to 1300 kg/m^3. The chemical composition of slag influences the reactivity considerably. There have been different compositional moduli suggested for the hydraulicity of GGBS among which the best correlation of strength was seen to be predicted by $[(CaO + MgO + Al_2O_3)/SiO_2]$, which would exceed 1.0 for good performance. ASTM C989 (1994) suggests the slag activity index as a percentage ratio of the average compressive strength of slag cement mortar (50%–50%) to the average compressive strength of the reference cement mortar tubes within a designated age for understanding its effectiveness.

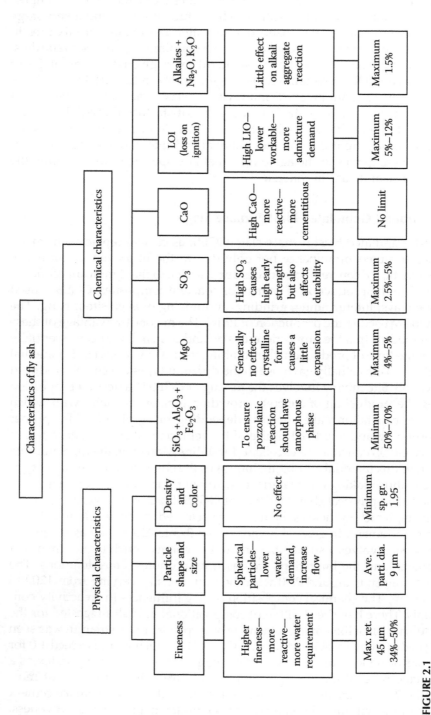

FIGURE 2.1
Influence of the characteristics of fly ash on concrete.

The slag activity index or the hydraulic index was used by many to characterize GGBS, but the values obtained could never be effectively related to the compressive strengths of concretes. Figure 2.2 presents a summary of the effects of the various constituents of GGBS in concrete. The limits available in the literature or in codal provisions are also incorporated in the figure.

2.4.3 Silica Fume

The necessity for the development of high-strength, high-performance concrete has forced researchers to examine various alternatives in terms of newer materials and admixtures along with improved production techniques. Silica fume as a mineral admixture in concrete (mostly combined with superplasticizers) has shown a possibility to develop concretes of over 100 MPa. Also, it is felt that the high efficiency of silica fume (in terms of its cementitious material property in strength) can lead to significant reductions in the basic cement content, which may affect the durability of these concretes adversely. Among the multitude of mineral admixtures available today, silica fume is the most promising and a uniquely placed material for the production of high-strength and high-performance concretes. Silica fume is a by-product resulting from the reduction of high-purity quartz with coal in the electric arc furnaces during the manufacture of silicon in ferrosilicon industries. It was reported (Sellevold and Nilson, 1987) that for application in concrete, silica fume should have an SiO_2 content in the range of 85%–98%, a mean particle size in the range of 0.1–0.2 microns, and a spherical shape with a number of primary agglomerates in the amorphous state. It is also known that the SiO_2 content depends on the end product of the industry. It was seen to be a highly pozzolanic material due to its high specific surface area (20–23 m^2/g) and amorphous structure. It was also reported that the incorporation of silica fume into concrete improves the properties in the green state as well as in the hardened state (Sellevold and Nilson, 1987).

2.4.4 Other Pozzolanic Admixtures

There are a host of other pozzolanic additives as already indicated earlier, which by virtue of being more easily available locally or from other considerations have all found their way into the concrete construction industry. The specific attributes, methodology, and limitations of these are obviously best discussed by presenting the production, processing, and utilization potential of these concretes, which will be seen later. To get an overview, Table 2.2 provides comprehensive information about the typical chemical and physical properties of many of the pozzolanic materials that have been used both in the production of normally vibrated concretes as well as in SCCs (after Kosmatka et al., 2003). Besides the outline that is presented in the table regarding the characteristics of the various pozzolanic admixtures, it is important to recognize a few methods by which they can be characterized in terms of

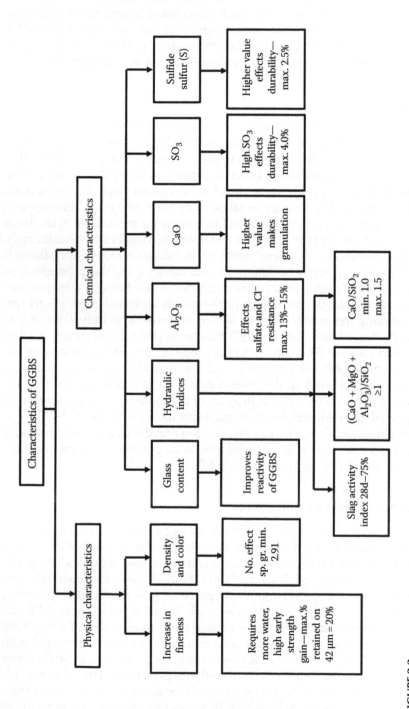

FIGURE 2.2
Influence of the characteristics of GGBS on concrete.

TABLE 2.2

Comprehensive View of the Chemical and Physical Properties of Mineral Admixtures

Pozzolan Charact.	Class F Fly Ash	Class C Fly Ash	GGBS	Silica Fume	Calc. Clay	Calc. Shale	Metakaolin	RHA	Zeolite
SiO_2 (%)	52	35	35	75–95	58	50	53	85	65
Al_2O_3 (%)	23	18	12	0.4	29	20	43	4	15
Fe_2O_3 (%)	11	6	1	0.4	4	8	0.5	0.5	0.5
S+A+F (%)	70	50	—	—	—	—	95	—	—
CaO (%)	5	21	40	1.6	1	8	0.1	0.5	—
Na_2O (%)	1.0	5.8	0.3	0.5	0.2	—	0.05	1	1.5
K_2O (%)	2.0	0.7	0.4	2.2	2	—	0.4	3	3
SO_3 (%)	0.8	4.1	9	0.4	0.5	0.4	0.1	0.5	0.1
LOI (%)	2.8	0.5	1.0	3.0	1.5	3.0	0.7	6	10
Sp. Gr.	2.38	2.65	2.94	2.40	2.50	2.63	2.50	2.05	2.15
Ave. part. size (μm)	20	20	20	1	10	15	2	7	20
Fineness (m²/kg)	420	420	400	20,000	990	730	12,000	35,000	460

Source: After Kosmatka, S.H. et al., *Design and Control of Concrete Mixtures*, 14th edition, Portland Cement Association, 2003.

their pozzolanic activity and appropriateness to be used as supplementary cementitious material. One simple yet reasonably acceptable methodology is to study them through their lime reactivity or even better the cement reactivity instead of lime. Another aspect that probably shows their potential is to look at the X-ray diffraction patterns of the different pozzolans that exhibit a clear diffusion band, as suggested by Mehta (1989). Another important factor that may have a bearing on both the pozzolanic activity and the water demand of the cementitious material combination is its average particle size and shape, and particle size distribution, which will be critical in the production of SCCs even with the addition of more modern superplasticizers like polycarboxylic esters and viscosity modifying agents of today. A more detailed discussion on the characteristics and influence on concrete in both the green and hardened states is provided later to ensure its efficient and appropriate utilization.

2.5 Superplasticizers and Other Chemical Admixtures

The fact that chemical admixtures are utilized to modulate the various characteristics of concrete in practice today is well known. Modifiers that impart specific characteristics like plasticizing effects, accelerators, retarders, air entrainers, and water proofers apart from a host of other minor modifiers constitute this group known as chemical admixtures. In fact in the present context it is difficult to find a concrete without at least a superplasticizer, which is almost an integral part of the ready mix concrete industry. While others like accelerators and retarders model at both the set and strength characteristics of concrete, air entrainers are referred for concretes in the marine environment.

2.5.1 Superplasticizers or High-Range Water Reducing Admixtures

Superplasticizers are admixtures that when added to normal concrete impart high workability and/or allow a reduction in water requirement at the same workability level as that of the concrete without the admixture. The first step towards flowing concretes started with the addition of superplasticizers, which could help in producing the flow without exhibiting either excessive bleeding or segregation, and with retardation and air entrainment being generally absent. Initially the modern superplasticizers were classified into four categories—sulfonated melamine formaldehyde (SMF) condensates; sulfonated naphthalene formaldehyde (SNF) condensates; modified lignosulphonates (MLS); and others. In recent years the introduction of the next generation of superplasticizers based on polycarboxylic esters has further improved their capability and is the most preferred one for the manufacture of SCCs. Several manufacturers are producing a range of superplasticizing admixtures tailored to specific user requirements, sometimes in combination with other admixtures that are needed. The main aim is to ensure the required water reduction and/or fluidity while maintaining their dispersing effect during the time required for transport and application. The requirement of this consistency retention will depend on the application, with precast concrete generally requiring a shorter retention time than for concrete that has to be transported to and placed on site.

2.5.2 Viscosity Modifying Admixtures

The production of SCCs can be categorized into three different approaches—powder type (high fines approach), viscosity modifying admixture (VMA) type, and combination type. It is obvious that ensuring the required amount of fines along with superplasticizers can certainly result in a SCC. VMAs can be useful in providing a wide range of possibilities to the designer, even in situations where modulation of the other constituents is a little difficult. VMAs impart viscosity and thixotropy of the mixture and help in reducing the bleeding and segregation problems often associated with SCCs—particularly due to variations in the grading, fineness, and moisture of aggregates—and cement and superplasticizer characteristics, which are indeed difficult to control while operating a ready mix plant. Also, in low-strength concretes with low water content the control of viscosity becomes difficult and VMAs will indeed be essential. Natural polymers such as guar gum or synthetic polymers like hydroxyl propyl methyl cellulose as well as water-soluble polysaccharides like welan gum have all been used as viscosity modifiers in concrete.

2.5.3 Other Admixtures

As already stated, there are several other types of chemical admixtures that are used for modifying the characteristics of concretes either in the green state or the hardened state—accelerators, retarders, shrinkage compensating admixtures, water proofers, and so on. Apart from these there are other additives like pigments that are used for special effects in the manufacture of concrete products.

2.5.4 Mixing Water

The specifications for mixing water are essentially the same as those prescribed for conventionally vibrated concretes. Most often a simple prescription for this would be to use potable water. The most important and probably significant limitation could be from the chlorides in the water, which could lead to corrosion of reinforcement, while in certain cases the sulfates may also have to be addressed. Specific limitations on some of these dissolved chemicals are already available in the various codal provisions.

2.6 Aggregate Characteristics

The aggregates in concrete occupy approximately three-fourth of the volume and do not participate in the complex chemical reactions (Mehta, 1997). In a properly designed concrete mix, all the particles of aggregate are expected to be bound by the cement paste. However, the cement content normally constitutes a low percentage (200–600 kg/m^3) of the total concrete with the aggregate cement ratios generally varying between 2.5 and 7.5. Thus, it is absolutely essential to have a proper grading of these aggregates to give a concrete with minimum voids resulting in high strength and economy. The fineness and the strength of cement will also influence the characteristics of concrete both in the green state and the hardened state. The design requirements of structural concretes, apart from strength and durability, are also based on element size (particularly its thickness and depth), reinforcement detailing (spacing), placement, and compaction procedures, all of which decide the maximum size of the aggregate and the workability requirements. A rational approach for the design of a concrete mix is to evaluate the properties of the constituent materials available—cement, aggregates, and water; based on this, for a required design strength and workability, the water–cement ratio and water content are determined. From these, the cement and total aggregate contents are worked out either by absolute volume method or by weight method. An appropriate ratio of the fine aggregate to total aggregate is calculated based on the maximum size of the coarse aggregate and grading of the fine aggregate (as in the case of BIS, ACI, and BS), while the combined grading of the total aggregate is ensured in the DIN standards.

The CEB-FIP model code (1994) clearly states that the aggregates consist of a mixture of particles of different sizes that are combined in accordance with specific requirements. The principal aspects concerning the overall aggregate grading can be summarized as follows:

1. To arrive at the maximum particle size depending on the dimensions of the member, thickness, concrete cover, spacing of reinforcement, and handling and placing conditions
2. To have sufficient workability for the compaction adopted with optimum content of fines for achieving maximum packing and closed textured surface

These two conditions will automatically result in the lowest water demand ensuring optimum cement content and thus lowest reactivity. At this stage, as already stated the grading of the aggregate is achieved by the combined all-in-aggregate grading in a few codes like CEB-FIP, DIN (1988), and Euro norms. However, the ACI, BS, and BIS look at grading of coarse and fines aggregate separately and mixing them in an appropriate proportion based on the type of finer material used (fineness modulus of sand as in ACI and particles below 600 microns in the latest BS specification). It is obvious that an increase in the finer particles (sand) can ensure a better packing and also better flexibility in the concrete. However, this increase in fines content will also mean a higher surface area required to be wetted and will need higher water contents resulting in higher cement contents at the same water–cement ratio.

CEB-FIP recognizes the effect of smaller fractions in fine aggregates and states the following:

- Too high an amount of ultrafines will reduce the workability of fresh concrete unless superplasticizers are used.
- A replacement of the cement for ultrafines beyond limits (through mineral admixtures or supplementary cementitious materials) can adversely affect the concrete durability.
- If the addition of fines allows a reduction in the cement content without an increase in the water–cement ratio, shrinkage and creep of concrete are reduced.

Aggregates constitute the bulk of the total volume of concrete. The characteristics of the coarse aggregate that influence the behavior of concrete are its type, size, shape, surface texture, and gradation (Ganesh Babu et al., 1992). The type of the aggregate is based on the nature of the parent rock, which influences its strength, porosity, and water absorption. Of particular interest in this regard are the chemical reactions between the aggregates and the cement paste, such as alkali–aggregate reaction. Most igneous aggregates do not show specific problems related to the alkali–aggregate reactivity. However, when other types of aggregates such as dolomite are used the increase in water requirement for the given workability has to be considered. Similarly, variation (possible increase) in strength is also an important factor.

The grading of aggregates has assumed a greater significance in recent years with the increased demand for higher strengths and more durable (high-performance) concretes. The characteristics of both coarse and fine aggregates in terms of the size, shape, and gradation affect the workability of the concrete mixes. It is known that if the maximum size of the coarse aggregate increases, the mix becomes economical as the cement content decreases (Neville, 1971). But in practice, the dimensions of structural members and the spacing and distribution of the reinforcement will have a restraint on the maximum size of the aggregate (IS 456, 1979). Moreover, mixes with smaller-size aggregates yield higher strengths in comparison to mixes with relatively larger-size aggregates (Neville, 1971, ACI 363, 1984). These things have to be borne in mind while designing high-strength concrete mixes that are essential for tall and long span structures and for structures in aggressive environments, typically the marine environment (Ganesh Babu and Raja, 1989). Though the angularity and gradation of aggregates affect workability, their effect on strength has not yet been exactly quantified.

The grading requirements of coarse and fine aggregates according to IS 383 (1979), ASTM C 33 (1986), and BS 882 (1973) have been presented in Tables 2.3 and 2.4, respectively. The combined grading requirements of DIN and the all-in-aggregate grading according to BS and BIS are presented in Table 2.5. The DIN code of practice also presents recommendations for gap graded aggregates, which are not shown as we are not discussing this topic here. It is clear that in most codes the allowable range of variation is of the order of 40% at some of the sizes. In the experience of the authors it is very

TABLE 2.3

Grading Requirements for Coarse Aggregates

Sieve Size (mm)	Code	Percentage Passing					
	BIS	40	20	16	—	12.5	10
	ASTM	37.5–4.75	19.0–4.75	—	—	12.5–4.75	—
	BS 84/73	40–5	20–5	—	14–5	—	—
75	BS 84/73	100	—		—		
63	BIS	100	—	—		—	—
50	ASTM	100	—			—	
50	BS 84	100	—		—		
40	BIS	85–100	100			—	—
37.5	ASTM	95–100				—	
37.5	BS 73	95–100	100		—		
37.5	BS 84	90–100	100		—		
25	ASTM	—	100			—	
20	BIS	0–20	85–100	100		—	—
19	ASTM	35–70	90–100			100	
20	BS 73	35–70	95–100		100		
20	BS 84	35–70	90–100		100		
16	BIS	—	—	85–100		100	—
14	BS 84/73	—	—		90–100		
12.5	BIS	—	—	—		85–100	100
12.5	ASTM	—	—			90–100	
10	BIS	0–5	0–20	0–30		0–45	85–100
9.5	ASTM	10–30	20–55			40–70	
10	BS 84/73	10–40	30–60		50–55		
4.75	BIS	—	0–5	0–5		0–10	0–20
5	BS 84/73	0–5	0–10		0–10		
4.75	ASTM	0–5	0–10			0–15	
2.36	ASTM	—	0–5			0–5	

TABLE 2.4

Grading Requirements of Fine Aggregates

Sieve Size (mm)	Code	Overall Limits	Percentage Passing			
			Zone I C[a]	Zone II M[a]	Zone III F[a]	Zone IV
10	BIS[b]		100	100	100	100
10	BS 84	100				
9.5	ASTM[c]	100				
9.5	BS 73		100	100	100	100
5	BS 84	89–100				
4.75	BIS		90–100	90–100	90–100	95–100
4.75	ASTM	95–100				
2.36	BIS		60–95	75–100	85–100	95–100
2.36	ASTM	80–100				
2.36	BS 84	60–100	60–100	65–100	80–100	
1.18	BIS		30–70	55–90	75–100	90–100
1.18	ASTM	50–85				
1.18	BS 84	30–100	30–90	45–100	70–100	
0.6	BIS		15–34	35–59	60–79	80–100
0.6	ASTM	25–60				
0.6	BS 84	15–100	15–54	25–80	5–100	
0.3	BIS		5–20	8–30	12–40	15–50
0.3	ASTM	10–30				
0.3	BS 84	5–70	5–40	5–48	5–70	
0.15	BIS		0–10	0–10	0–10	0–15
0.15	ASTM	2–10				
0.15	BS 84	0–15				

[a] As per BS 84 (C, coarse; M, medium; F, fine).

[b] Grading requirements of BIS and BS 73 are same but for 10 mm.

[c] The fine aggregate shall have not more than 45% passing any sieve and retained on the next consecutive sieve of those shown in the table, and its fineness modulus shall not be less than 2.3 nor more than 3.1.

difficult to maintain the required slumps with such a wide variation allowed in the grading for both coarse and fine aggregates. This problem is highly accentuated for high-strength concretes (of low water–cement ratios) with or without admixtures. At present, with the aggregates in weight batching plants, the required gradings for coarse and fine aggregates or even combined aggregates can be easily obtained within a really narrow range. As a corollary, consecutive sieve sizes for establishing the grading requirements of coarse aggregates can be chosen nearer. This obviously means that there is an urgent necessity for reconsidering this aspect.

The DIN specifications present the combined grading modulus for the all-in-aggregate grading as given in Table 2.5 for the selection of the required

TABLE 2.5

All in Aggregate Grading for Different Codal Provisions

Percentage Passing for Different Maximum Size of Aggregates

| | DIN | | | | | | | | | | | | BS 84 | | | BIS | |
| Sieve Size (mm) | 63 mm | | | 32 mm | | | 16 mm | | | 8 mm | | | 40 mm | 20 mm | 10 mm | 40 mm | 20 mm |
	A63	B63	C63	A32	B32	C32	A16	B16	C16	A8	B8	C8					
80	100															100	
63	100	100	100														
50													100				
40														100		95–100	100
37.5													95–100				
31.5	67	80	90	100	100	100											
20													45–80	95–100		45–75	95–100
16	46	64	80	62	80	89	100	100	100								
14															100		
10															95–100		
8	30	50	70	38	62	77	60	76	88	100	100	100					
4/5/4.75	19	38	59	23	47	65	36	56	74	61	74	85	25–50	35–55	30–65	25–45	30–50
2/2.36	11	30	49	14	37	53	21	42	62	36	57	71			20–50		
1/1.18	6	24	39	8	28	42	12	32	49	21	42	57			15–40		
0.5/0.6	4	15	26	5	18	29	8	20	34	14	26	39	8–30	10–35	10–30	8–30	10–35
0.25/0.3	2	7	14	2	8	15	3	8	18	5	11	21			5–15		
0.15													0–8	0–8	0–8	0–6	0–6

water content for this grading through charts or figures. This value is based on a total of 9 sieves starting from 63 to 0.25 mm. This is similar to the fineness modulus of the fine aggregate as considered by the ASTM C 136 (1995). The ASTM considers only the cumulative percentage of the material retained on the sieves, 0.15, 0.3, 0.6, 1.18, 2.36, and 4.75 mm, for calculating the fineness modulus of sand, which influences the water requirement most because this is the finer fractions in the mix. A detailed discussion on the comparisons between the various factors involved and the differences between the grading distributions have all been presented comprehensively by Ganesh Babu (1999). They also presented a methodology for the calculation of the combined grading modulus with a total of 20 sieves (150–0.15 mm) so that the effect of all the fractions involved in any concrete mix can be taken into consideration. These in a way are basically the methodologies applicable for the calculation of water requirements based on the aggregate grading, combined or otherwise, based on a methodology followed, for use in normally vibrated concretes but yet for achieving the different consistencies ranging from a stiff mix with their slump to a collapse slump concrete with slump values reaching 170 mm.

In the case of SCCs, these are excellent guidelines because the mix is appropriately proportioned with a good combined grading having sufficient fines, and experience shows the most concretes with even slump values of about 25 to 50 mm can be brought to their collapse slump through the addition of a small quantity of superplasticizers. In most cases, minute adjustments in terms of the powder content and probably sand content if required may ensure that these concretes are nearly self-compacting or self-compacting.

Apart from what was the status regarding grading of aggregates in concretes for the normally vibrated concretes hitherto, the new Euro norms (BS EN10260, 2008) recently specified a separate set of recommendations that are almost an offshoot of the earlier recommendations of the British standards mostly. It is only appropriate that these are also examined to have clarity in what the current recommendations are toward the grading of aggregates. Table 2.6 presents the recommendations for graded as well as single-sized coarse aggregates up to 40 maximum sizes for structural concretes. One important observation is that in doing so, the specifications recognize not only the maximum size but also an appropriate minimum for each of these, which were missing hitherto. Similar to these recommendations for coarse aggregates the fine aggregates were also defined in three separate categories of coarse medium and fine grades as can be seen (Table 2.7). The minimum particle size mentioned is 0.5 mm, but for all practical purposes it is to be taken that the 150 μm sieve is essentially the lowest particle size acceptable under the fine aggregate category. This is because the particles below this particular size have to be considered as powders, particularly while recognizing the category of powders in the SCC regime.

TABLE 2.6

Grading Limits for Coarse Aggregates

Sieve Size (mm)	Percentage of Coarse Aggregates (d/D)						
	Graded Aggregates			Single-Sized Aggregates			
	4/40	4/20	4/14	20/40	10/20	6.3/14	4/10
80	100	—	—	100	—	—	—
63	98–100	—	—	98–100	—	—	—
40	90–99	100	—	85–99	100	—	—
31.5	—	98–100	100	—	98–100	100	—
20	25–70	90–99	98–100	0–20	85–99	98–100	100
16	—	—	—	—	—	—	—
14	—	—	90–99	—	—	85–99	98–100
10	—	25–70	—	0–5	0–20	—	85–99
8	—	—	25–70	—	—	—	—
6.3	—	—	—	—	—	0–20	—
4	0–15	0–15	0–15	—	0–5	—	0–20
2.8	—	—	—	—	—	0–5	—
2	0–5	0–5	0–5	—	—	—	0–5

Source: BS EN 12620:2013, Aggregates for concrete, British Standards Institution, 2008.

TABLE 2.7

Grading Limits for Fine Aggregates

Sieve Size (mm)	Percentage Passing (Mass) for Fine Aggregate Size (d/D)				
	0/4 (CP)	0/4 (MP)	0/2 (MP)	0/2 (FP)	0/1 (FP)
8	100	100	—	—	—
6.3	95–100	95–100	—	—	—
4	85–99	85–99	100	100	—
2.8	—	—	95–100	95–100	—
2	—	—	85–99	85–99	100
1	—	—	—	—	85–99
0.5	5–45	30–70	30–70	55–100	55–100

Source: BS EN 12620:2013, Aggregates for concrete, British Standards Institution, 2008.

2.6.1 Reinforcing Fibers

The use of fibers for bridging the microcracks that occur in concrete during setting and hydration processes (the shrinkage of the heat of hydration to be more specific) is a major factor in substantially reducing the strength of concrete produced. A simple yet the most appropriate solution to this particular problem is an appropriate curing regime. Even so, particularly when concretes of very high strength involving higher cementitious material contents are required, avoiding these types of microcracks appears to be a major problem. One of the best solutions for this would be the use of fine, discontinuous fibers. These could be metallic (steel), polymer (polyvinyl, polyacrylic, and also carbon), or even ceramic based (glass, Basalt) depending on the application. One important factor that needs an appropriate understanding is the fact that these could significantly influence the flow characteristics of SCCs, particularly while using stiff and inflexible fibers like steel fibers. The fact that steel fibers are indeed considered as equivalent aggregates (accounting for the special gravity differential between aggregates and steel) in some of the mixture designs that could be seen later obviously justifies their inclusion in the discussions under aggregates. If the steel fibers used have surface indentations or more importantly hooked ends to ensure better transmission of the tensile stresses, their inclusion in a SCC mix could be a much more serious aspect that needs a specific understanding of the parameters that influence the flow characteristics with these fibers. Some of these assumptions are discussed later extensively to understand how the long fibrous inflexible steel fibers could influence the flow of SCC and to suggest methodologies for their inclusion appropriately. More flexible fibers other than steel may not pose that serious a problem but yet at higher percentages could have certain implications that need to be addressed.

2.7 Interactions and Compatibility

SCC being a highly complex composite with so many materials of varying sizes, shapes, and textures involved, the prediction of the strength and performance of the resultant material has always been a matter of serious concern, without stringent control on the constitutes. The large and ever expanding database of research findings with several different local constituents, be it aggregates, the cements produced, or the admixtures utilized, on these concrete composites also causes a certain amount of confusion in understanding the material.

One important parameter that is often emphasized is the compatibility between the cement and the superplasticizing admixture that is proposed to be used. A few simple initial tests through cement pastes and mortars can easily indicate the suitability (sometimes even quantitatively) of a particular admixture among the several alternatives. It is not yet fully understood though there are arguments about the chemical kinetics that influences these characteristics. Apart from this the need for a larger quantity in the presence of larger unburnt carbon particularly in fly ash is always observed.

There are several other interactions that can only be discussed at a later stage while looking at the parameter addressed like the alkali silica reactivity, efflorescence characteristics, leachability and chemical binding character-istics of heavy metals, and many more. These studies could lead to more specific domains such as nanotechnology in concrete, which have been coming into focus in recent years. A review of the present trends in these was reported by Sanchez (2010), discussing both the physical and chemi-cal characteristics of superfine or nanoengineered admixtures like precipi-tated silica and nanosilica in concrete beyond the realms of silica fume that is being discussed here. In a way even silica fume concretes are almost at the threshold of nano cementitious composites if only the material is properly chosen and the remaining cementitious materials are appropriately regraded to accommodate this material so that the characteristics that are exhibited fully qualify for it to be addressed as nano cementitious composites. One of the most important aspects in this regard is the particle size distribution typically determined through laser diffraction methods. A broad discussion on the characteristics of particle size distribution that is also quite useful in understanding the behavior of fly ash and silica fume concretes is presented by Mehta and Monteiro (2006). Particle packing methodologies and their use in ensuring the right composition of cementitious constituents are discussed later.

References

ACI 363R-84, State of the art report on high strength concrete, *ACI Journal*, 81(4), 1984, 363–410.

ACI Committee 226, Use of fly ash in concrete, *ACI Material Journal*, 84(5), 1987, 381–409.

ASTM C 136-95, *Standard Test Method for Sieve Analysis of Fine and Coarse Aggregates*, Annual Book of ASTM Standards, West Conshohocken, PA, 1995.

ASTM C 150, *Specification for Portland Cement*, Annual Book of ASTM Standards, West Conshohocken, PA, 2009.

ASTM C 33, *Standard Specification for Concrete Aggregates*, Annual Book of ASTM Standards, 1986.

ASTM C 618, *Specification for Coal Fly Ash and Raw or Calcined Natural Pozzolan for Use as a Mineral Admixture in Portland Cement Concrete*, Annual Book of ASTM Standards, West Conshohocken, PA, 1994, pp. 304–306.

ASTM C989, *Standard Specification for Ground Granulated Blast-Furnace Slag for Use in Concrete and Mortars*, Annual Book of ASTM Standards, West Conshohocken, PA, 1994, pp. 494–498.

BS 882, Aggregate from natural source for concrete, British Standard Institute, London, U.K., 1973.

BS EN 12620:2013, Aggregates for concrete, British Standards Institution, London, U.K., 2008.

CEB-FIP Model code, Design code 1994, Thomas Telford, London, U.K., 1994.

Dhir, R.K., Pulverised-fuel ash, in Swamy, R.N. (Ed.), *Concrete Technology and Design*, Vol. 3, Cement Replacement Materials, Surrey University Press, London, U.K., 1986, pp. 197–255.

DIN 1045, Beton und Stahlbeton, Beton Verlag GMBH, Koln, Germany, 1988.

EN 197-1, Cement—Part 1: Composition, specifications and conformity criteria for common cements, Brussels, Belgium, 2000, pp. 1–29.

Ganesh Babu, K., Kiran Kumar, A., and Pandu Ranga Prasad, R., Effect of grading on strength and performance of concrete, in *Fifth NCB International Conference on Concrete and Concrete Technology for Developing Countries*, New Delhi, India, November 1999.

Ganesh Babu, K. and Raja, G.L.V., Concrete mix specifications for marine environment: An overview, in *Third National Conference on Dock and Harbour Engineering*, K.R.E.C, Surathkal, India, 1989, pp. 873–877.

Ganesh Babu, K., Raja, G.L.V., and Rao, P.S., Concrete mix design: An appraisal of the current codal provisions, *The Indian Concrete Journal*, 66(2), 1992, 87–95, 102.

IS 383, *Specification for Coarse and Fine Aggregates from Natural Sources for Concrete*, Bureau of Indian Standards, New Delhi, India, 1979.

IS 456, *Code of Practice for Plain and Reinforced Concrete*, Bureau of Indian Standards, New Delhi, India, 1979.

Kosmatka, S.H. Kerkhoff, B., and Panarese, W.C., *Design and Control of Concrete Mixtures*, 14th edn., Portland Cement Association, Skokie, IL, 2003.

Mehta, P.K., Standard specifications for mineral admixtures: An over view, in *Proceedings Second International Conference on Fly Ash, Silica Fume, Slag and Natural Pozzolans in Concrete*, Madrid, Spain, 1, ACI SP-91, 1986, pp. 639–658.

Mehta, P.K., Pozzolanic and cementitious by-products in concrete—Another look, in *Third International Conference on Fly Ash, Silica Fume, Slag and Natural Pozzolans in Concrete*, Trondheim, Norway, ACI, SP-114, 1989, pp. 1–44.

Mehta, P.K., *Concrete: Microstructure, Properties and Materials*, First Indian Edition, Indian Concrete Institute, Chennai, India, June 1997.

Mehta, P.K. and Monteiro, P.J., *Concrete: Microstructure, Properties and Materials*. McGraw-Hill, New York, 2006.

Neville, A.M., *Hardened Concrete, Physical and Mechanical Aspects*, ACI Monograph No. 6, American Concrete Institute, Detroit, MI, 1971.

Sanchez, F. and Sobolev, K., Nanotechnology in concrete—A review, *Construction and Building Materials*, 24, 2010, 2060–2071.

Sellevold, E.J. and Nilsen, T., Condensed silica fume in concrete: A world review, in Malhotra, V.M. (Ed.), *Supplementary Cementing Materials for Concrete*, CANMET, Ottawa, Ontario, Canada, SP 86-8E, 1987, pp. 167–246.

Wesche, K. (Ed.), Fly ash in concrete, state of the art report, RILEM TC 67-FAB, 1990.

3

Insights into Standards and Specifications

3.1 Standardization Principles

The commercial trade practices of today's highly industrialized world need specific understanding and authenticity of any particular transaction to ensure a level playing field. The primary objective of standardization, in simple terms, is to provide a fundamental document containing the generic rules or guidelines to achieve a broad order in any given context. The International Organization for Standardization (ISO), a nongovernmental organization based in Geneva, Switzerland, looks into the development of the technical standards for products as well as processes for trade the world over. These can be broadly divided into two groups: product standards and process or service standards overall, which have local, regional, or global applicability. They can also be either voluntary or mandatory depending upon the regulatory policies of national or international bodies to which they are signatories and can be legally binding. In the context of construction and building materials, particularly concrete, some of the prominent standards that one looks at could be the Euro norms (BS, EN), the American standards (ASTM) apart from the various national standards (past and present) for an understanding of the requirements of the material.

The several possible advantages like not having to vibrate, ease of placement through pumping, and better surface finish, and if appropriately constituted a higher performance have drawn several professionals to adopt self-compacting concrete (SCCs) more extensively. Some of them have even modulated the concrete to be suitable for structural requirements, and others modified the structural configurations so that they could use the advantages of SCCs. The fact that the constituent materials of SCC are essentially the same as the ones used for making normally vibrated concretes and even the equipment and machinery in the production, transport, placing, and finishing are the same, there is very little need to look very far for the formulation of specific standards for that part. In fact, as considered by many, SCCs are only logical extensions of the collapse slump concretes that exist even today. Though the characteristics of SCC composites are not very different, the fine balance that exists between self-consolidating and potential segregation possibilities of these types of concretes makes it essential to have a relook at this particular scenario and to ensure that there are enough methods that can characterize the specific aspects that make these concretes self-compacting or self-consolidating. It is this aspect of segregation that actually mandates the need for a specific understanding of the viscosity, filling, and passing resistance without segregation. While it is possible to characterize the SCCs adequately through the simplest of tests that can be done like slump flow, which typifies its fluidity and passing ability, a trained eye can yet have an idea of the other characteristics of the mix from experience. However, to be able to quantify and prescribe limits on each of the parameters that are normally looked for in a SCC, there is a need for understanding the specific test methods that quantify these parameters accurately and with some ease and comfort.

3.2 Fundamental Characterization and Classification

There have been several methods for the assessment of the workability or the consistency of concrete over the last nearly 100 years. There were also several attempts to categorize them into different segments to understand their relevance in practice. Wesche (1975) presented a simple and yet a comprehensive picture of several of these concepts, which in a way shows that many of the present-day concepts proposed for even SCCs actually have their origins from a long time back. In the words of Wesche, "from the rheological point of view, for each testing method another type of loading is applied and thus another property of fresh concrete is measured (workability, compactability, transportability, etc.)". The characteristic values for the different consistency regimes were also presented by him. In line with what was discussed by him, a broad outline of the test methods and the consistencies that are generally relevant for these test methods to be adopted (as slump and Vebe time) are presented in Table 3.1. Table 3.1 not only presents a comprehensive picture of these methods of testing, but also explains the relevance of each of these for the different concrete consistency regimes as given by ACI 309 (1996). It also discusses the relevance of each of these methods along with their applicability and limitations. In fact, most of the workability and consistency assessment methods actually tried to simulate broadly the concepts and methodologies adopted in the different construction practices. It is the versatility of concrete emanating from its amenability to suit such diverse needs—starting from an almost semidry mass required for applications like roller-compacted concrete to the other extreme of flowing, or self-levelling status and even for applications underwater—that actually makes it the choicest material for construction. Naturally, the methods of compaction followed for each of these cases have to be changed from the simple rodding or

TABLE 3.1

Outline of the Test Methods for Different Consistency Ranges

Property	Test Method	Methodology	Applicable Consistency Range	
			Slump (mm)	Vebe (s)
Subsidence	Slump	Subsidence due to self-is weight	25–190	5–0
	German box	Subsidence after vibration	0–125	18–3
	Vebe	Time for remolding through vibration	0–125	30–3
	Remolding	Time for remolding through vibration and pressure	0–125	30–3
Density	Density ratio	Density at standard compaction	25–125	5–3
Spread	Slump cone	Free flow diameter of slump cone	125–250	5–0
	Inverted slump cone	Flow diameter of inverted slump cone	125–250	5–0
	Flow table (mm)	Flow diameter after a specific number of drops	25–125	5–3
	LCL flow time	Flow through a channel	125–250	5–0
	L-box	Flow through channel with reinf. obstructions	125–250	5–0
	U-box	Flow through a tube	125–250	5–0
Flow	Marsh cone	Time of flow through an orifice (for mortar)	NA	NA
	Inverted slump cone	Time of flow (for fibrous concretes)	75–190	3–0
	V-funnel	Time of flow	125–200	0–0
Pressure	Plate vibrator	Direct vibratory compaction	NA	32–10
Penetration	Kelly Ball	Penetration of the ball due to self-weight	25–125	5–0
	Standard pin	Penetration of a pin or a due to self-weight	25–125	5–0
	Standard pin	Penetration under standard pressure	25–125	5–0
	Standard pin	Penetration with number of drops of standard weight	25–125	5–0
Others	K-slump	As a material flow through openings	125–200	3–0

Sources: After Wesche, K., *Mater. Struct.*, 8(45), 1975; ACI 309R, *Guide for Consolidation of Concrete*, ACI Manual of Concrete Practice, American Concrete Institute, Farmington Hills, MI, 1996, pp. 1–39.

Vebe time/slump predefined for the consistency ranges by ACI 309 are the following: Extremely dry—NA/32–18; very stiff—NA/18–10; stiff—0–25/10–5; stiff plastic—25–75/5–3; plastic—75–125/3–0; highly plastic—125–190/NA; flowing—190+/NA.

needle vibrators to methods like application of pressure and vibro-pressing, high-frequency vibrators of needle, shutter and float types, vacuum dewatering, and vibro-vacuuming to name just a few. In a way the consistency measurement methods appear to have followed the principles involved in each of these systems through slump, compaction factor, Vebe time, remolding effort, flow table, penetration, and so on, as can be seen in Table 3.1. A closer look at this could show that the methods depicted and several of the methods that are being followed even for SCCs can be broadly classified into the following parameters, just to name a few:

- Height of subsidence due to self-weight
- Density at standard compaction
- Spread due to subsidence (flow) without or with jolting (number of drops)
- Subsidence due to vibratory compaction (varying frequencies)
- Subsidence due to applied pressure or pressure and vibration
- Penetration due to applied pressure or pressure and vibration

3.3 Methods of Consistency Measurement

One of the most important aspects of any program trying to establish authentic characteristics for a given entity is to ensure that it is representative of the whole and not of a sample and that it is repeatable within an accepted variation probability that is predefined. It is for this reason the nations while adopting the standard practices either send it for open wetting by competent organizations or put in place an extensive evaluation program through inter-laboratory or inter-institutional participation. It is obvious that these programs are openly advertised for an effective participation and are mandated to ensure competence in terms of both knowledge and practice. The fact that quality assurance is of paramount importance in ensuring the right performance of constructed facilities, the Euro norms (EN 206-1, 2000) delineate the specifications for the quality control of concrete. In terms of the standardization practices followed for sampling of fresh concrete, while the American standards (ASTM C172, 2004) prescribe specific regulations for sampling, there are also others from the different national bodies like the Euro standards (EN 12350-1, 2000).

It may be added that the international standards for each of these practices originate from the engineer at the site or maybe the academician in a lab giving a solution to problems that are faced by the industry. Many a time it becomes necessary to appropriately articulate the specific aspect that is to be investigated from the application perspective to find the composition of test methodologies including some that may need to be put together for the specific purpose on hand to ensure a proper appraisal of the aspects that are being investigated. One can easily talk about several such situations where the existing practices and methodologies do not lend themselves to be directly applied. It can be seen that most of what is talked about as consistency measurement is for concretes as we know and use them in general construction practice, and the moment we extend these philosophies to situations needing to understand the likes of foamed concretes that tend to settle or segregate even sand, lightweight aggregate concretes that tend to float the aggregate to the top, or heavy weight concretes that are used in nuclear containment applications, the present practices would simply not be adequate or even applicable.

For the samples to be representative of the materials, the batch, and even delay times, they are collected just before the concrete is placed at the site, in suitable molds required for testing. The number and collection methodology will depend on the batch size, and care is taken to protect the batches from open drying and not to disturb them in transit if any. It is ensured that the curing regimes are in consonance with the practices adopted for the construction so that they are an appropriate representation of the site practice and for being used in the structure. Occasionally, particularly when there is a large discrepancy between what was expected and what was observed, nondestructive assessments are undertaken. These should not be invasive to the structural component and should cause the least damage to the structural strength of the member in case there is a limited invasive practice to be adopted. Some of these tools are probably very useful, but have not been adequately investigated as yet to be in a position to enthuse confidence in their results for SCCs. Some of these aspects will be later discussed for a better understanding of the present state of the art in these practices.

The guidelines for self-compacting concrete, or self-consolidating concrete, require that it is a highly workable concrete that can flow through the highly congested reinforcements of complex structural elements under its own weight and adequately fill voids without segregation and/or bleeding under its own weight without any need for vibration to consolidate. It is probably most appropriate to revisit the definition proposed by ACI 237 (2007) that the self-consolidating concrete in a highly flowable, non-segregating concrete that can spread into place, fill the formwork and encapsulate the reinforcement without any mechanical compaction effort. This consistency is characterized by the following properties and the concrete mixture is considered self-consolidating if all the requirements below are satisfied.

- *Filling ability*: The ability to flow under its own weight (and also under pumping pressure) and fill completely all spaces of even an intricate formwork with congested reinforcement. The fact that an SCC should encapsulate the reinforcement completely is probably one of the most significant filling ability characteristics of an SCC that needs to be looked for.

- *Flowability*: Flow over significant distances and yet exhibit the filling ability above without the problems of segregation. The maximum flow distances also have a relation with the depth and maybe even the thickness of the member and should be considered in the design of the SCC mixture.
- *Passing ability*: The ability to flow through openings between reinforcement approaching the maximum size of the coarse aggregate without segregation and blocking. The fact that the maximum size of coarse aggregate governs the spacing of reinforcement should also be recognized and enforced.
- *Stability*: The ability to remain homogeneous without segregation during transport, placement, and initial hardening. The capacity to hold the water in its body without bleeding particularly during the initial hardening, which could significantly alter the strength of the top layer, is also a part of the stability. A very thin layer of water appearing on top could, however, be beneficial in certain dry conditions to compensate for the loss of hydration water in the top concrete.

It is also obvious that there have been many attempts to quantitatively assess the characteristics of SCCs by different researchers using several principles to establish the flowability, passing ability, and segregation resistance. A few reports are available presenting the complete scenario to have a comprehensive overview of all the consistency measurement methodologies used, particularly from the national and international bodies (EFNARC, 2002; ACI 237, 2007). Also, there have been a few research efforts investigating the characteristics of SCCs through some of the methodologies proposed in the literature to understand their efficacy and acceptability (Pade, 2005). Having looked at the fundamental properties that are used to characterize concrete composites in general earlier, it is only appropriate to discuss a few of the major test methods that go a long way in understanding SCCs. These presently include the test methods like the slump flow and T_{500} tests, J-ring, V-funnel, L-box, U-box, fill box, screen stability, and Orimet tests, and so on, which have been recommended for such assessments by the national recommendations discussed later on. It is also to be noted that most often it is required to perform not just one but a small group of tests to ensure an appropriate, quantitative, and clear understanding of the green state performance of SCCs. Table 3.2 presents the tests along with the range of the results for an SCC and also an outline of some of the effective combinations that may comprehensively address the needs of the industry for a few specific requirements. The details of the tests themselves are also presented hereunder.

TABLE 3.2

Characteristics and Test Methods for Performance

Characteristic	Method of Test	Preferable Range	Alternative Tests	Complementary Tests
Filling ability	Slump flow spread	650–800 mm	J-ring flow	L-box/stability
	Kajima fill box	90%–100%	Mesh box	—
Flowability/ viscosity	Slump flow time T_{500}	2–5 s	J-ring T_{500} time	L-box/stability
	V-funnel flow time	8–12 s	O-funnel	L-box/stability
	O-funnel (<12 mm)	5–15 s	Orimet	—
	Orimet flow time	0–5 s	O-funnel	J-ring/stability
Passing ability	L-box height ratio	0.8–1.0	J-ring	Stability
	U-box height diff.	30 mm max.	J-ring	Stability
	J-ring stoppage	0–10 mm	Kajima test	Stability
	Kajima fill box	90%–100%	J-ring	—
Segregation resistance	Penetration	0–2 mm	Stability	Slump flow/J-ring
	Screen stability	0%–15%	Penetration	Slump flow/J-ring
	Settlement column	0%–15%	Penetration	Slump flow/J-ring
	V-funnel – $T_{5\,min}$.	+3 s	Penetration	—
	K-slump	—	Penetration	—
	Bleeding	<2.5%	—	—

Sources: After EFNARC, *Specification and Guidelines for Self-Compacting Concrete*, EFNARC, Surrey, U.K., 2002, p. 32; ASTM C1611, *Standard Test Method for Slump Flow of Self-Consolidating Concrete*, Annual Book of ASTM Standards, ASTM, West Conshohocken, PA, 2009.

3.3.1 Slump Flow and T_{500} Tests

This test in a way is a logical extension to the slump test that is conventionally adopted in many countries for understanding the consistency of traditionally vibrated concretes. The test uses the conventional Abrams slump cone (a truncated conical mold of 300 mm height with base and top diameters of 200 and 100 mm) filled with concrete and measures in the horizontal spread of concrete on lifting instead of the vertical slump. The concept of using unconfined spread for assessing consistency of concrete using a much shorter conical mold for highly fluid mixes is already well known as the flow table test of yesteryear proposed by Graf (Wesche, 1975). In fact the flow table test existed both in the American (ASTM C124, 1971) as well as the British standards (BS 1881:105:84, 1984). The earlier American flow table consisted of a circular brass plate of 140 kg mounted on a camshaft arrangement at a height of 450 mm that was firmly bolted to a concrete base. The mold itself was the frustum of a cone made of smooth metal casting with base and top diameters of 250 and 170 mm and a height of 120 mm. The mold filled with two layers of concrete was compacted with 25 strokes of a round metal rod of 600 mm length. The camshaft allows a drop of 12.5 mm and the diameter of concrete spread after 15 drops was taken as a measure of the concrete flow consistency. This test is a very good indicator to check if the concrete is prone to segregation. Laboratory measurements at intervals of every five drops could also indicate the floor parameters of concretes. In an effort to quantify the flow characteristics of superplasticized concretes, the British standards advocated a flow table of 700 mm weighing 16 kg consisting of a wooden board covered by a steel plate hinges on one side and with a stop on the opposite side allowing for a drop of 40 mm. A mold of 200 mm height with a bottom diameter of 200 mm and top diameter of 130 mm is filled with concrete and the tabletop was lifted and allowed to drop freely 15 times in about 60 seconds. The flow or spread is an estimate of the consistency of superplasticized concretes which are actually thixotropic and need only very light vibration, generally using a shutter vibrator for deeper sections. In a way this is a near SCC and is still probably one of the best technologies for use in prefabrication wherein the concerns of segregation even with light tamping are remote. It is probably only appropriate to add at this stage that this concrete is the one which probably will need almost negligible additional quantity of either superplasticizer or water itself (sometimes both) that will render itself to be a total collapse slump, same as the SCC presently being discussed.

The slump flow test also uses the same Abrams slump cone filled with concrete placed on a flat, unyielding, and nonabsorbent surface with the 200 and 500 mm circles clearly marked (Figure 3.1). This should be filled without any compaction or even tamping to represent the field situation striking off the top to be level using a trowel. The highly workable SCC will exhibit a collapse slump and the resulting average diameter of the horizontal spread due to the unconfined flow on lifting the mold measured in two perpendicular

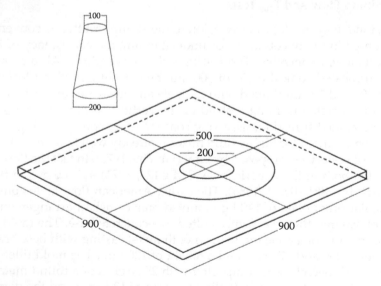

FIGURE 3.1
Slump flow test setup.

directions is the slump flow. It is necessary to ensure that the slump cone is held down during the filling of concrete to not allow the material from escaping from the bottom. The use of a weight ring (>9 kg) of 40 mm height and 225 mm diameter that sits on the top fitting flush with the top opening is advised for this purpose. Concretes with spread diameters of greater than 500 mm may be considered on self-compacting as can be seen from the various national recommendations and concretes with spread values of over 850 mm could have serious problems of stability leading to segregation. The time taken for the concrete to reach the marked 500 mm spread circle is T_{500}, which is an indicator of the viscosity of the SCC. As the cohesion or the thixotropy of the SCC increases, the time that it takes to reach the 500 mm circle is higher. The SCCs used in practice are generally characterized by T_{500} values ranging from 2 to 10 seconds. Because of its simplicity this is the most commonly used test and is quite useful to assess the consistency of ready mix concrete from load to load. A trained eye can also detect signs of segregation and bleeding through the concentration of the aggregates toward the center and the appearance of a halo of cement paste and/or bleed water at the outer periphery of the spread circle.

It can also be seen that some of the investigators have done a similar test with an inverted slump cone resulting in a minor variation in the slump flow values. This method of using an inverted slump cone was originally adopted to assess the workability of fibrous concretes and to see if the long, hard, and unyielding steel fibers in combination with the larger aggregates exhibited a blocking effect due to the constriction at the mouth of the cone. It is obvious that even in the absence of fibers the constriction at the mouth of the call could hamper the flow due to the arching effect of the aggregate matrix in the concrete. A limitation of the method is the fact that by lifting an inverted cone the concrete is allowed to fall freely from a height and the height of fall could influence the results substantially and could also lead to segregation effects.

In trying to alleviate this, some of these limitations specified that the mold be lifted by a standard distance of about 100 mm in some of the specifications using the inverted slump cone methodology. There have also been efforts to even standardize the test (DAfStb, 2012), the one additional advantage being the possibility of recognizing segregation and maybe even bleeding due to the fall. In fact it can be seen that the proposed method uses a cone of 390 mm height with top and bottom diameters of 194 and 63 mm which in a way pushes it from the Abrams cone to be nearer the marsh cone and V-funnel tests, obviously relevant for an understanding of all the characteristics of SCCs, filling, passing, and segregation resistances through the spread, spread rate, the time taken for emptying, and the observed segregation if any due to the fall (Figure 3.2). The concretes resulting in a flow diameter of about 640–750 mm and a slow time of about 12–23 seconds by this method were found to satisfy the consistency characteristics required for SCCs.

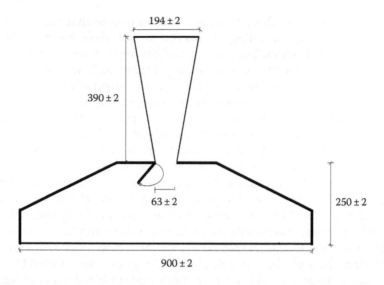

FIGURE 3.2
Proposed alternative to flow cones developed for SCCs. (From DAfStb, DAfStb-Richtlinie Selbstverdichtender Beton (SVB-Richtline)—Teile 1, 2 und 3, Entwurf, DAfStb, September 2012, pp. 1–18.)

3.3.2 J-Ring Test

The J-ring test is a logical extension to the slump flow test as it addresses the one important characteristic of the passing ability of the SCC that was clearly possible through the slump flow test itself. As the name indicates, it consists of a steel ring into which reinforcing bars of a specific diameter are threaded to form a circular comb representing the well-spaced reinforcement bars in the structural member. The diameter of the ring (with the comb of reinforcing bars of the required diameter and 100 mm height) was chosen to be 300 mm to accommodate the slump cone in the middle of it for the slump flow test without disturbing the ring. There have been variations in the diameter of the reinforcing bar chosen, ranging from 10 to 18 mm approximately, and the numbers also varying accordingly. As a broad principle the spacing of bars was retained as three times the maximum size of the aggregate, from passing ability considerations. It is possible to choose a much closer spacing to represent a specific field situation to have a more rigorous estimate of the passing ability of the material. Reinhardt (2001) suggested the number of bars in accordance with the maximum sizes of the aggregate proposed in the German code as 22 bars for 8 mm, 16 bars for 16 mm, and 10 bars for 32 mm spaced uniformly around the circumference of the ring. In contrast to these recommendations the precast/prestressed concrete Institute interim SCC guidelines suggest a uniform 16 mm reinforcing bar diameter with a number of bars varying with the maximum nominal size of aggregate from 31 rebars at a center to center spacing of 30 mm for the 8 mm aggregate, 27 rebars at a center to center spacing of 35 mm for the 10 mm aggregate, and 17 rebars at a center to center spacing of 55 mm for the 20 mm aggregate. It was also suggested that the number of rebars will have to be re-evaluated for other reinforcement dimensions.

Apart from using the J-ring in contention with the slump cone to determine the presently obstructed slump flow, the ring can also be used with other flow spread test methods like Orimet or even the inverted slump cone if need be. The test as in the case of the slump flow test is done on a flat unyielding nonabsorbent surface, a flat plate measuring not less than 900 mm^2, with circles marked from the center at 200, 300, and 500 mm for ease in locating the slump cone and grading and also to take other required measurements during the test.

The slump cone is placed in the middle of the ring locating it appropriately at the required markings on the flat plate. The slump cone as in the slump flow test is filled to the top without any type of compaction effort or even tamping and the concrete at the top is struck off with a trowel. Care should be taken to ensure that the slump cone is neither disturbed from its location nor lifts up to avoid leaking from the bottom. The cone is carefully lifted in one smooth motion without twisting so that the concrete is allowed to flow freely in all directions. It can be seen that the obstruction due to the reinforc-ing bars from the J-ring may result in a level difference between the concrete inside the ring and that which has passed through the reinforcement cage. After the flow process is complete this level difference between the concrete outside and inside the ring is measured, which indicates the passing ability of the particular concrete being investigated (Figure 3.3). The time taken for the advancing spread boundary to reach the 500 mm mark can also be a mea-sure of the viscosity of the SCC as in the case of the slump flow test discussed earlier. Any accumulation of coarse aggregate or larger fractions of coarse aggregate within the blocking ring should be specifically reported. All frac-tions of the coarse aggregate in representative proportions must have passed through the J-ring to present a uniform distribution. There should also not be any obviously detectable separation of water or paste at the edge of the concrete as already discussed in the slump flow test.

An important fact that should be remembered is that the results of the J-ring test are intimately linked to the characteristics of the ring itself, the number and diameter of bars, the spacing, and the maximum size of the aggregate that is being used for the construction. In general, most investigators have used only smooth bars to represent the reinforcement and it is the spacing, while it is certainly in order to use the deformed bars that are in practice as reinforcement. The recommendations presented in EFNARC (2002) sug-gested that the J-ring be fixed with 10 mm diameter smooth bars of 100 mm length at a spacing of 48 mm. But the same EFNARC recommendations (2005) though recognized the test later they did not explicitly present any specifica-tions as the different SCC classes were all defined based on the slump flow, T_{500}, V-funnel, L-box, and sieve segregation aspects primarily. The Norwegian

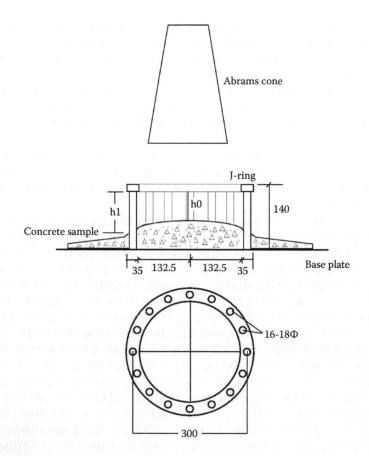

FIGURE 3.3
Typical arrangement of the J-ring test setup.

recommendations (Pade, 2005) have chosen to adopt 16 numbers of 18 mm plain steel bars around the 300 mm perimeter of the J-ring. The ASTM C1621 (2009) recommended 16 smooth bars of 16 mm diameter equally spaced for the J-ring.

3.3.3 V-Funnel Test

The V-funnel test setup primarily consists of a V-shaped flat funnel of a rectangular cross-section with the dimensions presented in Figure 3.4. At the very outset one should understand that the dimensions of the opening are essentially related to the aggregate size and for the 20 mm aggregate in the middle opening size of 65 mm is prescribed. This is the best and essentially is also very similar to the Marsh cone test used in the case of mortars. The test was developed in the University of Tokyo by Ozawa et al. (1994) who also used an O-funnel with a circular cross-section in their initial studies on SCCs.

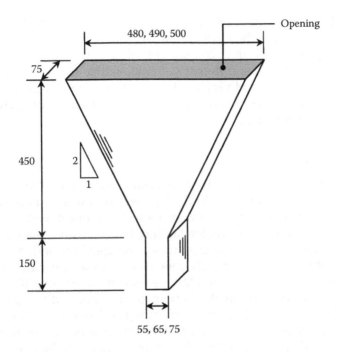

FIGURE 3.4
General V-funnel test apparatus.

The fact that the circular funnel with its three-dimensional flow could easily lead to arching of the coarse aggregate and block the flow, while the rectangular cross-section will only result in a two-dimensional flow representing a much nearer situation to the flow of concrete in structural members needs to be appreciated. This is the reason why the funnels with rectangle cross-section are indeed to be preferred for this test. The funnel is filled fully with the concrete without any tamping as in the case of slump flow, and the time taken for the entire concrete to empty is measured by opening the bottom completely almost immediately after the funnel is filled. The test while assessing the viscosity of the mixture in this way can also be used to understand the possibility of segregation. To understand the effects of segregation it is recommended that the funnel is refilled with concrete and the flow time is measured after a 5-minute waiting period to see if there is a significant variation in the flow times indicating segregation. An exit opening at three times the maximum aggregate size is recommended as is done for the spacing of bars in the J-ring. In an effort to study the influence of a coarse aggregate, Okamura (2003) observed that the flowability is not affected by the shape of the coarse aggregate as long as the ratio of the coarse aggregate content to its solid volume in concrete is the same. It was also found by them (using different funnel openings of 55, 65, and 75 mm as shown in Figure 3.4) that the flow time of concrete through the funnel with an outlet width of 55 mm was largely influenced by the grading of the coarse aggregate. The V-funnel with a 65 × 75 mm outlet is the most commonly adopted for concretes of maximum aggregate sizes of up to 20 mm. This obviously brings us to the situation that the results of the different self-compacting tests in devices having different opening sizes and varying the maximum size of aggregates cannot be compared without a serious consideration for the effects of aggregate size in relation to the opening size.

3.3.4 U-Box Test

This test is the one that was used to define the different ranks of SCCs (Ranks 1–3) by the Japanese code (JSCE, 1999) after it was used initially by Okamura (1997). It was reported that the test was developed by the Taisei group (Hayakawa, 1993) for understanding the behavior of super workable concretes and is still one of the most widely accepted for understanding the passing ability of SCCs. It actually consists of a rectangular tube separated by a central divider resulting in a U-type tube of dimensions 200 × 140 mm having a similar opening at the bottom of the tube with a gate to close (Figure 3.5). At the opening a grill consisting of 3 bars of 13 mm diameter at 50 mm center to center is fixed to form an obstruction for the concrete filled on one side passing to the other empty side when the gate is opened. The bottom of the box can be either semicircular or flat. The large difference between the two sites of the tube is a measure of the filling and passing ability of the SCC.

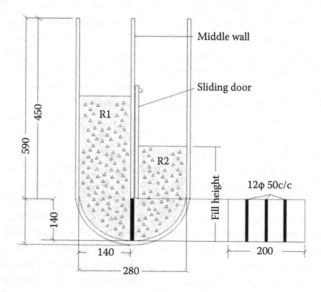

FIGURE 3.5
U-box test arrangement.

3.3.5 L-Box Test

The L-box test appears to be a modification of the old LCL flow test used to understand the plastic viscosity of concrete through a rectangular channel of 150 × 150 mm, which is 600 mm long. The time required for the flow of uncompacted concrete at one end to the other end due to vibration is a parameter that indicates the plastic viscosity of the concrete (Wasche, 1995). The present L-box (EFNARC, 2002) adopted from an original Japanese test is in fact a channel of 150 × 200 mm, 800 mm long containing a vertical column of 100 × 200 mm that is 600 mm high as shown in Figure 3.6. The column opening to the channel is closed initially and after filling the concrete the gate is opened to allow the concrete to flow through the channel. As the primary aim of this particular test is to characterize the passing ability of concrete, two or three smooth reinforcing bars of 12 mm diameter are placed vertically obstructing the flow resulting in gap of 59 or 41 mm between them. These reinforcing bars are located at a distance of 70 mm from the gate opening of the concrete column. Considering the fact that the spacing of reinforcement is generally fixed in relation to the maximum size of the aggregate being used (a spacing 3 times maximum aggregate size) the applicable maximum aggregate sizes for the concretes to be investigated will be 20 and 12 mm approximately. After the concrete flow attains equilibrium, the height difference between the opening end and the opposite end expressed as a ratio is an indicator of the passing ability characteristics of the SCC being investigated. Similar to that in the case of J-ring, the time for the concrete front to reach

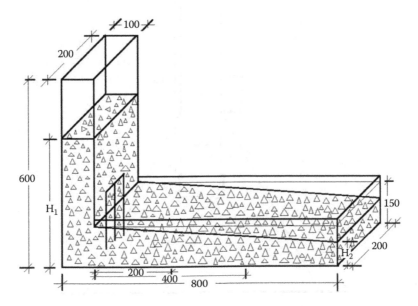

FIGURE 3.6
Outline of the L-box test apparatus.

the distance of 200 and 400 mm from the opening is also recorded to have a broad idea of the viscosity characteristics of the SCC under investigation. The time values recorded for the concrete front to reach 200 and 400 mm have to be judged with caution, looking for any obvious blocks restraining the larger-size aggregates at the reinforcement cage and thus not allowing a free flow of the material in question.

3.3.6 Orimet Test

The Orimet proposed by Bartos (1998) and later investigated by Sonebi (2002) was indeed a test that was specifically developed to understand the flowability of the SCC in a setting more akin to the site conditions of the practical application. The idea is that instead of measuring the flow characteristics of the concrete that is perfectly at rest in the slum cone through the slump flow test, the Orimet test actually measures the flowability in its dynamic state as it comes out of the pump at the site. The test equipment consists of a 120 mm diameter tube of 600 mm length, with the end reducing to an 80 mm orifice in the next 60 mm (Figure 3.7). The complete table arrangement is set on a stand to allow the concrete to flow and fall freely

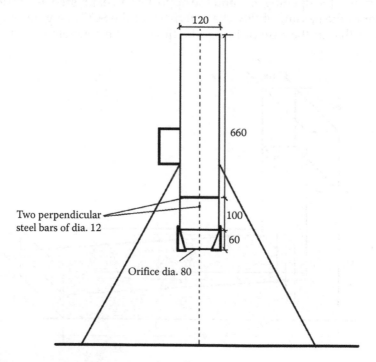

FIGURE 3.7
Orimet test apparatus. *Note*: All dimensions in mm.

on the bottom plate. The flow time of concrete through the orifice characterizes the flowability of concrete, particularly coupled with the difficulty
of overcoming the resistance due to the constriction introduced at the end.
Two particular steel bars of 12 mm diameter were also introduced later at
a point 500 mm from top to be able to evaluate the passing ability of SCCs.
As already discussed, a J-ring can also be introduced at the bottom of the
Orimet to evaluate the passing ability of the concrete in such a state through
the spirit diameter measured.

3.3.7 Mesh Box Test

This test was originally suggested for evaluating the passing ability of
concrete vertically through reinforcement (Ozawa, 1990). The test apparatus consists of a simple box fixed with a reinforcement mesh of 50 mm c/c
both ways filled with 30 liters of concrete. A minor pressure of 0.1 kg/cm^2
is applied before lifting the box. The volume of concrete flowing through
the reinforcement mesh is an index of the passing ability of the SCC being
investigated.

3.3.8 Fill Box (Kajima Box) Test

This test was used during the original development of SCC by Ozawa et al.
(1990) to understand the filling and self-levelling characteristics of an SCC
within the narrow gaps of a congested reinforcement location. The box
consisted of a tapering section with the 18 mm diameter reinforcing rods
arranged in decreasing levels of 4 rows to one (starting with $5 \times 4, 4 \times 3, 4 \times 2$
and 3 totaling 43 numbers at 50 mm spacing). One side of the box was fitted
with a transparent plate so that the flow of concrete dropping from a height
of 400 mm from bottom can be observed.

The test that was later modified to measure the filling ability of SCC of
20 mm maximum size of the aggregate is also known as the "Kajima test"
(EFNARC, 2002). The test is performed by filling one end of a rectangular container (300×500 mm and of height 300 mm containing 35 obstacles of 20 mm
PVC tubes at 50 mm c/c in in 5 rows from the other end) through a 100 mm
funnel with a stem of 500 mm length and 100 mm diameter (Figure 3.8). The
container is filled with fresh concrete through the funnel till the concrete just
covers the first top tube across in front. The ratio of the difference in heights
on the opposite side expressed as a percentage characterizes the filling ability
of the SCC. A higher percentage will obviously mean a better SCC. Though
the test is indeed a simulation of the actual conditions that could prevail at a
highly congested location, the possibility of arriving at an exact quantitative
understanding of any specific characteristic of the SCC is not possible. It is,
however, possible to say that the ratio of the filled area to the total area to the
top can be considered the filling ability.

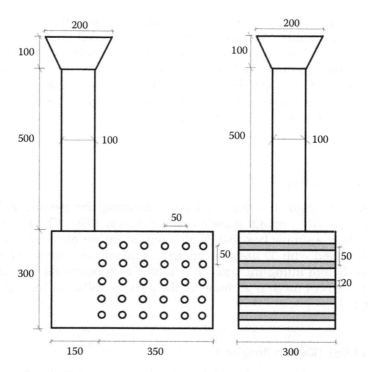

FIGURE 3.8
Typical fill box test (Kajima test).

3.3.9 Screen Stability Test

This is a very simple stability test for assessing the segregation resistance of SCCs. It is also known as the GTM screen stability test named after the French contractor who developed the test. Its simplicity lies in the fact that a 10-liter sample collected in a container is allowed to stand for about 15 minutes without disturbance to allow any spontaneous segregation. The top half of the sample is then poured into a 5 mm sieve of 350 mm diameter. The mortar that passed through the sieve in a couple of minutes, expressed as a percentage of the original sample, is considered the segregation index of the SCC. The test apparently is not highly scientific, can be influenced by the sample and a method of collection (at the start or the end of the pour from a truck mixture), the handling of the container itself, and so on, but yet gives a fair idea of segregation resistance. The inordinate length of time that the sample has to be kept undisturbed before testing makes it impractical to be applied in the field. Segregation ratios between 5% and 15% are considered appropriate, with values below 5% indicating a very stiff mix and those above 30% potentially segregating.

3.3.10 Column Settlement Test

There have been different approaches in the assessment of the segregation resistance of concrete. The settlement of the small acrylate plate placed on top of a cylindrical mold of 800 mm height and 200 mm in diameter is an indicator of the segregation characteristics of the SCC filled in the mold. In fact a similar test in a standard cylinder of 300 × 150 mm diameter is also used to quickly assess segregation characteristics of the concrete in the laboratory. There were also other segregation column tests that allow establishing the coarse aggregate content in the top and bottom segments leaving a much larger middle segment of about double these segments. The methods adopted varied from exposing the aggregate by split tension test of both the top and bottom segments and calculating the number of aggregate larger than 8 mm at the cracked face to washing off the mortar phase in its shiny hardened state and establishing the difference between the top and bottom segments as a percentage.

Another such approach is a method that can establish both the dynamic and static segregation possibilities of an SCC (Bartos, 2002). The equipment consists of a rectangular column of 150 × 100 mm with three-hinged doors permitting the extraction of samples from the top 150 mm, middle 220 mm, and the bottom 150 mm of the column as shown in Figure 3.9. The segregation column can be left undisturbed for about 5 minutes to understand the

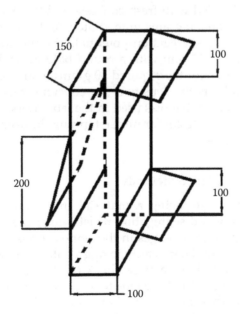

FIGURE 3.9
Column settlement test apparatus.

static segregation effects or can be subjected to the jolting effects (20 drops of the mortar flow table) by clamping it onto the top of a flow table to assess the effects of dynamic segregation. The samples collected in pans are washed clean to remove the mortar and the remaining weights of the coarse aggregate after drying as the ratio of top to bottom is an indicator of the level of segregation. Experimental investigations suggest that the aggregates ratios of 0.96, 0.95–0.88, 0.87–0.72, and below 0.71 are indicative of negligible, mild, notable, and the severe susceptibility levels of segregation of the SCCs studied.

ACI 237 (2007) also suggests similar test with a fresh concrete placed in a cylindrical mold. The mold itself in four segments in the concrete is placed in one lift without temping or vibration. After allowing for segregation, the concrete in the top and bottom segments are wet-washed on a 4.75 mm sieve to establish the coarse aggregate content in each. The difference as a percentage is an indicator of the segregation characteristics, and a value less than 10% is considered acceptable for SCCs.

3.3.11 Penetration Resistance Test

Another way of studying the segregation resistance is to look at the changing penetration resistance through the change in the coarse aggregate concentration due to settlement of the coarser particles. A container of 200 × 200 mm cross-section of about 600 mm depth (maybe something like the German box for consolidation) is filled with fresh concrete and the depth of penetration of a pin weighing about 50 grams immediately and after the lapse of about 5 minutes indicates the segregation potential. The pin could be a 200 mm long plastic tube of 10–12 mm with a round bottom, with a weight at the bottom end (making up for the required 50 grams total weight) so that it can sink vertically to indicate the density of the concrete at the top from time to time. Another alternative that was used in the laboratory with some success is the use of the K-slump tester for understanding the segregation resistance in a very similar manner.

3.3.12 Job Site Acceptance Methods

Self-compacting concrete, being a material that requires a fine balance between the viscosity and segregation resistance of the concrete, is highly susceptible to even minor changes in the cementitious material characteristics, the gradation and the moisture in aggregates, and so on. Given these facts it is imperative to have a close check on the quality of the different batches of SCCs being used at the jobsite, in particular, just before they are placed in position. This means that there is a need for continuous monitoring of the material as it goes into the structure which is impossible for all practical purposes without the system that can detect any variations directly online as the concrete is being supplied in position. It is obvious that monitoring self-compactability is not possible through the tests mentioned earlier, as nonconformance to SCC characteristics by concrete even a small part can lead to disastrous consequences or maybe even failure. This apparatus ensures that only if the concrete is able to pass through it gets pumped. In fact, the concrete pump in itself is a good piece of equipment that will show several consistency defects due to its dynamic pumping that could lead to segregation. Ouchi et al. (1996) proposed an additional attachment that is installed between agitator truck and pump at the job site. The concrete is poured into apparatus consisting of two vertical layers of 13 mm bars spaced 60 mm apart, the layers separated by 40 mm and the second layer situated at the center of the other layer, obstructing the flow of concrete. If the concrete is stopped by the apparatus, the concrete is considered as having insufficient self-compactability and mix-proportion has to be adjusted.

3.4 Japanese Recommendations

The first pioneering efforts to produce SCCs in their present form, even if they were to be not very different from the flowing concretes of yesteryear was in Japan (Okamura, 2003). The most important factor in this method is the lowering of the coarse aggregate in the granular skeleton and modulating the fine aggregate and powder contents along with the water and superplasticizer contents needed to achieve self-compactability. The Japanese recommendations primarily adopted the methodology suggested by Okamura to arrive at SCCs. It is obvious that in designing concrete structures there is a need to recognize the three aspects—function, performance, and quality—defined by service conditions, environmental conditions, and properties that can be quantified through testing, which are also broadly accepted by the Japanese code. In line with this the Japanese court proposes three types of SCCs:

1. *Powder type*: an SCC with reduced water-powder ratio (0.16–0.19) through increased powder content (0.85–1.15) and a high-range water-reducing admixture. The higher powder contents of this type will be helpful in achieving higher strength concretes, but the risk of higher shrinkage associated with the higher powder content should be recognized.
2. *Viscosity modifying agent type*: a SCC of lower powder content (300–500 kg/m³) that will naturally need the assistance of VMAs for ensuring the segregation resistance. While the plastic viscosity can be modulated through VMA, the compatibility of the various admixtures

with the cementitious materials in particular and the possible delay in setting and strength gain rate due to the addition could be a matter for concern.

3. *Combination type*: an SCC that uses the characteristics of both the above types for concretes with the intermediate range of powder content (>0.13).

Incidentally, the Japanese code proposes three different ranks of SCCs that are possible using coarse aggregates of 20–25 mm maximum size, and for the powder type SCC which are as follows:

Rank 1: 0.28–0.30 m^3/m^3

Rank 2: 0.30–0.33 m^3/m^3

Rank 3: 0.32–0.35 m^3/m^3

It is suggested that the water content should be minimized to be preferably in the range of 155–175 kg/m^3, with the water powder ratio ranging from 28% to 37% and the cement content of 0.85–1.15 (powder content of 0.16–0.19 as already stated). Similarly, the recommendations for the other two types of SCCs containing VMA are also available. All these different requirements for the various ranks of the SCCs along with the member considerations and their appropriate consistency requirements as recommended by the guide to construction of high-flowing concrete (JSCE, 1998) are presented in Table 3.3. It is easy to see that there are only very minor differences in the coarse aggregate volumes between the different types of SCCs and the strength and consistency requirements will probably be deciding the maximum water powder ratios and powder contents, and minimum water contents as can be seen from the table.

The recommendations essentially are in four parts that deal with a general introduction, the mixture proportioning for the various types, in the production of replacement aspects, and finally the test methods for conformity and quality assurance. While in the first part it is suggested that the verification of self-compactability be done through the U-box, the last part additionally suggests slump flow spread, V-funnel time, and L-box standard test methods for SCCs. The actual details of these tests along with a few others proposed by other codes and researchers are being discussed in detail later in the chapter.

3.5 Euro-EFNARC Guidelines

The Euro recommendations, essentially coming out as specifications and guidelines for SCC initially (EFNARC, 2002) followed and were in line with the then existing methodology of the recommendations suggested by

TABLE 3.3

Japanese Recommendations for SCCs of Different Ranks

| | Mix Requirements | | | | Member Considerations | | Consistency Requirements | | | |
| | | | | | | | Passing Ability | Filling Ability | | Segreg. Resist. |
Rank	Coarse[a] Aggregate (vol. %)	Water[b] Content (kg/m³)	w/p[b] (by wt.)	Powder (vol. %)	Min. Gap in Reinf. (mm)	Reinf. Amount (kg/m³)	U-Box (mm)	Slump Flow (mm)	V-Funn. (s)	T_{500} (s)
R1-Powder	0.28–0.30	155–175	0.28–0.37	16–19						
R1-VMA	0.28–0.31	Min. fresh	Max. hard	Sufficient	35–60	≥350	≥300 (5–10φ)	600–700	9–20	5–20
R1-Mixed	0.28–0.30	Sufficient	—	>13						
R2-Powder	0.30–0.33	155–175	0.28–0.37	16–19						
R2-VMA	0.30–0.33	Min. fresh	Max. hard	Sufficient	60–200	100–350	≥300 (3–13φ)	600–700	7–13	3–15
R2-Mixed	0.30–0.33	Sufficient	—	>13						
R3-Powder	0.32–0.35	155–175	0.28–0.37	16–19						
R3-VMA	0.30–0.36	Min. fresh	Max. hard	Sufficient	≥200	≤100	≥300 (none)	500–680	4–11	3–15
R3-Mixed	0.30–0.35	Sufficient	—	>13						

Sources: JSCE, Recommendation for self-compacting concrete, in T. Uomoto and K. Ozawa (Eds.), Concrete Engineering Series 31, JSCE, Tokyo, Japan, 1999, p. 77; JSCE, *Guide to Construction of High Flowing Concrete*, Gihoudou Pub., Tokyo, Japan, 1998.

a Maximum size of the coarse aggregate is 20 or 25 mm.

b The water content should be the minimum for required for consistency to ensure minimum powder, superplasticizer, and VMA, and maximum strength.

Okamura. With the primary aim of achieving satisfactory values for the filling ability, passing ability, and segregation resistance, the guidelines for the initial mix composition were to keep the relative volume proportions of the constituents as the following. The basic change in the recommendations is that the water powder ratio by volume was increased from the range 0.90–1.00 proposed by Okamura to 0.80–1.10, with the total powder content at 160–240 liters (400–600 kg/m³) and the coarse aggregate content at 28%–35%. It was also suggested that the water cement ratio was selected based on the available Euro norms for normally vibrated concrete to meet the strength and performance requirements and ensuring that the water content does not exceed 200 L/m³, probably a recommendation that is aimed at limiting the total powder content in the mix. The methodology for determining the optimum paste composition, water powder ratio, and superplasticizer dosage were also ensured through the method proposed by Okamura, through slump flow and V-funnel studies on the paste and mortar.

These recommendations were later revised to form the European guidelines for SCC (EFNARC, 2005) after detailed studies from several European countries essentially to take advantage of the lower manpower requirement and ease of construction apart from other factors. In principle, it has to naturally confirm to the existing strength and exposure class requirements in the Euro norms for the traditionally vibrated concrete. These revised Euro norms for SCCs as in the Japanese code now present them under three consistency classes, but the additional construction-related requirements are also projected under separate sets of classes.

- Flowability, defined by types—SF1, SF2, and SF3
- Viscosity or rate of flow, defined by T_{500} in slump flow—VS1 and VS2 or by V-funnel time—VF1 and VF2
- Passing ability, defined by L-box ratio or J-ring step—PA1 and PA2, and
- Segregation resistance—defined by sieve segregation—SR1 and SR2 classes

As in the earlier set of recommendations, the present guidelines also specify a broad range for the typical constituents both by weight and volume showing that the powder content be limited to a range of 300–600 kg/m³ arriving at the paste content of 0.30%–0.38% with the water powder ratio of 0.85–1.10, a minor change from the previous recommendation but yet almost close to the values originally proposed by Okamura. The water content recommended is between 150 and 210 kg/m³, the coarse aggregate content line between 750 and 1000 kg/m³ (0.27–0.36), and fine aggregate content constituting the balance (0.48–0.55). It is also stated that in general the slump flow class alone need be specified while the viscosity, passing ability, and segregation resistance requirements come into play only for specific requirements. It can be

seen that the qualifying tests and the expected range of values are also specified to ensure that the proposed concrete can be verified to see if it satisfies
the requirements. Complete details of the test for each of these consistency
parameters and their actual range for each of the class mentioned in the code
was tabulated and is presented in Table 3.4. The table also presents the possible and recommended applications for each of these classes and also the
limitations for each of these classes, which probably is not only informative,
but is also one with respect to which care should be taken to ensure a satisfactory functioning of the structure.

It is obvious that a complex interaction of all these factors is always at play
and the choices available for any given scenario of construction requirement
is indeed not so easy to put in place. Even so a broad outline of which of
these factors will have a primary say in each situation is outlined earlier by
Walraven (2003), indicating their prominence for applications in ramps, walls
and piles, and tall and slender structures and for floors and slabs, though
these are not correlated with the actual test values expected for these classes
as defined (EFNARC, 2005). Presently, an effort is being made to put them in
some perspective with reference to the classes they belong and the range of
test values they represent (Figure 3.10). While it may not be totally correct to
say that the correlations suggested are an exact and actual representation of
the correspondence of the different ranges of test methods, it is good to say
that these were arrived at considering the ranges of values that were suggested by the EFNARC recommendations itself. A more detailed discussion
on how these correlations were made will be later discussed while presenting the information on the correlating ability of the different parameters of
interest and the correlations of their test results from the literature.

The recommendations also prescribe the test methods for evaluating the
characteristics of SCCs along with the methodologies and ranges for establishing the different classes. It is noteworthy to see that the initial recommendations in 2002 present an extensive list of test methods for understanding
the characteristics of the SCCs (slump flow, T_{500}, J-ring, V-funnel, L-box,
U-box, fill box, screen stability, as well as Orimet tests), and the later recommendation in 2005 while recognizing them all concentrates mainly on the
few (slump flow, T_{500}, V-funnel, L-box, and sieve segregation tests) that are
used for the characterization of the SCCs in to the different classes. The methodology and interpretations through these test methods is also discussed in
the later part. Apart from these, a table outlining the possible defects, the reasons why they occur, and the avenues for their redressal are also appended
to the recommendations. In brief, the defects addressed were mainly surface
defects like blowholes, honeycombing, and layering at joints. However, the
other inherent problems like shrinkage, cracking in thin sections, inadequate
shear transfer through aggregate interlock, or bonding with reinforcement
have to be looked at separately. These aspects that have relevance to the functioning of SCCs have not discussed appropriately in the later parts while
discussing the limitations and remedial measures.

TABLE 3.4

Euro–EFNARC Guidelines for SCCs of Different Classes

Consistency	Test	Class	Range	Applications	Limitations
Filling ability	Slump flow (mm)	SF1	550–650	Housing slabs, tunnel linings, piles	Lightly reinforced, thin, top free
		SF2	660–750	Walls, columns	Normal applications
		SF3	760–850	Vertical, congested reinf., complex shapes	Segregation difficult to control
Viscosity (specified in special cases)	T_{500} (s)	VS1	≤2	Congested reinf., self levelling	Bleeding, segregation likely
		VS2	>2	Improved segregation resistance, thixotropic	Blow holes, stoppages between lifts
	V-funnel (s)	VF1	≤8	Congested reinf., self levelling	Bleeding, segregation likely
		VF2	9–25	Improved segregation resistance, thixotropic	Blow holes, stoppages between lifts
Passing ability	L-box (ht. ratio)	PA1	≥0.80 (2 bars)	Housing, vertical structures	Complex structures with a gap <60 mm, specific mock-up trials may be necessary
		PA2	≥0.80 (3 bars)	Civil engineering structures	
	J-ring step (mm)	PA1	≤15 (59 mm bar spacing)	Structures with a gap of 80–100 mm (e.g. housing, vertical structures)	Consider geometry, reinf. density, maximum aggregate size
		PA2	≤15 (41 mm bar spacing)	Structures with a gap of 60–80 mm (e.g. civil engineering structures)	Complex structures with a gap <60 mm, specific mock-up trials may be necessary
Segregation resistance	% Sieve segregation	SR1	≤20	Thin slabs, vertical applications, flow <5 m	—
		SR2	≤15	Vertical applications, flow >5 m	Segregation, strength at top critical.

Source: After EFNARC, *The European Guidelines for Self-Compacting Concrete*, EFNARC, 2005, p. 68.

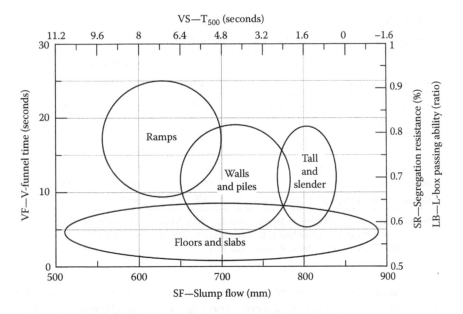

FIGURE 3.10
Potential application regimes for SCCs with different class characteristics. (After Walraven, J., Structural applications of self-compacting concrete, in *Third RILEM and International Symposium on Self-Compacting Concrete*, Reykjavik, Iceland, 2003, pp. 15–22.)

3.6 ACI Recommendations

The American Concrete Institute also presented a comprehensive report setting out guidelines for selecting the proportions along with specifications and test methods (ACI 237). Being one of the later ones the reports started with an outline of a few case studies which could illustrate the application potential of SCCs in general. The recommendations suggest evaluating the characteristics of SCC through slump flow for filling ability, J-ring, and L-box tests for passing ability like in the previous recommendations while the stability or segregation resistance was expected to be evaluated through the column segregation test. It can also be seen that the methodology was kept quite simple defining the requirements of the powder contents in the three recognized slump flow ranges, <550, 550–600, and >600. While accepting that the maximum size of coarse aggregate could be above 12.5 mm (maximum 25 mm), the ranges of the absolute volume of coarse aggregate, paste fraction, mortar fraction, cement, and water cement ratio are all broadly specified. Having had the benefit of several research efforts and international seminars and symposia and with a broad background of the normal strength concretes of ACI 211.1-4, these recommendations for SCCs appear to be the logical extensions to the already available methodologies. Table 3.5 presents

TABLE 3.5

Suggested Proportioning Possibilities for SCCs

Design Parameter	Slump Flow (mm)		
	<550	550–600	>600
Powder content (25 mm agg.) (as needed for strength)	355–385	385–445	458–475
Coarse aggregate (vol. %)[a]	<28	28–32	>32
Paste fraction (vol. %)	<34	34–40	>40
Mortar fraction (vol. %)	<68	68–72	>72
w/cm (typical)	<0.32	0.32–0.45	>0.45
Powder content (typical, 12 mm agg.)	355–385	386–475	>475

Source: ACI 237, Self-consolidating concrete, ACI Manual of Concrete Practice, American Concrete Institute, Farmington Hills, MI, 2007, 34pp.
[a] Up to 50% for 10 mm nominal maximum size.

a simple outline of these recommendations. It is also important to note that these recommendations present a combined aggregate grading for the different slump flow characteristics proposed to be achieved. Apart from this the recommendations also present the variables influencing the filling and passing ability as well as stability in different application scenarios. They also presented the effect of member characteristics on the slump flow requirements from an earlier report available, and in fact these will be discussed in detail a little later while looking at the same utilized by a different forum more comprehensively.

3.7 Other Perceptions

There have been a few other national bodies that looked at a few specific aspects—like the Nordic Innovation Centre reporting on test methods for SCC and the Danish Technological Institute discussing the mix design guidelines in general. There are also a few documents from academic institutes that had significant research programs leading to a better understanding of SCCs like the UCL method for proportioning of SCC by Domone (2009), a white paper by researchers at the Center of Advanced Cement Based Materials (ACBM) of the Northwestern University (2007), and a report on the development of SCC by the Goodier (2003) to name a few. Each one of these reports and some of the background papers that go into the making of these reports give an insight into the various aspects relating to SCCs. Probably one of the more explicit documents on the utilization of self-consolidating concretes in the industry is the interim guidelines available from Precast/Prestressed

Concrete Institute of North America (TR-6-03). These guidelines while discussing production issues clearly recognize that there are two main aspects that will affect the selection fresh SCC properties, namely the characteristics of the element being cast and the placement techniques. They also presented a list of seven basic member characteristics that have to be considered for the SCC mixture requirements which are listed hereunder.

1. *Reinforcement level*—as proposed by the JSCE for establishing the rank SCC
2. *Element shape intricacy*—helps determine the required fluidity level
3. *Element depth*—has greater segregation potential
4. *Surface finish*—for improved surface appearance
5. *Element length*—helps determine the required fluidity and dynamic stability
6. *Coarse aggregate content*—controls passing ability
7. *Wall thickness*—Thin walls usually require a higher fluidity

Based on this critical understanding of the controlling parameters associated with the element characteristics particularly in relation to the precast industry the appropriate rank or class for each of the flow characteristics has been indicated—slump flow, U-box, T_{500}, L-box, V-funnel, and J-ring—in line with the recommendations suggested by Daczko (2001). The enclosed Table 3.6 presents the complete matrix of these interactions collated into a single entity, thus facilitating how each of the member parameters above will demand the different consistency characteristics. This will help in deciding the set of consistency measurements needed to ensure a trouble-free casting operation. Even so it is necessary to keep in mind that although there is very little energy needed for the compaction effort which invariably is taken care of by the SCC itself, a significant amount of energy is needed for the transport and in particular the placement operations at site. This will indeed have a significant influence on the flow, passing, and segregation characteristics of the concrete as it is placed. In fact, it is necessary to keep the proposed SCC one step below the best to ensure that the construction-induced vibrations and other disturbances during the site operations will not lead to serious consequences like segregation and bleeding. The guidelines actually suggest that the relative energy delivered by the different placement techniques in particular be appropriately addressed to ensure a satisfactory website experience. The factors related to each of the site operations like placement, discharge rate, type, volume, and related energies are all summarized in these guidelines and need to be appropriately addressed. Maybe at this stage it is important to recapitulate the views expressed earlier that for the sake of the required quality assurance with the specific confidence levels most often is mandatory to perform not just one but a small group of tests. Table 3.6 also

TABLE 3.6

Guidelines on the Flow Property Requirements for the Different Member Characteristics

Controlling Parameters Associated with Element Characteristics			Slump Flow (cm)			U-Box (Rank)			T_{500} (s)			L-Box (%)			V-Funnel (s)			J-Ring (mm)[a]		
			<55	55–65	>65	3	2	1	<2	2–5	>5	<75	75–90	>90	<6	6–10	>10	<10	10–15	>15
Element characteristics	Reinforcement level	Low																		
		Med.																		
		High																		
	Intricacy of element shape	Low																		
		Med.																		
		High																		
	Element depth	Low																		
		Med.																		
		High																		
	Importance of surface finish	Low																		
		Med.																		
		High																		
	Element length	Low																		
		Med.																		
		High																		
	Wall thickness	Low																		
		Med.																		
		High																		
	Coarse aggregate content	Low																		
		Med.																		
		High																		
	Placement energy	Low																		
		Med.																		
		High																		

Source: PCI, *Interim Guidelines for the Use of Self-Consolidating Concrete in Precast/Prestressed Concrete in Precast/Prestressed Concrete Institute Member Plants, TR-6-03, Precast/Prestressed Concrete Institute,* Chicago, IL, 2003.

[a] With the inverted slump cone.

presents the test requirements and considered in conjunction with the paste and characteristics at the site will help in identifying which of the test parameters need to be addressed to ensure an appropriate functioning of the SCC for the construction operations at hand.

3.8 Summary and Suggestions

The various provisions discussed earlier provide a broad outline of the basic consistency characteristics of the cementitious composites that can be classified as self-consolidating concrete by nature. A broad panorama of the different consistency assessment methods along with the underlying principles was also discussed and summarized. The recommendations of the various national and international bodies that tend to characterize even the small group of self-consolidating composites into different regimes (termed as either ranks or classes) was also discussed.

Having looked at the consistency requirements including their assessments and classifications it is probably important to note two specific aspects. First, it may not be easy to enforce even the two specific consistency requirements proposed and continuously monitor them during construction unless a small yet highly trained workforce dedicated only for the quality assurance purposes are available at the site. Second, the effects of the placement mechanisms like pumping are not clearly appreciated and addressed.

Also, as was presented in the opening paragraphs on consistency (through the words of Wesche, 1975) that though from a rheological point of view each test method addresses a different property of the fresh concrete (filling, flowing, passing, segregation resistance, etc.), the general relationships between some of these have to be understood and delineated to present basic comfort to the construction site personnel who cannot be burdened with the intricacies of the effects of minor changes in the characteristics of the constituent materials. While there have been a few efforts of the particular type, the discussions and even suggestions are few and far between. Essentially, this is because of the fact that though each individual research effort naturally uses different setup materials available, locally making the concretes is very difficult to compare between investigations, even the test methods (be it the opening size of a V-funnel, as also the number, diameter, and spacing of the bars in a J-ring, and their relationship with the maximum size of the aggregate) makes any correlation almost impossible. As a corollary the type of the material that is investigated in each case is so different that it becomes impossible to even confidently predict the various strength relationships (like modulus, tensile, shear, and bond strengths for a concrete of a specific strength). It is indeed these limitations if not inhibitions that have to be addressed appropriately to ensure the confidence of the field engineer. There are a few other aspects that probably show more relevance to the constituent materials and their appropriate utilization in realizing the maximum potential out of the materials being used, which will be discussed a little later.

References

ACI 237, Self-consolidating concrete, ACI Manual of Concrete Practice, American Concrete Institute, Farmington Hills, MI, 2007, 34pp.

ACI 309, *Guide for Consolidation of Concrete*, ACI Manual of Concrete Practice, American Concrete Institute, Farmington Hills, MI, 1996, pp. 1–39.

ASTM C124, *Method of Test for Flow of Portland-Cement Concrete by Use of the Flow Table* (withdrawn 1973), ASTM, West Conshohocken, PA, 1971.

ASTM C1611, *Standard Test Method for Slump Flow of Self-Consolidating Concrete*, Annual Book of ASTM Standards, ASTM, West Conshohocken, PA, 2009.

ASTM C1621, *Standard Test Method for Passing Ability of Self-Consolidating Concrete by J-Ring*, Annual Book of ASTM Standards, ASTM, West Conshohocken, PA, 2009.

ASTM C172, *Standard Practice for Sampling Freshly Mixed Concrete*, Annual Book of ASTM Standards, ASTM, West Conshohocken, PA, 2004.

Bartos, P.J.M., An appraisal of the Orimet Test as a method for on-site assessment of fresh SCC concrete, in *International Workshop on Self-Compacting Concrete*, Japan, 1998, pp. 121–135.

Bartos, P.J.M., Sonebi, M., and Tamimi, A.K., *Workability and Rheology of Fresh Concrete: Compendium of Tests*, RILEM, Cachan, France, 2002.

BS 1881:105:84, *Testing Concrete. Method for Determination of Flow*, British Standards Institution, London, U.K., 1984, 10pp.

Daczko, J.A. and Constantiner, D., Rheodynamic concrete, in *43rd Congresso Brasileiro do Concreto*, Brazil, 2001.

DAfStb, DAfStb-Richtlinie Selbstverdichtender Beton (SVB-Richtline)—Teile 1, 2 und 3, Entwurf, DAfStb, Berlin, Germany, September 2012, pp. 1–18.

Domone, P., Proportioning of self-compacting concrete—The UCL method, UCL, 2009, p. 30.

EFNARC, *Specification and Guidelines for Self-Compacting Concrete*, EFNARC, Surrey, U.K., 2002, p. 32.

EFNARC, *The European Guidelines for Self-Compacting Concrete*, EFNARC, Surrey, U.K., 2005, p. 68.

EN 12350-1, Testing fresh concrete, Part 1. Sampling fresh concrete, 2000, European Committee for Standardization, Brussels, Belgium, pp. 1–6.

EN 206-1, Concrete—Part 1: Definitions, specifications and quality control, European Committee for Standardization, Brussels, Belgium, 2000, pp. 1–74.

Goodier, C.I., Development of self-compacting concrete, *ICE-Structures and Buildings*, 156(4), 2003, 405–414.

Hayakawa, M., Matsuoka, Y., and Shindoh, T., Development and application of super-workable concrete, in *RILEM Workshop on Special Concretes Workability and Mixing*, 1993, pp. 183–190.

JSCE, *Guide to Construction of High Flowing Concrete*, Gihoudou Pub., Tokyo, Japan, 1998.

JSCE, Recommendation for self-compacting concrete, in T. Uomoto and K. Ozawa (Eds.), Concrete Engineering Series 31, JSCE, Tokyo, Japan, 1999, p. 77.

Lange, D.A., Self-consolidating concrete. ACBM, Northwestern University, Evanston, IL, 2007, p. 42.

Okamura, H., Self-compacting high performance concrete, Ferguson Lecture, *Concrete International*, 19(7), 1997, 50–54.

Okamura, H. and Ouchi, M., Self-compacting concrete, *Journal of Advanced Concrete Technology*, 1(1), 2003, 5–15.

Ouchi, M., Ozawa, K., and Okamura, H., Development of simple self-compactability testing method for acceptance at job site, in *First International Conference on Concrete Structures*, Cairo, Egypt, 1996, pp. 9.11–9.20.

Ozawa, K., Maekawa, K., and Okamura, H., High performance concrete with high filling capacity, Admixtures for Concrete: Improvement of Properties, *Proceedings of the International Symposium held by RILEM*, May 14–17, 1990, Barcelona, Spain, E. Vazquez (Ed.), Chapman and Hall, London, U.K., 1990, pp. 51–62.

Ozawa, K., Sakata, N., and Okamura, H., Evaluation of self-compactability of fresh concrete using the funnel test, *JSCE*, 23(490), 1994, 71–80.

Pade, C., Test methods for SCC, NORDEN Nordic Innovation Centre, Oslo, Norway, 2005, p. 56.

PCI, *Interim Guidelines for the Use of Self-Consolidating Concrete in Precast/Prestressed Concrete Institute Member Plants*, TR-6-03, Precast/Prestressed Concrete Institute, Chicago, IL, 2003.

Reinhardt, H.W., DAfStb guideline on self-compacting concrete, *Betonwerk und Fertigteil- Technik/Concrete Precasting Plant and Technology*, 67(12), 2001, 54–58 and 60–62.

Sonebi, M. and Bartos, P.J.M., Filling ability and plastic settlement of self-compacting concrete, *Materials and Structures*, 35, 2002, 462–469.

Walraven, J., Structural applications of self-compacting concrete, in *Third RILEM and International Symposium on Self-Compacting Concrete*, Reykjavik, Iceland, 2003, pp. 15–22.

Wesche, K., Comparison of different consistency values, RILEM TC 14-CPC, *Materials and Structures*, 8(45), 1975, 255.

4

Methodologies for the Proportioning of SCC Mixtures

4.1 Introduction

The need for a specialized and skilled labor force trained specifically to the continuously changing needs of the increasingly urbanized societies of recent years has forced the construction industry to look for both alternative materials and methods of construction, even if they were slightly more expensive. The necessity for a speedier construction process in each facet of the construction activity is also of paramount importance to ensure economy apart from making sure that a quicker turnaround period will help in reducing the expenditure on support facilities like scaffolding and shattering in the project site as well as in capital investment interest. Also, the growing emphasis on the performance of materials and structures, particularly to perform in highly stressed environmental regions near the oceans and in situations where deicing salts are in need, necessitated a closer look into the complete picture of the construction scenario. The need for higher workability of the concrete for ensuring appropriate level of competition to be able to produce a defect-free system with a reasonable guarantee is also well recognized. While it is not always an advantage to have highly fluid systems like self-compacting concretes (SCCs), which originally started as flowing or collapse slump concretes of yesteryears, their advantages in terms of the limited skilled labor requirements as well as substantially reduced noise pollution levels at the construction site have ensured their popularization world over.

However, the overarching simplification in feeling that a few simple modifications in the traditional mixture proportions would result in a SCC did not go very far. Starting from the concrete mixture that was designed through an earlier delineated procedure for traditionally vibrated concretes with a few simple modifications never really enforced a deeper understanding into the making of SCCs of the desired consistency and more specifically strength and performance. Though there were a lot of precursors for an understanding of the specific aspects of these SCCs, a comprehensive relook to develop an appropriate methodology eluded the industry. In fact concrete practitioners were playing with several different parameters to look for an appropriate methodology to achieve this goal. Even the fact that the original proponent of this particular philosophy, Okamura (1995), has chosen to modify the existing normally vibrated concrete mix design through an increase in the powder content and a corresponding reduction in coarse aggregate content, even if empirically, has not made the later efforts to look for an appropriate methodology that might be in line with the design of normally vibrated concretes. Also, the fact that free-flowing, collapse-slump, nonsegregating concrete composites did exist even at that particular point has not made much of a difference to their research efforts that were directed towards the development of the design procedures for SCCs. The most important reason for almost everybody choosing such a path is that highly fluid concretes tend to bleed and segregate, and suppression of these two phenomena is an extremely difficult task even with the adoption of additional fines as well as viscosity modifying agents (VMAs). For this reason initial semi-empirical efforts were suitably modulated essentially in two ways—to see that the water–cementitious materials ratio of the concrete composite is reduced through the addition of fines, which means that the modifications are in the paste and mortar phase, or alternatively the aggregate phase is modulated to ensure the loosening effect through appropriate grading norms. Once again under each of these subdivisions there are several methods of either modifying the fines content or the water, superplasticizer, and VMA contents on the one side and on the other side modulating the coarse aggregate packing factors or adopting an appropriate continuous grading curve for both the coarse and fine aggregates. These different approaches have led to different propositions for designing SCC composites. The later part of the chapter deals with some of the more prominent and hypothetically different approaches that are used in these methodologies for the proportioning of SCC mixtures.

At this stage it is important to observe that though it is believed by the community to be a novel invention in the area of concrete technology, SCC is yet another concrete composite with only a few distinctive qualities in terms of its green state and is expected to perform like the conventional normally vibrated concretes in practice. Although some of these assumptions and expectations may be questioned, the fact is that the concrete in a structural member is always designed for a particular strength and maybe in certain cases even for specific performance requirements. As stated earlier, the emphasis being on the development of self-compactable concretes, the strength characteristics of the resulting SCCs did not receive as much attention, though some of the methodologies have indeed started with the strength requirement itself. Another fact that is obvious and should not be forgotten is that most efforts for SCC were all directed toward the making of concretes for conventional applications that require strengths of only around 30–50 MPa, a strength level that can be achieved with almost any combination of the aggregates and materials accessible owing to the high-strength cements and high-end superplasticizers available in the market today. Also, it should be recognized that the modulation of powder content need not only be limited to the use of inert mineral powders or even the readily available simple low-end pozzolans like fly ash and ground granulated blast furnace slag (GGBS), while the significantly higher-performance pozzolans, like silica fume, metakaolin, and so on, available do present excellent possibilities for achieving very high-strength high-performance SCCs over a very wide range. It is also obvious that none of the mix design methodologies proposed to date have adequately addressed the effect of the pozzolanic efficiency of the powders used for the increased fines requirement of SCCs, whether it is the addition of low-end fly ash and GGBS or the more efficient ones like silica fume and metakaolin, but for the fact that some of them recognize their beneficial effects in terms of performance. It is in this context that one needs to understand that some of the mix designs proposed resulted in strengths much higher than those envisaged originally by their designs. This situation naturally requires an appropriate understanding of the performance of pozzolans in these cementitious composites. The primary reason for calling these concretes, cementitious composites, originates from the fact that one has to understand this aspect of the pozzolanic contribution to the cementitious strength and performance characteristics of the materials (maybe even as composite cements) involved, supplemented with the advantages of using the latest polycarboxylate based superplasticizers essentially to have a handle on the water requirements of the fines compositions. Also, in concretes with lower powder contents requiring the high-slump flow levels of SCCs, the obvious need for a VMA need not be overstressed to ensure the required stability and segregation resistance.

4.2 Design Viewpoints

The concept of SCC, first proposed by Okamura (1995), was mainly to harness the advantages like the elimination of vibration and decreasing time and labor costs along with a reduction in noise pollution during compaction. It is presumed that it could ensure better durability due to the compaction effectiveness and additional fines that generally are of a pozzolanic nature. The basic difference of SCCs from conventional concretes is in ensuring the three main attributes of filling, passing, and segregation of these concretes through modifications in the proportioning of the constituents, like increasing the fines or by the addition of a VMA that imparts the required segregation resistance.

The available procedures for the proportioning of self-consolidating or SCCs could be categorized broadly into three main types as originally defined by the Japanese recommendations (JSCE, 1999). The first is the powder type—essentially increasing the powder content while reducing the water–powder ratio to ensure that the aggregates are supported in such a viscous medium of paste and mortar. Alternatively, felt possible to generate a similar effect by adding a viscosity modifying admixture to provide the same segregation resistance to a concrete of high fluidity. Last, a combination of both these methodologies could be used to ensure an optimal utilization of the effects of both.

In an effort to design the concrete for the massive end-anchorages of the Akashi-Kaikyo Bridge, Okamura (1995, 2003) proposed a methodology to achieve a highly workable and thixotropic mix through augmented fines and superplasticizers while reducing the coarse aggregate content. After this initial semi-empirical approach of modifying the conventional concrete design, researchers appear to have looked for avenues to achieve self-compactability through several different approaches like ensuring the exact water requirement for the constituents to control fluidity, packing characteristics of the aggregates, utilization of different mineral admixtures, and even extending the same for the design of fibrous SCC composites. There have also been efforts to find governing relations by a mathematical analysis of the experimental results using fuzzy logic and neural network methods. It is to be seen that most of them, in their interest to ensure self-compactability, have not looked at strength as the criterion though some presented the strength results of the concretes produced with these designs. Also, the different methodologies proposed often appear to concentrate on specific parameters such as the granular skeleton and its optimum packing characteristics, the fluidity and its enhancement through additional fines and water, or addition of mineral and chemical admixtures without addressing comprehensively all the requirements together to ensure an optimal solution. Apart from this, aspects like the reduction of the maximum size of the aggregate and the continuous distribution in the grading of the aggregate in these mixes were also employed by some researchers, though these parameters were not explicitly discussed as a requirement for the mixture to be self-compacting. Felicoglu (2007) echoed almost the same view in stating that while in traditional concrete the water–cement ratio is kept constant to obtain the required strength and durability, of the several methods proposed for SCCs, no single method or combination of methods has achieved universal approval and most of them have their adherents. No single method has been found to be characterizing all the relevant workability aspects.

Naturally, the simplest method of categorizing the hitherto proposed methodologies will be through a comparison with the generic method of mix design for conventionally vibrated concretes. It is known that conventionally vibrated concretes essentially start with the strength required, be it the design or the target strength, which is fundamentally defined by the strength to water–cement (or cementitious materials) ratio relationship. Without getting into the discussions regarding the different relations defining these, in some cases depending on the age, in others maybe the strength of the constituents involved, and so on, the fact that the strength is related to the water–cementitious materials ratio cannot be disputed and is accepted universally. Once this parameter of the water–cementitious materials ratio required for the specific strength is established as for the particular cementitious material combination proposed, the workability requirements that essentially depend on the type and methodology of construction will present a specific water content requirement for the appropriate type of constitutes. In some cases this water requirement is defined through only the maximum size of the aggregate, and in other cases it is defined through a parameter established to represent the surface area of all the constituents. However a detailed discussion on these methodologies and their efficacy or applicability in different situations is outside the purview of the present discussions. Having established the above parameters, the quantities of water and the cementitious materials (the binder component) are available to the designer. It now remains to have an appropriate modulation of the filler component through either the individual grading of the different aggregate fractions or a combined aggregate grading that is specified in a few cases. Most of these concepts and interrelations between the constituents for ensuring an appropriate mix design methodology have all been developed over a period of time and are based on extensive investigations reported on such concretes. They have withstood the test of time and have served the construction engineering industry well over the past hundred years, to say the least. Based on this understanding, it is possible to categorize the available mix design methodologies for SCCs reported into the following categories.

- *Empirical methods*—these only attempt to transform a relatively well-proportioned conventionally vibrated concrete design mix of a specific strength by simply enhancing the fine powder quantities to increase the paste content. The increasing volume of fines is compensated by reducing the coarse aggregate content generally to values below 50%. To control the possibility of segregation of the larger fractions in the increased mortar volumes, sometimes viscosity modifying admixtures are used. These methods incidentally require a perfect understanding of the various factors associated with the design of conventionally vibrated concretes to modify them suitably.
- *Strength requirement based designs*—these essentially follow the basic methodology similar to the conventionally vibrated concretes earlier, but have sometimes digressed in some ways to ensure the other requirements of self-compactability. This method requires a specific understanding of the strength contribution of the additional fines added to the system, in terms of both the pore filling and pozzolanic effects if any from them. Obviously an exact understanding of the strength to water–cementitious materials ratio relationships of the total fines that may include pozzolans is necessary, which is the same as the first step in conventionally vibrated concretes. This category

lays emphasis on the primary characteristic, the specific strength requirement for the application.

- *Matrix fluidity–based designs*—these look at the requirements of water for wetting the surface of the different ingredients, which will be critical for mixtures dominated by the finer fractions in the system, which include the cement and other powder additives in specific. This essentially is the second step of conventionally vibrated concretes, so that the water content essential for the required fluidity is established, though the strength of the system cannot be directly related.

- *Packing density and granular skeleton–based designs*—these enforce the creation of a certain loosening in the packing of the aggregates aiding free flow of the aggregates. This looks at the most appropriate grading of the filler ingredients, which will ensure a free flow of these aggregates, similar in the flow of solids in bunkers in silos, particularly when they are coated with the fine cementitious paste system wherein the required paste component is taken care of differently.

It is obvious that a perfect handle on all the parameters is required to ensure a justifiably high-performance concrete for any particular application. However, as already stated researchers attempting the design of SCCs have all followed independent paths starting at different points, recognizing the overarching need to self-compactability, though in principle they still attempt to ensure the definition of all the three parameters either directly or indirectly. It is thus possible that all these different mix design procedures advanced by different researchers can be grouped into the categories mentioned here. Attempts at such a categorization presently appear to show that some of the methodologies have inadvertently not paid enough attention to one parameter or the other, such as looking for a specific strength but arriving at a strength far different from the one that they have proposed, increasing the superplasticizer and VMA contents significantly while substantially reducing the water contents, or not recognizing the additional cementitious efficiencies of high-end pozzolans like silica fume.

A critical look at the micromechanisms in the manifestation of each of these design procedures enunciated shows several individual traits and variations, probably depending on the specific viewpoint that each of these procedures addressed. In a way, going back to the earlier parameters discussed for the design of SCCs, it can be seen that they were essentially attempting the same, maybe starting at a different point in view of the specific requirement of self-compactability, as perceived by the individual author. It may be appropriate to understand this broad spectrum listed here in some detail first before making any specific attempts to discuss them appropriately.

- Semi-empirical methods
- Compositions based on wetting water requirements of the constituents

- Methods based on aggregate distribution and packing factors
- Methods of limiting the cementitious materials through water content
- Methods of incorporating the cementitious efficiency of pozzolans
- Procedures for incorporating different pozzolans
- Approaches for a specified compressive strength
- Methods based on rheometer tests
- Methods based on the rheological paste model
- Methods based on the rheological paste model incorporating fibrous materials
- Guidelines based on statistical evaluations

The above list may be an indication of the complexity of the processes under which the different methodologies hitherto proposed can be put into. Even with such a diverse representation in the different methodologies proposed by the authors in their approach, it is obvious that each of these have to ultimately ensure the quantification of all the constituents. Naturally, this would pave the way to locate them in the proper perspective and later look for a well-directed approach to arrive at for ensuring a high-performance SCC.

Even before attempting to have a relook at the methods for proportioning of SCCs, it is probably appropriate to recognize and appreciate the need for the stage when there have already been several that seem to exist and were already categorized as noted earlier. It may not be out of place to recount the words of Kheder (2010) while reviewing 14 methods for proportioning of SCCs that most of these give only general guidelines and ranges of quantities of materials to be used in the proportioning of SCCs and also that the emphasis is on properties of fresh concrete and not on concrete strength (other than Su (2001), whose method also requires primarily a specific quantity of the water content to be compatible to ACI as proposed) as the case of conventionally vibrated concretes. They also presented a comprehensive experimental investigation on SCCs incorporating limestone powder for the additional fines required and the resulting strength to water–cement ratio relationships considering these fine powders only as inert additives to concrete. Some of these aspects will be discussed more in detail later while discussing the actual mix design methodologies proposed.

Another aspect that probably comes to mind is the fact that the simplest way adopted to supplement the finer fractions required in most cases was to use the ubiquitously available pozzolanic materials like fly ash and GGBS, as they are available in such large quantities as needed and are also relatively, if not almost totally, inexpensive. However, it is also to be noted that the minimum quantities of cement required in the various environmental zones to ensure an appropriate and adequate performance throughout the life of the structure are broadly codified in the various national standards, which is still mandatory but was never even mentioned by anyone. Apart from this it is obvious from all the discussions here that there is a need for a minimum paste in the system to increase the fines content. While the discussions on this will be continued later while the performance characteristics are being discussed, it is obvious to say that even at the 30% addition of these pozzolans, with an appropriate control on the water and superplasticizer to cementitious materials ratios, the strengths of concrete that can be produced (which can easily be assessed through the presently existing ACI or Euro norms) appeared to have not been realized in the reported literature. Also the possibility of achieving very high strength and significantly higher performance SCCs through the use of the higher-end pozzolans has not even been suggested much less addressed, though these have indeed been adopted in the construction of many very tall structures. It is for this reason that the following chapters address ways to ensure synergy that could result in SCCs of high performance, while realizing the highest strength possible with the cementitious system adopted.

An attempt is made in this chapter to look deeper into the different methodologies proposed by the various researchers, categorize them based on the major thrust in their design philosophy, and also go through some of their designs in specific to get a comprehensive picture of the resulting SCC composites. To be consistent in making a comparison of the concretes designed by the various methods available, a specific gravity of 3.15, 2.20, and 2.60 gm/cc for the cement, fly ash, and both the aggregates, respectively, was adopted in all the design calculations. All the other parameters were kept the same as reported in the different methodologies. Incidentally, most researchers have employed only 30% fly ash in the binder, probably to limit the cement content while increasing the fines content. An important fact as already stated is that most methodologies do not attempt to design these concretes for a specific strength, self-compactability being the primary target. However, with only 30% fly ash content being advocated for the cementitious materials by most researchers, the strength of these could have been reasonably predicted without much difficulty through the available water–cement ratio strength relations of either the American or Euro regulations. This prompted the evaluation of the possible strengths of the concretes designed with these methods based on both the American and Euro regulations for an effective comparison. If the methods have reported the strengths from their own experimental studies, these were also presented along with the strengths assessed for a comprehensive understanding. Further, in each of the methods the details of the concrete mixes obtained were presented in a graphical form to facilitate not only a comparison but also to enable obtaining a concrete mix, starting at any specific parameter of interest.

4.3 Semi-empirical Methods

Some of the early attempts to produce SCCs have relied heavily on the experience of the researchers with conventionally vibrated concretes and their perception of how they could be modified to achieve the desired result, making these efforts essentially semi-empirical. The fact that the available ACI recommendations present methods to produce normal concretes of different workabilities ranging from no slump to almost a collapse even without the addition of superplasticizers (ACI 211.1, 1998), as well as high-strength concretes through the use of the high-range water reducing admixtures in conjunction with fly ash as pozzolans (ACI 211.4, 1998), is ample testimony to the fact that the precursors to the so-called SCCs were always available. Most other national codes also had similar methods for arriving at such concretes. The use of other chemical admixtures to control the cohesiveness wherever required was also available in the domain of concrete technology. The terms flowing concrete, underwater concretes, and self-levelling concretes were all common terminology earlier to the postulation, if it can be termed so, of the present-day SCCs.

Okamura (1995, 1997, 2003), who pioneered this quest for what he termed as SCCs, appears to have modified the generally accepted normal high-slump or flowing concrete mix by modulating the fines and coarse aggregate contents along with superplasticizers to ensure the required thixotropy or consistency. Okamura (1995) opines that though there is a broad similarity between SCC and anti-washout underwater concrete, the need for particularly larger-sized aggregates to pass through the reinforcement cage makes its design that much more complicated. He considers that though the manipulation of the water–cement ratio and superplasticizer content could lead to improved flowability, the passing ability of the coarse aggregate through the narrow spaces limits its content to around 50% of its solid volume. Also at higher fine aggregate contents above 40%, the direct contact between sand particles could result in a decreased deformability. Thus, he proposes a simple mix design system to the following.

- The coarse aggregate content is fixed at 50% of the solid volume.
- The fine aggregate content is fixed at 40% of the mortar volume.
- The water–cement ratio in volume terms could vary between 0.9 and 1.0 depending on the properties and cement.
- The superplasticizer dosage and the final water–cement ratio should be modulated to ensure self-compactability.

In fact the use of VMA is considered only if it is absolutely necessary, maybe to suppress the bleeding and segregation tendencies if exhibited by the resulting fluid mix. The most important factor in this method, after fixing the coarse aggregate content at 50% of the granular skeleton and the fine aggregate content at 40% of the mortar volume, is that the water–powder ratio was assumed to be varying between 0.9 and 1.0 depending on the cementitious materials. As can be seen the method recognizes the need for a higher paste content to support the aggregates in suspension and suggested the use of additional powder content through fly ash at 30% of the total powder content. Figure 4.1 is a graphic representation of the concretes resulting from this procedure, paving the way to directly arrive at the compositions of the mixtures without going through any of the steps required by the method. Assuming that the requirement is to design a SCC containing 50% coarse aggregate with 40% fine aggregate in the mortar volume, the volume of the total powder at any water–powder ratio (by volume) can be seen in Figure 4.1. The water–powder ratio by weight can now be asserted from the figure, which defines the expectant compressive strength of concrete as per the recommendations of either ACI or Euro norms. It is enough to say that the superplasticizer dosage

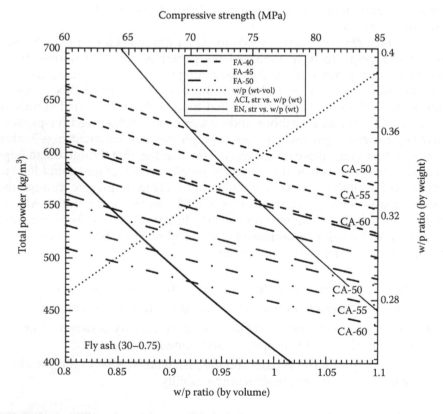

FIGURE 4.1
Composition of SCC mixes by semi-empirical method. (After Okamura, H. and Ozawa, K., *JSCE*, 25, 107, 1995.)

and water content are adjusted to ensure a SCC mix. However, Figure 4.1 will not define the superplasticizer content, which in fact is dependent on its type and composition apart from its compatibility to the cement and powder proposed to be used.

It is seen that the Japanese (1999) and Euro recommendations (2002, 2005) were all based on the original recommendations of Okamura, though the limits suggested for some of the parameters varied occasionally. The Japanese recommendations suggest the following limits for concretes with a maximum coarse aggregate size of 20 or 25 mm.

- The absolute volume of coarse aggregate for the different ranks defined are
  - Rank 1—0.28-0.30 m³/m³
  - Rank 2—0.30-0.33 m³/m³
  - Rank 3—0.32-0.35 m³/m³
- The water content should be in the range of 155–175 kg/m³.
- The water–powder ratio should be in the range of 28%–37% by mass of cement or 0.85–1.15 by volume ratio to cement.
- The powder content should normally be in the range of 0.16–0.19 m³/m³.
- The remaining fraction will be contributed to the fine aggregate while the dosage of the chemical adventure is determined from trial mixes.

The European guidelines for SCCs present the typical range of constituents both in weight and by volume as given in Table 4.1, by suggesting that these are in no way restrictive and could also contain other constitutes.

TABLE 4.1

Typical Range of SCC Mix Composition

Constituent	Typical Range by Mass (kg/m³)	Typical Range by Volume (Liters/m³)
Powder	380–600	
Paste		300–380
Water	150–210	150–210
Coarse aggregate		
Fine aggregate (sand)	Remaining volume (typically 48%–55% of total aggregate weight)	
Water–powder ratio by vol.		0.85–1.10

Source: EFNARC, Specification and Guidelines for Self-Compacting Concrete, European Federation of Producers and Applicators of Specialist Products for Structures, 2002.

The addition of a viscosity-modifying admixture is considered a useful tool for compensating the stability requirements of the concrete in all these methodologies. Incidentally, it can be seen that Figure 4.1 was generated for this broader range of specifications proposed by all these methods to accommodate them. Figure 4.1 also presents the probable strengths in accordance with the American and Euro norms at the corresponding water–cementitious materials ratios.

There have been several other attempts to arrive at optimal solutions for SCCs very similar to the semi-empirical method presented earlier over the years. In fact as already presented, Kheder (2010) reviewing some of these methods for proportioning of SCCs clearly stated that most give only general guidelines. Even so it is important to realize that some investigations particularly have suggested methods based on comprehensive experimental investigations on several of the parameters related to the various aspects of SCCs in the laboratory. Nevertheless, these specific investigations on SCCs also had the benefit of experience in the laboratory with several of the concretes designed for varied applications over a long period of time.

Among these, one of the most comprehensive and most discussed in the literature is the one by Domone (2009), which in fact was based on several years of experience on such concretes (Domone, 1999, 2002, 2006). The method, popularly known as the University College London (UCL) method, initially attempts an optimization of the mortar characteristics through V-funnel tests for the mortar to have a grip on the wide variety of parameters that influence the flow characteristics of the paste. This is then integrated with the investigations on concretes containing 20 and 16 mm aggregates through studies on slump flow, V-funnel, J-ring, and segregation resistance tests. The effect of parameters like coarse aggregate volume and water–powder ratio on the green state characteristics are graphically presented to allow the designer to choose the most appropriate combination that is required for the task at hand. The design methodology suggested involves the following steps:

- Define the specifications of the SCC desired, particularly with the details regarding the passing ability in addition to slump flow and viscosity.
- The constituent materials and their characteristics are checked for conformity to be relevant national standards.
- An initial coarse aggregate content is proposed for the various slump flow, viscosity, and passing ability characteristics as recommended by the EFNARC.
- The fine aggregate volume is set at 45% of the remaining mortar volume.
- The paste volume and composition in terms of the water–powder ratio and admixture dosage are then obtained from the experimental mortar test results available from the laboratory.
- The resulting trial mixes are tested for conformity, and suitable adjustments are made to ensure a satisfactory performance.

A close look at the development of the method and the steps followed for arriving at the final SCC mix appears to be primarily dictated by the characteristics of the mortar and the capacity to support the aggregate in the matrix. In a general way this could be a good starting point for the design of such concrete mixtures in the ready mix concrete or prefabrication industry scenario, where the different constituents of the SCC mixture are generally consistent over long periods of the manufacture, and thus the necessary modifications would be restricted to the minimum.

4.4 Compositions Based on Wetting Water Requirements of the Constituents

It is well recognized that while fine materials help lubricate the matrix and fine powders help in giving the paste the robustness or thixotropy and also help through pore filling effects, the need for additional wetting water for a substantially increased surface area due to their fineness is well established and cannot be neglected, even with the use of the more modern polycarboxylate-based superplasticizers. Marquardt (2002) and Marquardt (2002b) proposed a method for the design of SCCs through studies on the wetting water required for the different constituents in concrete. The fundamental principle of his design is that "self-compacting concrete requires a precise knowledge of the water demands of the constant materials and in particular those of the fines contents, the cement and additives, whose water requirements add up to approximately 90% of the overall water content." The basic methodology is to assume a suitable combined aggregate volume with the corresponding cement content. After the requirements of water for these aggregates and cement are established, the fly ash content is calculated to make up for the remaining volume along with the additional volume of water required for the fly ash.

- As a first step, the granular matter composition of the aggregate mix with sand contents of 30%–40% and a grading curve between standard DIN grading curves A 16 and B 16 are found to be suitable.
- The aggregate volume should lie between 640 and 615 dm³/m³, with an inter-grain volume of about 220 dm³/m³. This helps in arriving at a granular composition with a sufficiently high sand content.
- This quantity is selected based on the cement type and strength class.
- At this stage the water demand for wetting the surfaces of the aggregates and cement is calculated.

- The remaining volume is supplemented through additives like fly ash, limestone, or quartz powder, and the wetting water requirement for these is also evaluated.
- The superplasticizer and stabilizer volumes are also then calculated at 3.5% and 2.5% of the weight of the binder as established by Domone (2009).

Though it is not directly based on the strength to water–cementitious materials ratio relation as in normally vibrated concretes, the resulting strengths of a few concretes from an experimental investigation were also reported. One very important factor that should be recognized from this investigation is that this trend achieved at all the water–cementitious materials ratios ranging from about 0.35 to 0.50 shows that it far exceeded the expected strengths (by almost 50% or to be more specific, almost 20–25 MPa, most of them ranging from 65 to 75 MPa) even with a cement of only 32.5 strength grade of the Euro standards. This is ample testimony to the fact that by using

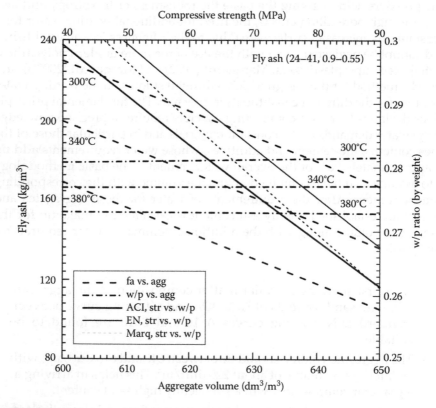

FIGURE 4.2
SCC mix compositions from wetting water requirements. (After Marquardt, I., *Betonwerk + Fertigteiltechnik BFT*, 11, 22, 2002.)

an appropriate validation of aggregates with suitable cementitious materials and water content, one can achieve successfully a reasonably high-strength SCC without much difficulty.

It is important to recognize that they propose a combined aggregate grading between DIN-A and DIN-B, which will be discussed later. As was done earlier the authors have gone through the exercise of calculating the typical mixes based on the proposed methodology and presented the results graphically in Figure 4.2 to make it possible for anyone to directly arrive at an SCC mix as proposed by Marquardt. The strengths reported by them and those calculated based on the American and Euro norms were also superimposed on Figure 4.2. Considering an aggregate volume and cement content, the fly ash content is assessed from Figure 4.2, and for this combination the water–powder ratio is ascertained. Given the water–powder ratio, the strength of the SCC according to the ACI or Euro norms can be obtained.

It is important to note that several of the other methods get a fix on the water requirements based on the recommendations from ACI and British recommendations of earlier years, proposed for conventional vibrated concretes. The fact that these have served well for a reasonably wide spectrum of cementitious materials contents till date, coupled with the fact that most general-purpose SCCs may not have significantly larger proportions of fines in the system, these assessments could still be reasonably appropriate. However, this water content may be considerably inadequate if superfine materials like silica fume, nanosilica, and metakaolin are to be involved in the system.

4.5 Methods Based on Aggregate Distribution and Packing Factors

Su (2001) proposed a method to produce an SCC based on the packing factors of both fine and coarse aggregates in terms of the volume ratio of fine aggregate to total aggregate. The packing factor of aggregate is defined as the ratio of the mass of aggregate of tightly packed state to that of loosely packed state. The method tries to fill the voids in a loosely packed aggregate matrix with a paste of binders that was said to be leading to the use of a smaller amount of binder. The facts that a higher packing will hinder self-compactability due to the higher aggregate content and a lower packing of the aggregates will require higher paste content that could lead to higher shrinkage and higher cost are well recognized. The SCCs investigated showed that as the aggregate packing factor increases from 1.12 to 1.18, the compressive strengths at all ages decrease by almost 15 MPa, a reduction of more than 25%. The inflection point of trend reversal, which certainly must exist, is not obvious from the investigation.

The steps involved in the design can be listed as follows.

- The fine aggregate–to–total aggregate ratio is taken to be ranging from 50% to 57% while the packing factors could vary from 1.12 to 1.18.
- The cement content is calculated as a function of strength.
- The water content is established on the water–cement ratio as was proposed by ACI or any other method in previous studies.
- The remaining volume, made up of fly ash (70%) and GGBS (30%), is calculated from the absolute volume method.
- The additional water requirements for these are calculated based on empirical relations to ensure good fluidity for the mixture of cementitious materials.

Figures 4.3 and 4.4 present the proportions of the different constituents obtained through the above methodology with the addition of only fly ash. It may also be seen that Figure 4.3 presents the possible mixtures and the corresponding strengths calculated by the authors as per the ACI recommendations, while Figure 4.4 presents the same for strengths calculated as per the Euro norms separately, essentially for clarity in presentation. The packing factors and the aggregate ratios for which they were calculated, as proposed by Su, can also be seen in the figures. The actual strengths obtained in his investigations were also presented in both the figures for comparison. In Figure 4.3, considering the specific strength required, the cement content is first established. The water–powder ratio and powder content for the assumed packing factor and volume ratio of fine aggregate to total aggregate (s/a ratio) are then ascertained. However, using Figure 4.3 or 4.4, one can obtain the expected strength as per ACI or Euro norms. It is obvious that the expected strengths from either of these methods are significantly higher than that proposed in the design originally to start with.

It can be seen that in this method the fine aggregate–to–total aggregate ratios at both the extremes of packing factors suggested did not show any major influence. However, a very important observation is that the additional pozzolanic materials content (70% fly ash + 30% GGBS) proposed for the different strength grades of concrete was not a constant and surprisingly varies from 35–50% for the proposed design strength regime of 27.5–48.0 MPa (or his obtained strength regime of about 40–55 MPa). It can also be seen that by extending the limits of the water-binder ratio (w/b) from his 0.31–0.42 to the probable 0.25–0.55 (as shown in Figure 4.3), the additional pozzolan content by this method will vary from 10% to 70%. Incidentally, it may be noted that Ganesh Babu et al. (1993) have presented earlier a method for incorporating different pozzolans in concrete through the efficiency concept, which shows that there is a specific maximum percentage of fly ash that can be accommodated in a concrete of a particular strength. This was

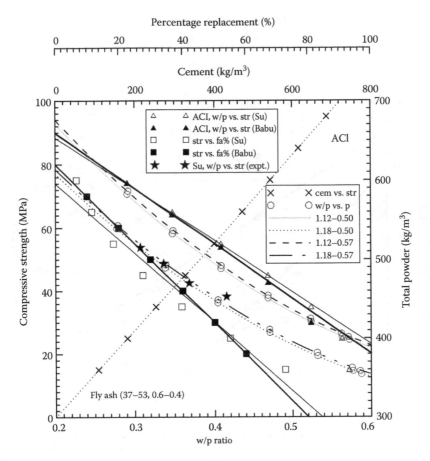

FIGURE 4.3
Composition of SCC mixes by packing factor method—ACI. (After Su, N. et al., *Cement Concrete Res.*, 31, 1799, 2001.)

reported to be varying from 10% to 60% for fly ash with strengths varying from 70 to 20 MPa, respectively, which was actually a very highly conservative limit and could safely be revised to an additional 10% fly ash, yet remaining to be conservative. It is indeed surprising to note that the proposed additional pozzolan contents by Su coincide almost exactly with the proposals by Ganesh Babu (2000) though the strengths predicted were indeed much higher, after appropriately accounting for the changing efficiencies of fly ash at the different percentages. However, it can easily be recognized that the method of accommodating fly ash as an additional component at any percentage through the efficiency concept, limited to the maximum possible suggested here for obtaining a concrete of desired strength, is indeed comprehensive and versatile for appropriately modulating the setting and strength gain rates as well as the durability criteria of the resulting concretes.

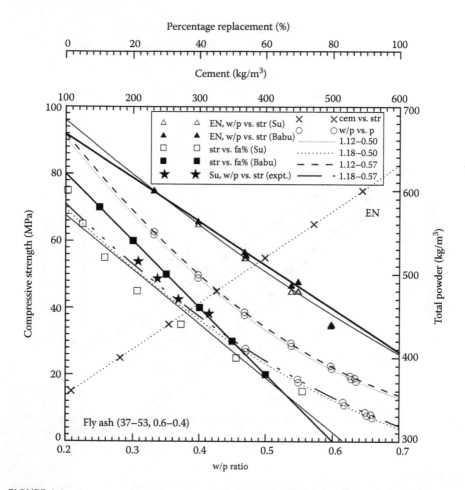

FIGURE 4.4
Composition of SCC mixes by packing factor method—EN. (After Su, N. et al., *Cement Concrete Res.*, 31, 1799, 2001.)

The reason for limiting the range and for adopting the two different pozzolans suggested by Su appears to be to ensure that the problems associated with concrete setting and the strength gain rates do not get seriously affected, though this cannot be guaranteed. Both the design strengths and the strengths obtained are far below those possible by either the ACI or EN methods as can be seen from Figures 4.3 and 4.4. A much more detailed discussion on the efficiencies concept and its potential to realize SCCs of different characteristics through the various pozzolans is relegated to the following segment as it is more relevant there.

4.6 Methods of Limiting the Cementitious Materials through Water Content

An experimental program on SCCs made with high volumes of fly ash (40%–60%) was presented by Bouzoubaa (2001) showing the way for achieving concretes of strengths ranging from about 25 to 50 MPa with relatively low cement contents, by keeping the cementitious materials constant at 400 kg/m³, which is in fact the best approach to achieve normal concretes of lower grade. This will limit the need for additional water and also the corresponding superplasticizer required with higher cementitious materials and lower fly ash contents. Though it is not explicitly stated, the earlier method of Su (2001) also presents a case where the pozzolan content varies substantially. The use of higher cement contents and the more expensive higher-end pozzolans, like silica fume, metakaolin, and so on, is in fact required only for achieving high-strength concretes, where their additional costs can be justified.

An earlier proposed method by Petersson and Billberg (1999) also appears to be based on the concept of minimization of the paste content (also known as the CBI method). The fundamental concepts involved are the two basic aspects of maximizing the aggregate content ensuring that there is no problem of blocking to have a minimum paste content, yet providing for sufficient paste content to ensure both fluidity and strength. The method proposes the use of overall grading of the aggregates, which are assessed for their blocking characteristics through their blocking risk factor similar to the Miner's rule for fatigue. The paste content that makes up the coating and lubricating layer and fills the voids is essentially based on the strength characteristics. The method appears to derive its strength from the grading distribution adopted for the combined aggregate, which has been subjected to similar discussions and modifications in the later years and which is also the cornerstone of the propositions advanced in the later chapters for the design of SCCs of high strength and performance.

4.7 Methods of Incorporating the Cementitious Efficiency of Pozzolans

In an effort to investigate the characteristics of SCCs at various levels of cement replacement with fly ash, while keeping their filling ability constant through an increase in water–powder ratio and reduction in superplasticizer dosage, Liu (2009, 2010) was able to arrive at concretes ranging from about

80 to 40 MPa for replacements ranging from 0% to 80% by volume. It can be seen that the powder content has also reduced from about 539 to 439 kg/m^3 for these concretes, making the evaluation very difficult. The efficiency of the fly ash at these various levels of replacement was evaluated and compared with the efficiencies predicted through the methodology proposed by Ganesh Babu (1996) at different ages, showing a perfect agreement between the proposed relations and the relations obtained experimentally.

One important factor that has to be noted is that in the investigation reported by Liu (2009), the fly ash replacement attempted was by volume percentage while that used in the efficiencies predicted by Ganesh Babu (1996) was actually by the weight percentage and also contained superplasticizers. However, only those up to the limit of 2% were chosen as appropriate for economic considerations. The values proposed by Ganesh Babu (1996) were actually the reduced factors for practical application. While keeping these differences in mind, without looking too deep at some of these variations regarding the actual mix constituents and proportioning, it can still be seen that the earlier efficiencies proposed by Ganesh Babu (1996) were indeed found to be reasonably accurate for the SCCs investigated by Liu (2009, 2010). A similar conclusion was also made in a different context earlier while studying the SCC mix design methodology suggested by Su (2001). These discussions and the conclusions that are derived from them are used in proposing the methodology for the design of SCCs later in this book.

4.8 Procedures for Incorporating Different Pozzolans

The additional fines required for the production of SCCs was, in general, achieved through the use of fly ash, because of its availability in abundance and also its negligible cost. Most researchers have used fly ash at a 30% level, a level that will not seriously affect the setting times and the strength gain characteristics. However, the possibility of using fly ash at the different levels or the use of higher-end pozzolans like silica fume (with much higher cementitious efficiencies) requires a better understanding of the effects of not just the powder content alone but also the effects of the very high fineness associated with some of these materials. In a way as is already known, the use of these materials, if appropriately proportioned, could lead to higher strengths and also higher performance required for specific applications.

Sebaibi (2013) used a compressible packing model similar to that proposed by Su (Domone, 2006) for the design of SCCs using silica fume at 10% as a material for partial replacement of cement. The mix design is similar to that of Su with the same relation for the cement content, while the water–cement ratio strength relation used was from the Euro norms. The water content is then obtained from this, and the superplasticizer dosage required is

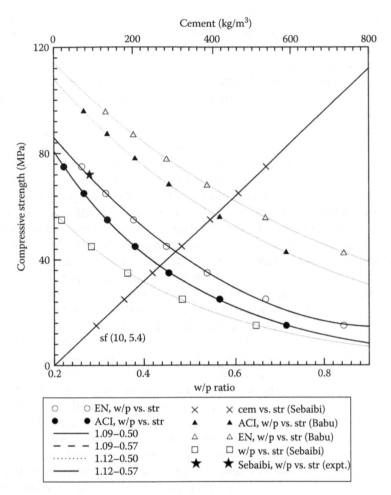

FIGURE 4.5
SCC mix compositions for incorporating different pozzolans. (After Sebaibi, N. et al., *Construct. Build. Mater.*, 43, 382, 2013.)

established through a Marsh cone test. The fact that a concrete containing 400 kg/m³ with the 10% silica fume achieved a strength of about 72 MPa amply testifies to the earlier discussions. Figure 4.5 presents the relationships between the various parameters of the resulting concrete composites through this particular methodology. The figure shows the strength of concrete as observed by the authors and as calculated through relationships proposed by the Euro norms as well as the ACI recommendations. Sebaibi did not consider the efficiency of such a high-end pozzolan like silica fume, and this resulted in the concrete of 50 MPa design strength resulting in a 72 MPa concrete. He also limited the water–binder ratios to 0.45. Also to be noted is the fact that a super fine pozzolan-like silica fume will certainly require additional wetting water, which was not looked into, and this will lead to very high superplasticizer contents. In these assessments by the codes, the efficiency values established earlier (1995) were used. In this method the compositions of the mixtures with the different packing factors and aggregate ratios do not vary significantly and the plotted relations overlap, facilitating the mixtures to be presented in a single graph, unlike in the case of the methodology proposed by Su et al. (2001).

As in the case of the design method by Su, considering the strength required, the cement content and the water–powder ratio (with 10% silica fume) are established from Figure 4.5. It was observed that according to Sebaibi, the water–powder ratio is a constant for any ratio of aggregate proportions (gravel–sand ratio). Also it appears that the method suggests a maximum strength possibility of 80 MPa and a minimum of 40 MPa because of the restricted water–powder ratio of 0.48. The superplasticizer content (with its 32.3% solids) is significant, and it is only appropriate to consider the water in it in the calculations for water–powder ratio to have a more realistic estimate of the strength as was suggested in the superplasticized concretes of yesteryears, though limited similar research efforts to incorporate other pozzolans were made by a few other researchers. Kheder (2010) reported an investigation on the design of SCCs containing limestone powder as almost an inert powder in SCCs. Khaleel (2014) reported a work on metakaolin indicating that the other high-end pozzolans can also be incorporated in SCCs with a suitable understanding of their effect on the hardened state characteristics.

At this point it is appropriate to recall some of the earlier efforts in our laboratory elucidating the cementitious efficiency of the different pozzolans like fly ash, silica fume, GGBS, and limestone powder (Ganesh Babu, 1995, 1996, 2000, 2012). The concretes investigated adopted the available water–cementitious materials ratio relations of normally vibrated concretes with due consideration of the efficiency of the particular pozzolan at the percentage adopted. It was also noted that, particularly at the very low water–cementitious materials ratios required for high-strength concretes, both the water and the plasticizer contents have to be addressed to achieve an adequate workability of concrete. In fact it is always seen that in these concretes of very high strengths of about 100–120 MPa, having high cementitious materials contents, there is a very thin line between a highly thixotropic mass of concrete (almost rubbery), which will need a considerable compaction energy, and its becoming a concrete with collapse slump and free flow without segregation (due to the high paste content). This transition can be realized during mixing, only with the use of adequate but a minimum amount of both water and superplasticizer. A water content in excess of the amount required will not only reduce strength but will also lead to bleeding and segregation, though at lower levels this can be addressed by the addition of a thixotropy agent like a VMA. However, without this minimum water required any further addition of superplasticizer is not really effective. In fact many research efforts utilizing very high superplasticizer contents in such cases do not realize that it is probably the water in the superplasticizer that is primarily responsible for the fluidity rather than the superplasticizer itself. These studies on efficiency of the different pozzolans also contain summary tables of the characteristics of the concrete mixes considered for evaluation, which clearly shows beyond any doubt that there is indeed a certain maximum percentage that can be accommodated in a concrete of a particular strength depending upon the efficiency in specific and also maybe the other characteristics of the pozzolan in consideration. It was also observed that superfine materials like silica fume require additional wetting water for fluidity, but will introduce better cohesion and thixotropy to the concrete reducing or avoiding the need for VMAs, making them essentially self-compacting.

4.9 Approaches for a Specified Compressive Strength

This is an approach that is similar to the ones that are normally adopted in the design of conventionally vibrated concretes world over in which the compressive strength of concrete is related to its water–cement ratio. The water required for a specified workability is then chosen based on the characteristics of the aggregates. The aggregates themselves are then defined in terms of either the coarse and fine aggregates or as a combined aggregate. However as was stated by Felekoglu (2007), unlike this method of proportioning of traditional concrete in which the water–cement ratio is kept constant in order to obtain the required strength, in SCCs the water–powder ratio that is in fact sensitive to self-compactability is the one that needs to be addressed. Even so it would only be appropriate to look at the possibility of starting with the necessary strength requirement and adjust the proportions to arrive at self-compactability.

Yu (2005) proposed a method for calculating the proportions in SCC based on an experimental study on concretes with strength grades of 30–60 MPa. The method involves the calculation of the water–binder ratio for a given strength of concrete through constants evaluated from the experimental study. It can also be seen that given the characteristics of the aggregates, they remain a constant for all compressive strengths of the SCCs designed by this method. It is reported that fly ash replacements of 30%–40% are in order, while the water–cementitious materials ratio should be below 0.4. The water content can now be calculated, and the plasticizer content is adjusted based on the requirements of SCC. A typical variation of the parameters designed according to the method suggested is presented in Figure 4.6. In generating this figure, the aggregate characteristics were assumed to be the same as those given in the article as can be seen. Also, the calculations in generating the figure were all done for 40% fly ash, and the figure presents the parameters with "w/p ratio" as investigated by Yu. The American and Euro

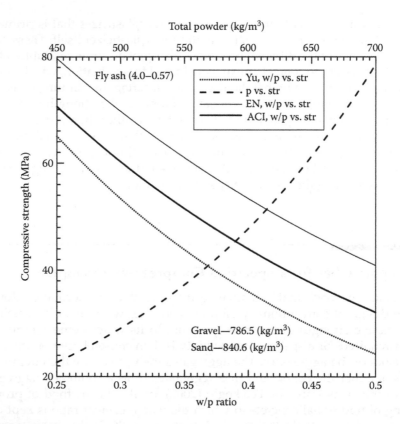

FIGURE 4.6

SCC mixture compositions for a specified compressive strength. (After Yu, Z. et al., Optimal mixture design of high performance self-compacting concrete, in *First International Symposium on Design, Performance and Use of Self-Consolidating Concrete*, 2005, pp. 181–190.)

relations presented are also corrected for the 40% fly ash the mixes contain, so that they can be used directly.

It can be seen that the designs proposed by some of the earlier studies have also been using the water–cement ratio strength relationships proposed by the American or the Euro standards as part of their design procedure (Su, 2001; Marquardt, 2002; Sebaibi, 2013). In some of these, an indirect relation between the strength and cement content was used, indirect as the water content required for a specific consistency is a constant as was proposed by Lyse (1932) a long time ago and was also discussed by Hansen (1992) recently. Interestingly, investigating the effect of water–cement ratio on the fresh and hardened state characteristics of SCCs containing limestone powder, Felekoglu (2007) studied five different mixtures with the same cement dosage (~377 kg/m³) by reducing the free water content from 227 to 140 kg/m³ while increasing the superplasticizer dosage from 3.7 to 13.0 L/m³ in order to obtain a slump-flow value in the range of 65–80 cm. Naturally, with the different water–powder ratios, the strengths, depending on the efficiency, were different, but the fact remains that they were all SCCs with different attributes of consistency. It is also noteworthy that with only a third as solids (35.7%, generally limited to 40% maximum by most producers), one should wonder if it is the plasticizer or the water in it that is responsible for the plasticizing effect. A better approach will be to account for this water in the calculation of the water–cement ratio as often advocated. Overall from his results on fresh concrete tests (slump flow, T_{50}, V-flow, L-box, air content), we can see that there is a marked change in the fresh concrete behavior at a water content of 180 kg/m³ in all the tests. Maybe one can say that a water content of about 160–200 kg/m³ is the most appropriate for producing SCCs with relative ease depending on the powder content and the type of plasticizer. The design involves establishing the powder content and the water–powder ratio for a specific strength. The fly ash content presently was assumed to be 40%. The aggregate contents were taken to be the same as was used by Yu (2005), assuming that their characteristics are the same.

4.10 Methods Based on Rheometer Tests

The Laboratoire Central des Ponts et Chaussees (LCPC), method proposed by Sedran (1996) uses a rheometer to characterize the shear yield stress (related to the slump) of fresh concrete along with software based on a mathematical model for optimizing the aggregate–binder granular skeleton. The mathematical model software used is based on what is termed as a compressible packing model. The software appears to help in determining the optimum packing density using less water for the same workability or improved workability at constant water content. The need for well-graded aggregates and an appropriate combination of cement (at about 70%); mineral admixture (at about 30%), varying with the strength requirements; and a compatible superplasticizer has been advocated.

4.11 Methods Based on the Rheological Paste Model

Self-compactability requirement envisages imparting additional fluidity to the concrete matrix while ensuring no segregation of the constituents particularly the larger fractions of coarse aggregate. One of the approaches followed in ensuring a nonsegregating mix is to control the viscosity of the paste phase, as attempted by Saak (2001), extended by Bui (2002), and used by Ferrara (2007) in their efforts to extend it further to accommodate fibers. Saak, recognizing the fact that the production of a cohesive fluid matrix in itself is not difficult but to keep particularly the larger particles of coarse aggregate in suspension in such a fluid is more difficult, tried to analyze the dynamics of segregation control and defined the concept of self-flow zone (SFZ). It is recognized that the exact values for the minimum yield stress and viscosity for the optimal performance of an SCC are dependent on the density difference between the paste and aggregates, and at higher viscosity, though segregation is avoided, workability will be lower and vice versa. Bui expanded the concept to include the effect of aggregate interaction. The effect of grouping and assemblage of aggregate masses that may include a part of the paste matrix were specifically addressed. In fact these problems can occur during the initial high thixotropy of the paste, which can be avoided with better continuous distribution of the aggregate and slightly larger sand content apart from proper mixing. Also, more importantly, inadequate water content particularly with higher superplasticizer content, inappropriate mixing sequence, and insufficient mixing time should also be addressed. Ferrara's concept included the effects of aggregate particle size distribution and the fine aggregate–coarse aggregate ratio. In a way they were specifically looking at the yield stress and the minimum viscosity required to avoid segregation, which probably is the most appropriate for mixes with very high fines content or with larger maximum size of the coarse aggregate. The importance of limiting excess paste layer on the aggregate to about twice the thickness and ensuring only the minimum cementitious paste required to coat the aggregate was recognized. This may ensure that the suspension of aggregate in paste matrix is restricted to the minimum required paste, ensuring segregation resistance while resulting in economy in terms of the binder required. In principle, the reduction in the coarse aggregate content and the additional fines proposed by many in all the methods is essentially to deal with this very specific problem. Also the addition of a VMA, which was hitherto not discussed, is also to ensure the same property in the mortar. It can be seen that even the very first proposal by Okamura (2003) also attempts the same effect. Ferrara (2007) presented an extension of this rheology of the paste model with suspended aggregates to fiber-reinforced concretes. As a part of this exercise, he also reported self-compacting mixes without the fibrous addition. Figure 4.7 presents the relations between the various parameters of the mixes designed at the three different water–binder ratios of 0.32, 0.36, and 0.40 adopted by him. The figure not only contains the estimated strengths based on the ACI and the Euro norms as already discussed, but also shows the strengths experimentally obtained during his experimental investigations. For the required compressive strength, the total powder and water–binder ratio are assessed. Now assuming the nearest water–binder ratio (say 0.36), the fine aggregate and total aggregate quantities are established for the total powder being used. Now the strength expected as per the ACI or Euro norms can also be seen from the figure.

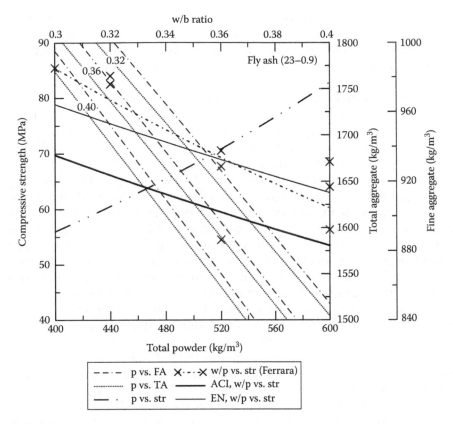

FIGURE 4.7
SCC mixtures resulting from the rheological paste model. (After Ferrara, L. et al., *Cement Concrete Res.*, 37(6), 957, 2007.)

4.12 Methods Based on the Rheological Paste Model Incorporating Fibrous Materials

After having looked at some of the methodologies by which SCCs could be produced, it is not out of place to see how to incorporate fibers into the system. In fact during the early stages of superplasticizers in concrete, the need for additional fines (through fine aggregates) was recognized to avoid bleeding and segregation. Later it was also seen that these additional fines could be quite advantageous while incorporating fibers also, and most designs for fiber-reinforced concretes increased the fine aggregate content by a small margin. It is only appropriate to think, knowing the facts that there are additional fines not only in terms of fine aggregate but also in terms of the cementitious paste, that the incorporation of fibers into SCCs should not be a serious problem, with the fact that their total quantity is limited to about 2% for the appropriate length of the fibers. The specific factors and the discussions on his investigations on fiber-reinforced concretes are actually discussed in detail later while looking at the problems associated with fibers in SCCs.

As already stated Ferrara (2007) presented an extension of the rheology of the paste method to fiber-reinforced concretes. In this method the fibers have been treated as equivalent aggregate particles, defined through the surface area equivalence and the density ratio. It is to be recognized that the method does not recognize the effects of the length of the fiber in relation to the max size and aggregate, which could be a good parameter that may be of significance in the design of fiber-reinforced concretes. It can also be observed that while the specific surface methodology will indeed cater to the needs of estimating the amount of paste required for coating them, it may not reflect adequately the interactions between the fibers (particularly in the case of longer fibers) and their effect on the workability and passing ability as desired in the case of SCCs. Probably considering the fiber interactions of a triangular mesh of some dimension related to the length of the fiber and size of the aggregate that could pass through it could be an appropriate method to define the fiber inclusions in concrete. A similar effort was earlier reported by Grünewald (2009), wherein the fiber length and aggregate diameter were connected to the risk of blocking in fibrous SCCs. Even so the limitations on fiber content as was the case for normally vibrated concretes may be appropriate to start with. Similar to the concretes without fibers investigated, the relations between the various parameters of the mixes designed with fibers at the different water–binder ratios are presented in Figure 4.8. It can also be seen that the fiber content was kept constant at 50 kg/m³ as was proposed by the authors. The strengths experimentally obtained during their investigations were also presented as was done in the previous figure. For the required compressive strength, the total powder and water–binder ratio are assessed. The methodology of arriving at the mixture composition is very similar to that described for the plain concretes that were discussed earlier. Assuming the nearest water–binder ratio (say 0.36), the fine aggregate and total aggregate quantities are established for the total powder used. Now the strength expected as per the ACI or Euro norms can also be seen from the figure.

4.13 Guidelines Based on Statistical Evaluations

At this stage it is important to point out that there have been several attempts to arrive at concrete mixtures based on statistical evaluations, be it through expert systems, fuzzy methods, or neural networks, and the SCCs were no exception. However, for these methods to be successful, first a large database

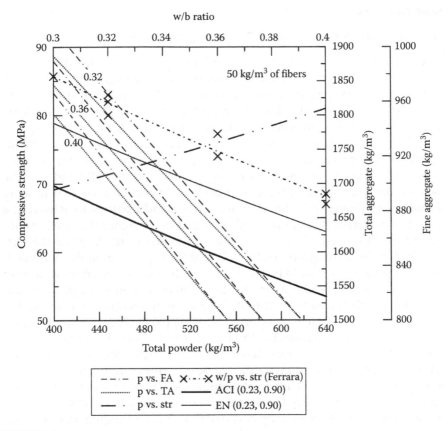

FIGURE 4.8
SCCs for incorporating fibers in a rheological paste model. (After Ferrara, L. et al., *Cement Concrete Res.*, 37(6), 957, 2007.)

that covers the effects of at least the most important parameters, which are by themselves far too many, is required. Also, most of the attempts are based on limited experimental investigations of their own or a few others from the literature that are available. Apart from this for these methods to be successful, a fundamental insight into the various aspects of the behavior constituents, mixture details, and the parameters that govern the characteristics of the designed mixes is required to be able to predict appropriately. For this reason it is not proposed to specifically examine any method that is based on this concept, yet recognizing the fact that these could still be excellent tools to arrive at fairly optimal solutions, specifically to solve the problems locally, like in the case of ready mix concrete plants that have limited variations in the concrete constituents and other production-related parameters. Khayat (1999) presented a factorial design model for proportioning SCCs based on this philosophy of design optimization with a comprehensive insight into the making of SCCs of high performance.

4.14 Need for a Relook and Proposed Methodology

Conventional concrete mix design incidentally suggests that the water–cement ratio or water–to–effective cementitious materials ratio in the case of concretes containing mineral admixtures will be most appropriate to ensure that the concrete of a desired strength is achieved. The water contents required for specific workability levels, even up to the level of flowing concretes, have already been established in these procedures taking into consideration the maximum size and the grading distribution of aggregates. The required grading characteristics of the aggregates individually and in some cases as combined grading were also defined. It can be seen that flowing concretes containing both chemical and mineral admixtures to achieve the desired characteristics of fluidity and segregation resistance that are the hallmark of the SCC were already available in the literature. Another fact that is evident in these design procedures is that for higher strengths, the maximum size of aggregate could be limited to much lower levels, from a value of 20–25 mm to about 10–12 mm or even lower to ensure effective packing and good compatibility, and the fact that these were all mostly self-compactible was not specifically stated as reported by Collepardi (2001). A similar need for anti-washout admixtures for the thixotropy of underwater concretes is also well known. The authors, having had the experience of working on high-strength high-performance pozzolanic concretes ranging up to 120 MPa with fly ash, silica fume, GGBS, metakaolin, and so on, always had to ensure that particularly at the very low water–cementitious materials ratios, a very high workability (collapse slump and free flow without segregation) is absolutely essential to avoid the critical problem of inadequate compaction of these concretes to achieve the highest strength and performance. The same is also true in our recent investigations on ultra-high performance concretes of about 180 MPa with very large cementitious contents along with fibers that were not subjected to either pressure or thermal regimes.

One look at the methodologies presented earlier for the design of SCCs shows that each of these concentrated on the effectiveness of a specific concept, while ensuring that the other factors are also modulated to realize the ultimate goal. It is enough to say that all the authors agree on the principle that the fines content be increased to ensure adequate flowability, while correspondingly reducing the coarse aggregate content. However, in doing so the fact that the actual gradation of aggregates, and their appropriate combination, could ensure the flowability with much less fines is recognized but not adequately addressed. Hu (2011) have shown that well-graded aggregate could considerably reduce the yield stress and consequently the viscosity of concrete. Some authors have also discussed the loosening effect in having a continuous gradation. Also, the increased fines suggested should not require large quantities of additional water for wetting, which could increase the water–binder ratios, thus decreasing the strengths possible. This being so, it is also important to note that a certain amount of minimum fines would be required for adequately coating all the aggregates with the paste and also fill the pores left in the system. One needs to understand that in the particle packing density, specifications should take into account this additional paste layer on the aggregates so that the aggregates can slide over each other most effectively. All these factors are not new to a designer attempting the SCC or even concrete, but the difficulty in addressing all these conflicting parameters in one go is the reason why there is some divergence in the methodologies proposed by the different investigators. Finally, it is obvious that a mixture in which the granular skeleton is of the highest packing density while having the grading characteristics that could facilitate free flow through the loosening effect (similar to the granular flow in silos) and aided by sufficient paste content (with superplasticizers and VMA) that will ensure a reasonable coating thickness to facilitate the sliding of the aggregates further resulting in an optimum cement content is the way forward for achieving a SCC.

At this stage it is appropriate to look at the broad picture of the concretes resulting through the various mix design procedures presented earlier. Table 4.2 presents the major parameters that govern the behavior of SCCs and the range of the values resulting from each of them. As some of them do not explicitly suggest the strengths of the resulting concretes, Table 4.2 also contains the probable strengths that these mixes could have based on the strength to water–binder ratio relationship proposed in the Euro norms as already stated. This comparison will throw light not only on the proposed methodologies but will also be able to help in understanding the makeup of constituents and their effects for ensuring self-compactability in concretes. Based on this understanding and also a comprehensive evaluation of the SCC mixtures reported in the literature, the authors attempted to suggest guidelines for producing SCCs with reasonable confidence.

After considering all the methodologies discussed here for the design of SCCs, it looks obvious that in an effort to ensure self-compactability, on the one side the fines content was revised by lowering the coarse aggregate proportion, while on the other side they looked for an effective packing methodology with the paste materials coating a loosely packed aggregate skeleton with an appropriate thickness to ensure sliding fluidity. In both these efforts, the resultant strength of the concrete that could be realized through these methods was primarily established by testing the resulting concretes, though some methods have chosen the powder content as a factor of the compressive strength. Incidentally, we have already gone through the broad recommendations by the Japanese, Euro, and American national bodies for arriving at the required self-compactability of concrete through general and overarching guidelines that ensure not only self-compactability but also a reasonable stability of the resulting mix. In fact these by themselves were guidelines put together by a select committee after looking at the available information at that point of time and to be consistent with the tradition and the construction practices prevailing in that country. Many of these have had the background

TABLE 4.2

Overview of the Recommendations for Self-Compacting Concrete

Reference	(p) Total Powder (kg/m³)	(f—Fly Ash) Pozzolan (%)	(w/p) Ratio	(w) Water (kg/m³)	Paste Volume (%)	(D) CA max. size (mm)	(CA) vol. (%)	CA in TA (%)	FA in Mortar (%)	Strength (exp./est.) (MPa)
					Parameters					
Okamura (1995) and (2003)	—	Fly ash (30%)	0.9–1.0 (vol.)	—	—	40	—	50	40	65–85 (est.)
Japanese (1999)	16%–19% (vol.)	Fly ash (30%)	0.28–0.37 0.85–1.15	155–175	—	20–25	28–35	—	—	65–85 (est.)
EFNARC (2002)	16%–24% 400–600	Fly ash (30%)	0.8–1.10 (vol.)	<200	—	<20	28–35	—	—	65–85 (est.)
European (2005)	380–600	Fly ash (30%)	0.85–1.10 (vol.)	150–210	30–38	12–20	27–36 750–1000	45–52 from FA	—	65–85 (est.)
Marquardt (2002)	500–600	Fly ash (24%–41%)	0.26–0.28	135–145	—	8–16	—	45	—	50–90 (est.)
Su et al. (2001)	425–553	Fly ash/ GGBS	0.3–0.41	170–176	34–37	25	26–28	45–52	48–50	15–75 (exp.)
Yu et al. (2005)	500–650	Fly ash (30%–40)%	<0.41	178–228	41	—	33	48	—	25–60
ACI 237 (2007)	400–475	Fly ash, SF, GGBS, LSP	0.32–0.45	0	34–40	<12.5	28–32	—	—	—
Sebaibi (2013)	250–650	SF	0.30–0.9	130–170	—	4–10	—	46–53	—	15–75 (exp.)
Ferrara et al. (2007)	420–605	Fly ash (30%)	0.32–0.40	150–215	32–44	10	—	42–44	—	65–85 (exp.)

Note: At fly ash 10%–70%, silica fume 5%–35%, GGBS 10%–70%, LSP 10%–50%.

of extensive laboratory testing with the types of materials and processes that are common in practice. It will be extremely difficult to go through each of these in detail, along with the background information that is available in making these recommendations. Many of them once again are also related to the fundamental concepts and processes that define the making of a SCC.

Segregation being the most important problem associated with the production of SCC, the primary aspects that needed a reasonable understanding obviously are the following.

- A reasonably fluid paste that is yet viscous enough to support particularly the coarser aggregate particles from settling, requiring an assessment of the fluidity of the paste or the mortar and even the settling velocity of some of the heavier aggregates
- A reasonable quantity of the paste itself to coat all the aggregate particles with a sufficiently thick layer, inducing fluidity and mobility to the mix
- An appropriate fix on the pozzolanic and pore filling effects of the additional fines included in the mixture along with the water–cementitious materials ratio to strength relation
- An appropriate grading of aggregates that permits a loosening effect, which enhances the fluidity and mobility of not just the paste but also the aggregate mass
- A restriction on the maximum size of the aggregate and its grading, to have the required passing ability through even a congested reinforcement cage
- A complex optimization of the physical processes involved in the above, through a mathematical methodology like fuzzy logic, neural network, or even simple factorial design

These aspects of the design for ensuring not only an optimal design of the SCC but also an SCC of higher performance will be addressed in a cogent and comprehensive manner in Chapter 5.

References

ACI 211.1, *Standard Practice for Selecting Proportions for Normal, Heavyweight and Mass Concrete*, ACI Manual of Concrete Practice, American Concrete Institute, Farmington Hills, MI, 1991, 38pp.

ACI 211.4, *Guide for Selecting Proportions for High-Strength Concrete with Portland Cement and Fly Ash*, ACI Manual of Concrete Practice, American Concrete Institute, Farmington Hills, MI, 1993, 38pp.

ACI 237R-07, *Self-Consolidating Concrete,* American Concrete Institute, Farmington Hills, MI, 2007, p. 30pp.

Bouzoubaa, N. and Lachemi, M., Self-compacting concrete incorporating high volumes of class F fly ash Preliminary results, *Cement and Concrete Research,* 31, 2001, 413–420.

Bui, V.K., Akkaya, J., and Shah, S.P., Rheological model for self-consolidating concrete, *ACI Materials Journal,* 99(6), 2002, 549–559.

Collepardi, M., A very close precursor of self-compacting concrete (SCC), in *Symposium on Sustainable Development and Concrete Technology,* San Francisco, CA, pp. 431–450, Suppl. Vol., 2001.

Domone, P., Proportioning of self-compacting concrete—The UCL method, University College London, London, 2009.

Domone, P.L., Relationships between the fresh properties of SCC and its mortar component, *Proceedings of the First North American Conference on the Design and Use of Self-consolidating Concrete,* Chicago, IL, 2002, pp. 37–42.

Domone, P.L., Self-compacting concrete; An analysis of 11 years of case studies, *Cement and Concrete Composites,* 28(2), 2006, 197–208.

Domone, P.L. and Jin, J., Properties of mortar for self-compacting concrete, in *RILEM International Symposium on Self-Compacting Concrete,* Stockholm, Sweden, 1999, pp. 109–120.

EFNARC, *Specification and Guidelines for Self-Compacting Concrete,* European Federation of Producers and Applicators of Specialist Products for Structures, Surrey, U.K., 2002.

Euro-EFNARC, *Specification and Guidelines for Self-Compacting Concrete,* European Federation of Producers and Applicators of Specialist Products for Structures, Surrey, U.K., 2005.

Felekoglu, B., Turkel, S., and Baradan, B., Effect of water/cement ratio on the fresh and hardened properties of self-compacting concrete, *Building and Environment,* 42(4), 2007, 1795–1802.

Ferrara, L., Park, Y., and Shah, S.P., A method for mix-design of fiber-reinforced self-compacting concrete, *Cement and Concrete Research,* 37(6), 2007, 957–971.

Ganesh Babu, K. and Chandra Sekhar, B., Efficiency of limestone powder in SCC, *MRS Proceedings,* 1488, 2012 , pp. 39–42.

Ganesh Babu, K., Rao, G.S.N., and Prakash, P.V.S., Efficiency of pozzolans in cement composites, in *Concrete 2000,* Dundee, U.K., Vol. 1, 1993, pp. 497–509.

Ganesh Babu, K. and Siva Nageswara Rao, G., Efficiency of fly ash in concrete with age, *Cement and Concrete Research Journal,* 26(3), 1996, 465–474.

Ganesh Babu, K. and Sree Rama Kumar, V., Efficiency of GGBS in concrete, *Cement and Concrete Research Journal,* 30(7), 2000, 1031–1036.

Ganesh Babu, K. and Surya Prakash, P.V., Efficiency of silica fume in concretes, *Cement and Concrete Research Journal,* 25(6), 1995, 1273–1283.

Grünewald, S. and Walraven, J.C., Transporting fibres as reinforcement in self-compacting concrete, *HERON,* 54(2/3), 2009, 101–125.

Hansen, T.C. and Hedegaard, S.E., Modified rule of constant water content for constant consistency of fresh fly ash concrete mixes, *Materials and Structures,* 25(6), 1992, 347–354.

Hu, J. and Wang, K., Effect of coarse aggregate characteristics on concrete rheology, *Construction and Building Materials,* 25(3), 2011, 1196–1204.

JSCE, *Recommendation for Self-Compacting Concrete*, Guidelines for Concrete, No. 31, JSCE, 1999.

Khaleel, O.R. and Abdul, R.H., Mix design method for self-compacting metakaolin concrete with different properties of coarse aggregate, *Materials and Design*, 53, 2014, 691–700.

Khayat, K.H., Ghezal, A., and Hadriche, M.S., Factorial design model for proportioning self-consolidating concrete, *Materials and Structures*, 1999, 32(9), 679–686.

Kheder, G.F. and Al Jaidiri, R.S., New method for proportioning self-consolidating concrete based on compressive strength requirements, *ACI Materials*, 107(5), 2010, 490–497.

Liu, M., Wider application of additions in self-compacting concrete, Doctoral thesis, submitted to University College London, 2009, 392pp.

Liu, M., Self-compacting concrete with different levels of pulverized fuel ash, *Construction and Building Materials*, 24, 2010, 1245–1252.

Lyse, I., Tests on consistency and strength of concrete having constant water content, *ASTM*, 32(2), 1932, 629–636.

Marquardt, I., Determination of the composition of self-compacting concretes on the basis of the water requirements of the constituent materials: Presentation of a new mix concept, *Betonwerk + Fertigteiltechnik BFT*, 11, 2002, 22–30.

Marquardt, I., Diederichs, U., and Vala, J., Determination of the optimum water content of SCC mixes, in *First North American Conference on the Design and Use of Self-Consolidating Concrete*, ACBM, North Western University, Chicago, IL, 2002, pp. 81–88.

Okamura, H., Self-compacting high-performance concrete, *Concrete International*, 19, July 1997, pp. 50–54.

Okamura, H. and Ouchi, M., Self-compacting concrete, Japan Concrete Institute, *Journal of Advanced Concrete Technology*, 1(1), 2003, 5–15.

Okamura, H. and Ozawa, K., Mix-design for self-compacting concrete, *Concrete Library, JSCE*, 25, 1995, 107–120.

Petersson, O. and Billberg, P., Investigation on blocking of self-compacting concrete with different maximum aggregate size and use of viscosity agent instead of filler, *Proceedings of First International RILEM Symposium on Self-Compacting Concrete*, Stockholm, Sweden, 1999, pp. 333–344.

Saak, A.W., Jennings, H.M., and Shah, S.P., New methodology for designing self-compacting concrete, *ACI Materials Journal*, 98(6), 2001, 429–439.

Sebaibi, N., Benzerzour, M., Sebaibi, Y., and Abriak, N., Composition of self-compacting concrete using the compressible packing model, the Chinese method and the European standard, *Construction and Building Materials*, 43, 2013, 382–388.

Sedran, T., de Larrard, F., Hourst, F., and Contamines, C., Mix design of self-compacting concrete, *Proceedings of the International RILEM Conference on Production Methods and Workability of Concrete*, Paisley, Scotland, 1996, pp. 439–450.

Su, N., Hsu, K.C., and Chai, H.W., A simple mix design method for self-compacting concrete, *Cement and Concrete Research*, 31, 2001, 1799–1807.

Yu, Z., Pan, Z., and Liu, X., Optimal mixture design of high performance self-compacting concrete, in *First International Symposium on Design, Performance and Use of Self-Consolidating Concrete*, Changsha, China, 2005, pp. 181–190.

5

Concepts and Criteria for High-Performance Self-Compacting Concretes

5.1 Introduction

The inadequate performance of the concrete used in constructed facilities is a cause for serious concern today, even for the most advanced of nations. This has led to the need for research and development of materials and technologies that could perform better in the increasingly hostile environments resulted due to industrialization. Also, it is important to realize that the durability and performance of structures should be the primary criteria even from the conceptual design stage. The planning, design, execution, quality assurance, and maintenance should all be an integral part of the strategy for achieving high-performance structures with a low life-cycle cost. The advent of computers in recent years has facilitated the structural designer to solve highly complex problems much faster. But it is seen that the increasing dependence on these packages and the associated computational problems have forced the designer to neglect the fundamental understanding of the behavior of structures. Apart from this, it should be observed that the performance of structures is primarily dependent on the behavior of the materials and the construction quality once the design is finalized. It is also obvious that the extent of savings possible with an appropriate choice of materials will be far more significant, while the possibilities of cost saving through alternate designs without the inputs from possible alternative and appropriate materials are at best marginal. Also, the newer and innovative methods of construction, including prefabrication in part or in whole, that have come into practice in recent years, coupled with the advancements in materials suitable for specific environmental zones in which the structure has to perform, can result in a high performance as well as a more cost-effective solution.

The most important ingredient in concrete, which actually binds the aggregate (filler), is the cement, and it is also the most expensive of all, because of the pyro-processing involved. There are several different varieties of cements, developed over the years, to suit the specific needs of the different environments. It is necessary that the behavior of these cements and cementitious composites of today be understood and evaluated for their performance. Also, it is recognized that the use of admixtures—chemical admixtures (to modify the green or hardened state characteristics of the concrete composites) and mineral or pozzolanic admixtures (to achieve better performance levels in terms of durability)—can significantly modulate or improve the various characteristics of the cements and concretes to suit the different types of construction and performance requirements. This has led to numerous studies related to these cementitious composites. Concrete has undergone several modifications to suit the needs of the construction industry. One such modification that has clearly attracted the attention of many researchers in recent times is the concept of self-compacting concrete (SCC), a concrete that consolidates under its own weight. Though the concept of flowing concretes, mostly for the grouts of yesteryear, is not totally new, the need to collate the methods of achieving this type of collapse slump concrete that is not prone to either segregation or bleeding even at that high slump has been addressed by several researchers as discussed earlier. At this stage it is imperative to have a comprehensive review of these efforts, at least the most prominent ones, among them which are generally accepted, and to see how a more comprehensive and effective solution can be put in place particularly to achieve high-performance SCC composites, an urgent need in the present industrial scenario. As has already been observed, most of the theories or design procedures have concentrated on the different proportions in the mix constituents to ensure self-compactability but have not focused on arriving at a concrete of a specific strength, which is often needed by the construction industry, let alone a concrete of a specific performance. Thus, the following part deals with the various parameters that influence the mix composition and attempts to propose effective guidelines for arriving at SCCs of specific strength with an assured performance level.

5.2 Fundamental Concepts of Performance

Structural concretes presently used in the construction industry can be broadly classified into three categories based on their strength, namely, normal-strength, high-strength, and very-high-strength concretes. Though there are no well-defined limits for this classification, the corresponding representative strength values can be taken to be 30–60, 60–90, and 90–120 MPa, which can all be achieved with the presently available materials, equipment, and construction practices. The specifications for concrete mix design are available for both normal and high-strength concretes, and, in general, high-strength concretes require the use of chemical and/or mineral admixtures. The general procedure for concrete mix design revolves around the following few well-accepted fundamental principles. First, the properties of the constituent materials available—cement, aggregates, and water—are to be evaluated.

Based on the above, for a required strength and workability, the water–cement ratio and water content are determined. From these, the cement content is obtained and the total aggregate content is worked out either by the absolute volume method or by the weight method. Then an appropriate ratio of fine aggregate content to the total aggregate content is calculated based on the maximum size of the coarse aggregate and the grading of the fine aggregate as in some codes, while the grading of the combined aggregate is ensured in the others. In addition, durability is ensured through recommendations in terms of the type of cement, minimum and maximum cement contents, and the maximum water–cement ratio depending upon the environmental conditions. Incidentally, it is felt that the increase in strength if any, due to the limitations on maximum water–cement ratio and minimum cement content, could be made use of in the structural design calculations to achieve economy.

Evaluation of the performances of the constructed facilities requires a specific and appropriate understanding of the three different domains of influence, namely:

1. Parameters related to environmental structure interactions
2. Parameters related to materials of construction
3. Parameters related to the construction and maintenance

A broad outline of how these parameters are to be accounted for in the design of high-strength SCCs was earlier presented by Okamura (2003). However, in the present context the specific details regarding the constituents and design of SCCs alone are being discussed, while the methodology of assessing the structural performance will be considered later.

5.3 Environmental Parameters

The exposure conditions of structures to the environmental forces and the interaction of the members of the structure with these forces will decide the performance of the structure. For this reason various national codes have suggested several indirect methods, ensuring that the materials that are used are resistant to these environmental forces, while the designer attempts to ensure that the structure is resistant to the loadings imposed. It will be enough to say that these provisions from the various national codes only suggest the basic minimum requirements and in several cases the industry has to go well beyond these guidelines to have a reasonable cushion to ensure that the performance of constructed facilities is not adversely affected even if there were to be some unforeseen and unprecedented acts of nature.

The CEB-FIP recommendations broadly present the environmental conditions under the categories of dry, humid, seawater, and industrial environments, which are further subdivided appropriately. Table 5.1 presents

TABLE 5.1

Durability Recommendations Related to Environmental Exposure

Env. Class	Environmental Conditions	Liquids			Soil	Reinforced Concrete						
		pH	CO_2 (mg/L)	SO_4^{-2} (mg/L)	SO_4^{-2} (mg/L)	Strength (MPa)	Max. w/c Normal	Max. w/c With SO_4^{-2a}	Min. Cement (kg/m³) Normal	Min. Cement (kg/m³) With SO_4^{-2a}	Max. Water Penetra. (mm)[b]	Cover (mm)
1.	Dry	—	—	—	—	≥C16/20 (25)	0.65	—	270	—	—	15
2. a	Humid	—	—	—	—	≥C20/25 (30)	0.60	—	300	—	—	30
b	Humid + frost	—	—	—	—	≥C20/25 (30)	0.55 (0.45)	—	300	—	50	30
3.	Humid + frost + deicing salts	—	—	—	—	≥C20/25 (35)	0.55	(0.50)	300	—	50	40
4. a	Seawater	—	—	—	—	≥C25/30 (35)	0.55 (0.40)	(0.45)	300	—	30	40
b	Seawater + frost	—	—	—	—	≥C25/30 (35)	0.50	—	300	—	30	40
5. a	Slightly aggressive chemical env.	6.5–5.5	15–30	200–600	2,000–6,000	≥C20/25 (35)	0.55	0.60[a]	300	300	50	—
b	Moderately aggressive chemical env.	5.5–4.5	30–60	600–3000	6,000–12,000	≥C25/30 (40)	0.50	0.50[a]	300	330	30	—
c	Highly aggressive chemical env.	4.5–4.0	60–100	3000–6000	12,000+	≥C30/35 (40)	0.45	0.40[a]	300	370	50	—

Source: After CEB-FIP, *Durable Concrete Structures: Design Guide*, Thomas Thelford, London, U.K., 1992, p. 128.

[a] Aggressive environments in which sulfate-resistant cement is used.

[b] Maximum water penetration.

a summary of the performance-based specifications for durability as proposed by the CEB-FIP recommendations for durable concrete structures (CEB-FIP, 1992), which is in fact the basis for the recommendations of the present Euro standards. Similar recommendations are also available from various other national codal provisions as well. The use of minimum quantity of cement for achieving the required strength and workability while satisfying the requirements of durability helps in producing concretes of lower environmental reactivity and lower shrinkage and creep problems. The standards stipulate that the design procedures recommended should be used only as a guide, and the final mix design should be obtained only by trial mixes. Table 5.1 clearly defines the required strength, maximum water–cement ratio, and minimum cement content for each of the environmental exposure classes. In addition, the code also specifies durability in terms of the maximum permeability acceptable in the different environmental classes, as it is well known that it is permeability that allows the environment to penetrate the concrete. The minimum strength recommendations proposed were actually revised in the light of the minimum cement content and maximum water–cement ratio as per the Euro standards and were presented in the strength column within brackets, showing clearly that the strength specifications have indeed been underestimated and the designer should ensure that all the conditions are satisfied.

The Eurocode classifies the exposure classes related to environmental conditions into a total of six different categories—no risk of corrosion or attack, corrosion induced by carbonation, corrosion induced by chlorides, corrosion induced by chlorides from seawater, freeze–thaw, and chemical attack—and also presents informative examples where these exposure classes may occur. It can be clearly seen that the performance is basically controlled by defining limitations on the minimum strength, maximum water–cement ratio, minimum cement content, and maximum permeability as was done earlier. A slightly different portrayal of these environmental exposure classes of the Euro code and the concrete technology–based remedial measures envisaged for ensuring satisfactory performance of structures was presented by Grube (2001) as given in Table 5.2. The addition of wear resistance as a further class of exposures in Table 5.2 is indeed noteworthy. They also explicitly stated how more than one class of exposure may be involved in the different locations of a structural facility in the field, a fact that is often forgotten and one that needs to be the primary target throughout the planning, design, construction, and also maintenance of the constructed facility.

It is expected that these recommendations are indeed necessary and sufficient for the designer to specify the concrete requirements for any structure, though the various parameters that are involved in the performance of concrete in a given environment are far too complex to be discussed individually in the present context. However, it will only be appropriate to at least put a picture of the interrelations between the causes and the possible remedial measures for constructed facilities as presented in Table 5.3.

TABLE 5.2

Environmental Exposure Classes and Remedial Measures

Class		Exposure Class (Environmental Effects, "Attacks")		Concrete Technology Measures ("Resistances")		
		Stress	Effect	Max. w/c	Min. c	f_{ck} Cube
XO		No attack—(moisture is still a concern)	No concrete attack	No requirement	No requirement	C8/10
XC	1	Carbonation—(presence of moisture accelerates but not immersion)	Dry	0.75	240	C16/20
	2		Constantly wet	0.75	240	C16/20
	3		Moderately moist	0.65	260	C20/25
	4		Wet/dry	0.60	280	C23/30
XD/XS	1	Chloride—(presence of moisture accelerates)	Moderately moist	0.55	300	C30/37
	2		Constantly wet	0.50	320	C35/45
	3		Wet/dry	0.45	320	C35/45
XF	1	Frost/+salt—(wetting/drying, marine env. seawater, splash zone, estuarine env. promotes attack)	Mod. water saturation (wo/ds)	0.60	280	C25/30
	2		Mod. water saturation (w/ds)	0.55 + LP	300	C25/30
				0.50	320	C35/45
	3		High water saturation (wo/ds)	0.55 + LP	300	C25/30
				0.50	320	C35/45
	4		High water saturation (wo/ds)	0.50 + LP	320	C30/37
XA	1	Chem. attack—(moist solids, liquids, gasses)	Weakly corrosive	0.60	280	C25/30
	2		Moderately corrosive	0.50	320	C35/45
	3		Strongly corrosive	0.45	320	C35/45
XM	1	Wear—(weight, traction wheel/track surfaces effect)	Moderate wear	0.55	300	C30/37
	2		Severe wear	0.45	320	C35/45
	3		Very severe wear	0.45	320	C35/45

Source: After Grube, H. et al., *Beton*, 51(3), 19, 2001.

Note: w—with; wo—without; ds—deicing salts. A structural component can be influenced by more than one of these 20 different classes in practice depending upon the location.

TABLE 5.3

Interrelations between Causes for Failures and Remedial Measures in Constructed Facilities

Causative Factors	Constituent Materials	Concrete Production	Concrete Strength	Structural Loading	Concrete Deterioration	Corrosion of Steel in Concrete
Effecting parameter	Cement Water Aggregates Admixtures	Design Mixing Compaction Curing temp.	Water content Cement content Admixtures	Early age Later age Overloading	Pore size Micro-cracking Porosity Alkali-silica reactivity (ASR)	Type of steel Creep, relaxation Stress, bond
Resultant effects	Plastic settlement Plastic shrinkage Heat of hydration ASR	Durability Segregation Bleeding Drying shrinkage	Water–cement ratio Porosity Micro-cracking	Micro-cracking Cracking	Erosion, abrasion Acid attack Sulfate attack Temp. effects	Carbonation Chloride ion ingress Cracking
Remedial measures	Aggregate grading Cement content Admixtures	Agg. grading Cooling/heating Curing compound Admixtures	Min. cement content Min. strength Admixtures Agg. grading	Age of stripping Age of loading	Cement type Cement content Maximum w/c Pozzolans	Admixtures, cover, coatings Inhibitors CP systems

It is, however, to be kept in mind that these are by no means complete and there are several situations particularly in the severe environmental conditions of oceans and the industrial atmospheric conditions where there is a need for further understanding and clarity. Table 5.3 presents the interrelations between the different aspects—the various causes, the parameters related to their effects, and the possible remedial measures. These are organized in the sequence of their actual utility in terms of the construction and its performance, namely the materials, production, strength, loading, deterioration, and corrosion. A brief glimpse at these will give an idea of the complexity of the problem on hand.

5.4 Practical Approach for High-Performance Design

At this stage, maybe, it is also appropriate to understand the significance of high-performance concretes that were the thrust for some time now. The term high-performance concrete was applied in recent years to concretes of high strength, which in most cases present a better performance. However, if the increased reactivity due to the higher cementitious content is not addressed specifically in terms of the environmental degradation possibilities, this higher strength need not necessarily lead to higher performance. Also, the primary objective in developing these was mostly to have adequate resistance to aggressive environments. In this regard, "High Performance Concrete" was defined as that meeting the following requirements—the first one regarding the strength and the second one enforcing the durability criteria (Zia, 1991)

(a) A maximum water–cement ratio of 0.35

(b) A minimum durability factor of 80%, as determined by ASTM C666, Method A

(c) A minimum strength criteria of either
  - 3,000 psi (21 MPa) within 4 hours after placement
    - Very Early Strength
  - 5,000 psi (34 MPa) within 24 hours
    - High Early Strength
  - 10,000 psi (69 MPa) within 28 days
    - Very High Strength

Even in these requirements, (a) and (c) look at only strength while (b) alone looks at a very specific aspect of durability. However, the freeze–thaw test of

ASTM C666 need not necessarily represent the performance of concrete in the environment it is supposed to perform. Even so, it should be conceded these recommendations were certainly justifiable for constructed facilities of national and strategic importance involving a very high cost or those having a high public utility value. However, at this stage, one should consider whether the larger proportion of the conventional and commonly constructed facilities like small private housing should be burdened with such demanding specifications. Also, these requirements obviously appear to stem from the evolution of the concept of performance that at water–cement ratios below 0.45 the pore structure of concrete will become increasingly discontinuous and is recommended for concretes in marine environments by many codes of practice. Thus, the need of the hour is to recognize that the losses due to the very commonly constructed facilities not functioning up to the required levels of expectation appear to be phenomenal by any national or international standards. Thus, naturally, it can be easily seen that this high performance need not necessarily be limited to only the high-strength concretes as given earlier.

In general, "high-performance concrete" can be defined as that concrete which has the highest durability (for any given strength class needed) for a particular application, of course, with the understanding that it should be produced at an economical cost. This means that, with the available knowledge, one can always strive to achieve the most durable concrete required for a particular application. At this point the method for the assessment of durability may also have to be well defined, keeping in view the performance requirements of the concrete for the specific application and the environment in which it is expected to perform. Also, the necessity for design mixes even for low-strength concretes in the construction industry has been recognized in recent years, due to the inadequate performance of concrete making and concreting practices of yesteryear. In fact, it is felt that the use of nominal mixes should be prohibited for any structural application.

5.4.1 Concrete Production Practice

High-performance concrete can only be obtained by adopting proper methods at the various stages of concrete making, apart from the choice of the constituents and design, namely mixing, placing, compaction, and curing. At the outset, weight batching should be recommended to avoid normally occurring errors. When admixtures are used, it is advisable to specify the time of mixing, which could be higher than that required for normal concrete. Also, proper care should be taken to avoid segregation during transportation or placement.

Different curing methods are used depending upon the site conditions, size and shape of the member, and to meet some special requirements as in the case of precast members. Effective and efficient curing is essential for obtaining concretes with less permeability. The curing should be continued for a minimum period of 7 days for greater durability. Seawater should not be used for initial curing of either reinforced or prestressed concrete. The current Norwegian practice, however, allows the submergence of even high-strength concrete members in seawater after 3 days of normal curing as the concrete is assumed to have become impermeable by that time, particularly for marine structures.

5.5 Performance Evaluation Methodologies

The performance of the concrete is presently evaluated based on the strength criteria (mechanical characteristics) only. However, there is a need to evaluate the concrete (particularly with admixtures) for its durability in general and the performance of the concrete under severe environmental exposure conditions in particular. There are no standard tests prescribed in this regard (other than ASTM C666). It can be suggested that suitable accelerated tests be performed to estimate the relative performance of different concretes, particularly for very large structures wherein an estimate after 28 days will not help in ensuring at least the curative measures in time.

There are several accelerated tests available in the literature that try to simulate the environmental degradation of concrete and corrosion of steel in concrete that can be used for the evaluation of durability/deterioration of concrete and the corrosion of reinforcement in the concrete. Some of the accelerated durability tests used are the wetting and drying test, acid attack, and sulfate attack. Similarly, chloride diffusion and accelerated electrolytic corrosion tests were performed for assessing the rebar corrosion resistance of different concretes. In addition, suitable methods to estimate the resistivity of the concrete, a parameter influencing the corrosion of rebars, and the potential of steel in concrete can also be used. Some of these are discussed in greater detail in the later chapters while presenting the results of the few experimental investigations reported in the literature for the SCCs studied.

5.6 Concept of Pozzolanic Efficiency and Strength Relations

The utilization of fly ash as a pozzolanic replacement material for cement is a well-accepted practice over the past several years. In spite of the numerous investigations reported on many of the aspects related to the various properties of the concretes containing fly ash, there appears to be a lack of a comprehensive understanding, particularly through systematic theoretical and experimental investigations even regarding the proportioning of these mixes at the various levels of replacement. In particular, an understanding of the exact quantitative contribution of fly ash to the strength of the concrete is not clear to many. It has been recognized that this contribution is not a constant and is determined mainly by its physical and chemical characteristics like cementitious compounds, fineness, and so on. It can also vary depending on the nature of cement, water–cement ratio, and so on. Nevertheless, it appears that the factors that maximize the contribution of fly ash are not known (Popovics, 1986). There have been a few reports on high-volume fly ash concretes (even up to 75% replacement) but there are only a few practical guidelines for the optimum utilization of fly ash in concrete under various circumstances.

Also, the ashes of yesteryear, because of the process of burning coal and the collection procedures of a generally much lower efficiency, have resulted in a larger fraction of the coarser particles apart from a larger proportion of unburnt coal (reflected through its loss on ignition). In contrast, the fly ashes available today are resulted from burning of powdered/pulverized coals and from the improved collection systems like the electrostatic precipitators of higher efficiency. The thrust in recent years has been to ensure a higher durability while achieving the required strength. However, fly ash concretes, in general, suffer from the difficulties arising out of the higher fine content, requiring a higher water content to wet the materials, resulting in lower strengths.

5.6.1 Efficiency Concept

The efficiency of fly ash is defined in terms of its strength characteristics with the control concrete as the reference. However, the improvements in durability upon the addition of fly ash being known, it is possible that other characteristics like durability factors can also be used for such an evaluation, though the exact methodology of the durability test has to be defined. Naturally, it is thus possible to define more than one durability factor (sulfate, chloride, freeze–thaw, etc.), and it would be difficult to compare or specify such a factor in codal provisions. However, it is accepted that the strength of concrete is a realistic indicator of the durability for at least normal concretes without any chemical modifications to improve other specific aspects and thus the efficiency of fly ash in concrete is always defined with respect to the strength of its control concrete. At this stage it is important to realize that the relationship for the strength to the water–cement ratio of the normal concretes without fly ash plays a significant role in the estimation of the efficiencies of fly ash at their different percentages.

The simple and modified replacement techniques of earlier times appear to have limited the replacement levels to a maximum of about 25%–35%, which is still adopted by the ACI committee 226 (1987) recommendations. The rational methods were expected to take into account the characteristic of fly ash that influences the workability and strength characteristic of concrete. Smith (1967) first proposed a factor known as cementing efficiency (k) such that a weight "f" of fly ash could be equivalent to a weight "k·f" of the cement. The strength and workability of this concrete with fly ash was comparable to the normal concrete with a water–cement ratio of $[w/(c + k·f)]$. The value of the cementing efficiency factor "k" was assessed to be 0.25 for fly ash contents of up to 25% by Smith. This method was observed to be insensitive to the cement used, curing conditions, and so on, and was also not suitable for richer mixes (Munday, 1983). Later, the German standard DIN 1045 adopted a value of 0.3 for replacement percentages of 10%–25%. The British code, however, recommended a value of 0.3 for replacement percentages of up to 50%. The CEB-FIP model code proposes an efficiency value of 0.4 for replacements between 15% and 40%. It was also reported that the Danish standards stipulate an efficiency value of 0.5 and can allow even higher efficiency factors if supported by an appropriate study (Bijen, 1993).

Schiessl (1991) pointed out that there could be variations in the efficiency factors with percentage replacement and age. Hansen (1992) suggested an efficiency of 0.25 for lean concretes. All these methods suggest only a single efficiency factor at all the ages and for all percentages of replacements adopted. A summary of the information available regarding the efficiency of fly ash has been presented earlier in Table 5.4 (Rao, 1996). Presently, the use

TABLE 5.4

Summary of the Efficiencies Reported for Fly Ash in Concrete

S.No.	Reference	Method	% Maximum Replacement	"k" Value	Remarks
Research					
1.	Smith (1967)	$w/(c + k·f)$	12–50	0.25	—
2.	Schiessl (1991)	$w/(c + k·f)$	28	0.50	All cements
3.	Hansen (1992)	$w/(c + k·f)$	—	0.25	Lean concretes
Standards					
4.	ACI 226 (1987)	Simple	15–25	(1.0)	Class F
	ASTM C 595-86	Replacement	15–35	(1.0)	Class C
			(15–40)		PPC
5.	DIN 1045 (1988)	$w/(c + k·f)$	10–25	0.30	Class F
			(10–32.5)		PPC
6.	BRE (1988)	$w/(c + k·f)$	10–50	0.30	PPC
	BS 6588 (1985)		(15–35)		PPC with PFA
	BS 6610 (1985)		(35–40)		—
7.	CEB-FIP	$w/(c + k·f)$	10–40	0.50	—
	CEB-FIP Model Code (1994)				
8.	Denmark	$w/(c + k·f)$	—	0.50	
9.	IS 1489 (1976)	—	10–25	—	PPC

Source: Rao, G.S.N., Effective utilization of fly ash in concretes for aggressive environment, PhD thesis, submitted to IIT, Madras, India, May 1996.

of higher volumes of fly ash in mass concretes and lower volumes in high-strength concretes is gaining importance, particularly with the understandable need for higher fines in the making of SCCs. It can also be seen that the cementitious efficiencies established by the earlier investigations will indeed be applicable for any type of concrete, including SCCs being discussed presently, as these are only chemical reactivity parameters at normal temperatures. In any case, the acceptability will be verified again later in the ensuing chapters to make sure that they are indeed valid.

Schiessl (1991) studied concretes containing different cements and fly ashes and showed that a value of 0.5 is more appropriate for this efficiency factor. The study indicated that with increasing fly ash content the efficiency of the fly ash tends to diminish and that the efficiency of fly ash could increase with decreasing water–cement ratio. It was also observed that the differences in the fineness properties of fly ashes used influenced the compressive strengths only marginally. The efficiency of fly ash has also not shown any significant variation in the range of 20%–28% replacements. An important contribution of this research was that it defined the reduction in water–cementitious materials ratio accounting for the efficiency of fly ash concrete compared to the water–cement ratio of the reference concrete through the Δw concept.

$$\Delta w = (w/c_o) - \{w/(c+f)\} \tag{5.1}$$

$$= \{w/(c+k \cdot f)\} - \{w/(c+f)\} \tag{5.2}$$

$$= (w/c)\left[1/\{1+k(f/c)\} - 1/\{1+(f/c)\}\right] \tag{5.3}$$

Equation 5.3 indicates that this reduction depends not only on the size of the efficiency factor (k), but also on the water–cement ratio and, more significantly, the fly ash content in the concrete mix. All of the above information (Table 5.4) clearly shows that a single efficiency value is recommended generally, while limiting the replacement level to a specified maximum. However, as was also suggested by Schiessl (1991), the cementing efficiency of fly ash is not only dependent on the water–cementitious materials ratio but is also dependent on the replacement level. Recent research efforts have also shown ways of incorporating high volumes of fly ash in concrete, with percentage replacement levels of up to 75%. Considering the above facts, it is necessary to understand the cementing efficiency of fly ash in concrete at the different ages of 7, 28, and 90 days and at percentage replacement levels ranging from 15% to 75% by reevaluating the results of earlier investigators and comparing them with the appropriate relation for normal concretes without fly ash.

A study of this type needs to first look at the quality and the quantity of the concretes from each individual reference and ensure that their important variations in the factors being looked at are understood. A set of more than hundred concretes from some of the references containing larger groups of concretes with complete data were chosen for evaluation. The concretes with larger-size aggregates (above 20 mm), higher superplasticizer content (higher than 2%), or different curing conditions (higher temperatures), and so on, were all deleted to ensure a representative group of concretes. All these concretes also contain minimum cementitious materials as well as the required slump to ensure ease of compaction. However, though some trend could be seen in the comparison of the strength to $[w/(c + f)]$ ratio or cement content, there is a significant scattering of results for acceptable relationships. Finally, it is felt that this direct evaluation was leading to only confusion, and as such the results were reevaluated through the efficiency concept. The data available over the past few years was now reevaluated through the proposed methodology.

5.6.2 Evaluation of Efficiency

The cementing efficiency of fly ash is defined as the number of parts of cement that could be replaced by one part of fly ash without changing the property being studied. Presently the efficiency of fly ash has been estimated by trying to bring together the water–cementitious material ratio to strength relations for both normal and fly ash concrete. This was done by using the "Δw" concept, which attempts to bring the water–cementitious materials ratio $[w/(c + f)]$ nearer to the water–cement ratio of the normal concrete (w/c_o) through the cementitious efficiency of fly ash "k." However, the attempt to bring the water–cementitious materials ratios to strength relations together through a single value (general efficiency factor, k_e) did not lead to an acceptable correlation at all percentages of replacement. In view of this, the remaining differences were reevaluated through the "percentage efficiency factor" (k_p). This will mean that the "overall cementing efficiency" (k) of fly ash is the sum total of "k_e" and "k_p." This makes the water–cement ratio (w/c_o) to strength relation of normal concretes be the same for fly ash concretes, by considering the "water to effective cementitious materials ratio" $[w/(c + k \cdot f) = w/(c + k_e \cdot f + k_p \cdot f)]$. A detailed presentation of this method for evaluation of the efficiency of fly ash in concrete has been discussed in detail earlier (Ganesh Babu, 1993, 1995, 1996, 2000) and is also presented hereunder briefly.

The evaluation of efficiency was attempted through the "Δw" concept explained earlier. Figure 5.1 is the conceptual diagram showing the relation between the compressive strengths of concrete and the water–cementitious materials ratio for both normal and fly ash–replaced concretes. The normal concrete relations were also verified to check if they were in line with the recommendations of the present national codes, which is explained later. It can be seen that the lower replacement percentages (may be up to 20%) show strengths higher than that of control concrete while the higher levels result in concretes of lower strengths. "A" and "B" are the two typical water–cementitious material ratios of concretes at the higher and lower percentages of replacement, with that of the normal concrete being at "N" for the same strength. The method tries to bring the $[w/(c + f)]$ ratios nearer to that of

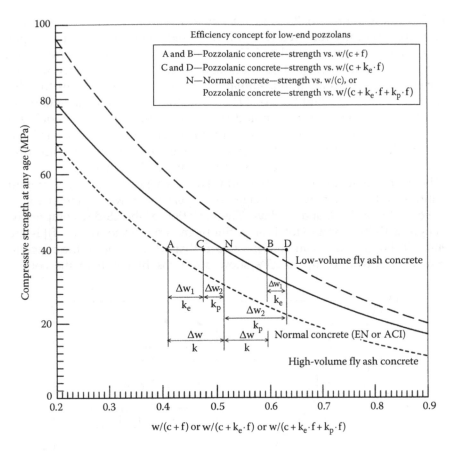

FIGURE 5.1

Conceptual diagram showing the effects of efficiency factors. (From Rao, G.S.N., Effective utilization of fly ash in concretes for aggressive environment, PhD thesis, submitted to IIT, Madras, India, May 1996.)

the control concrete by applying the cementitious efficiency of fly ash (k). Now the figure is replotted to check whether a unique value of "k" can help in bringing both "A" and "B" to "N." This means that the correction (Δw) required can be achieved by a single cementitious efficiency (k) at all percentages of fly ash replacement. This was not possible as the percentages of replacements were far too different, and the result was that at the higher replacements, the control values due to this correction approach the control concretes, and at the lower replacement, percentages resulted in a more conservative estimate. For this reason, the value of "k," which is generally applicable for all the replacement percentages, defined as the general efficiency factor (k_e), was adopted. This means that the points "A" and "B" now shift to their revised locations "C" and "D" due to the application of the general efficiency factor (k_e) with the axis as [w/(c + $k_e \cdot$ f)]. The original points "A" and

"B" have now shifted to "C" and "D," by a distance of "Δw_1." The revised correction presently required (Δw_2), considered to be the effect of the percentage replacement, is calculated. This is done to counteract this effect of the percentage replacement, and the additional factor "k_p" is evaluated for each percentage of replacement in a similar way. These two corrections together bring the points "A" and "B" to "N" so that the water–cement ratio of the normal concrete without fly ash and the water–cementitious material ratio $[\{w/(c + k_e \cdot f + k_p \cdot f)\} = \{w/(c + k \cdot f)\}]$ will be the same for any particular strength. It is important to state that the normal concrete relationships used were seen to be in line with the present Euro and ACI norms.

The water–cementitious material ratios $[w/(c + f)]$ to compressive strength relations at the different percentages of replacement were plotted for all the concretes at 7, 28, and 90 days. The variation of the 28-day compressive strength with the water to total cementitious materials ratio $[w/(c + f)]$ is presented in Figure 5.2. It can be clearly seen that while some of the concretes with replacements of up to 25% show strengths higher than the control,

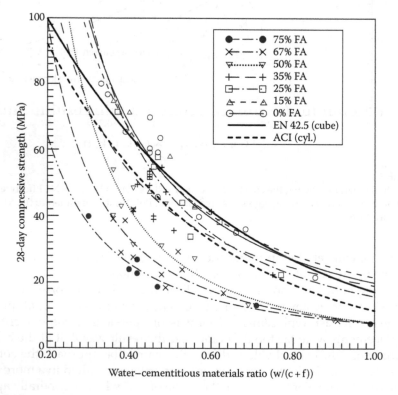

FIGURE 5.2
Compressive strength variation with $[w/(c + f)]$. (From Rao, G.S.N., Effective utilization of fly ash in concretes for aggressive environment, PhD thesis, submitted to IIT, Madras, India, May 1996.)

the higher-percentage replacements show a reduction corresponding to the percentage replaced and the $[w/(c + f)]$. It was also evident that while concretes above 70 MPa could be produced with replacements of up to 25% fly ash, replacements of about 75% can still lead to concretes of 40 MPa through suitable adjustments in the water–cement ratio and other concrete constituents.

After some trials with "k_e" values varying from 0.1 to 0.8, the final "k_e" values were found to be 0.3, 0.5, and 0.6 for the 7-, 28-, and 90-day strengths of these concretes. It can be observed from Figure 5.3 that the strength values come closer to the control with this corresponding "k_e" correction. This resulted in the concretes containing replacement percentages of up to 35% showing strengths higher than the corresponding control and those containing replacements between 50% and 75% showing lower strength with respect to the parameter $[w/(c + k_e \cdot f)]$ at all the above ages. It is evident that this general efficiency factor "k_e" could not bring the $[w/(c + k_e \cdot f)]$ to strength relations close to the water–cement ratio of normal concrete $[w/c_o]$ at all percentages.

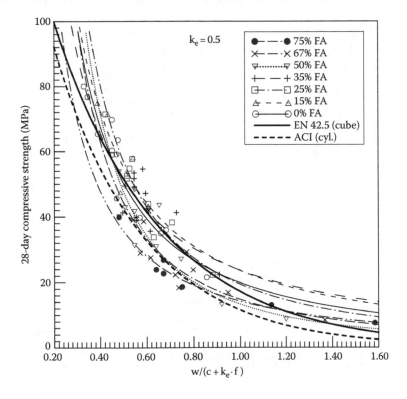

FIGURE 5.3
Compressive strength variation with $[w/(c + k_e \cdot f)]$. (From Rao, G.S.N., *Effective utilization of fly ash in concretes for aggressive environment*, PhD thesis, submitted to IIT, Madras, India, May 1996.)

At this juncture, the effect of percentage replacement on efficiency, which can bring the fly ash–concrete strength values closer to that of normal concrete, was evaluated through the remaining difference as the "percentage efficiency factor" (k_p). The variation of strengths with $[w/(c + k_e \cdot f + k_p \cdot f)]$ at 28 days is presented in Figure 5.4. This shows that by adopting the two efficiency factors "k_e" and "k_p" the strengths of fly ash concretes at different ages and at different percentages could be brought close to that of the normal concrete.

The variation of the "percentage efficiency factor" (k_p) with the fly ash–replacement percentage at the different ages of 7, 28, and 90 days is presented in Figure 5.5. This shows that the value of "k_p" was almost the same at all the ages studied. Also, the percentage efficiency factor decreased with increasing replacement level, and at about 45% replacement the value of k_p was almost zero for fly ash.

Thus, it is confirmed that only general efficiency factor is varying with age and is also a single value at all percentages of replacement. It can be seen that the k_e values were increasing with the age but rate of increase was lower at higher ages. Figure 5.6 presents a comparison of the overall predictions at 7-,

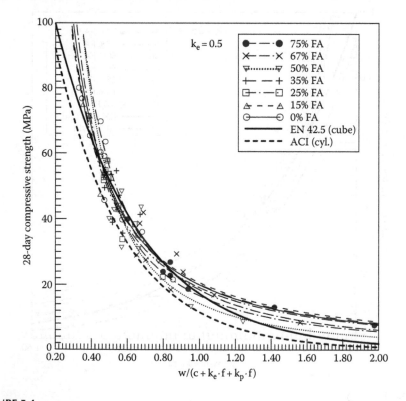

FIGURE 5.4
Compressive strength variation with $[w/(c + k_e \cdot f + k_p \cdot f)]$. (From Rao, G.S.N., Effective utilization of fly ash in concretes for aggressive environment, PhD thesis, submitted to IIT, Madras, India, May 1996.)

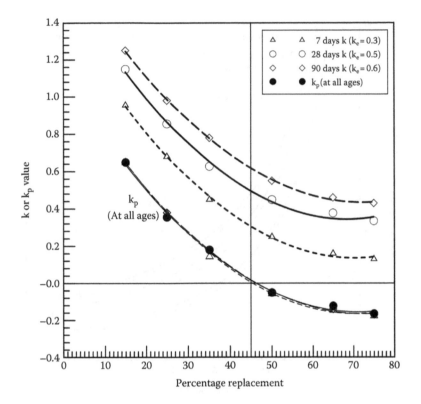

FIGURE 5.5
Variation of the efficiency factors with fly ash percentage. (From Rao, G.S.N., Effective utilization of fly ash in concretes for aggressive environment, PhD thesis, submitted to IIT, Madras, India, May 1996.)

28-, and 90-day strengths after modifying with "k_e" and "k_p." The regression coefficients for this prediction of the strengths of fly ash concretes at these ages were found to be 0.93, 0.95, and 0.94 for 7, 28, and 90 days, respectively, while the same for normal concretes was about 0.95 at all the ages.

The average "overall cementing efficiencies" ($k = k_e + k_p$) at the different ages studied are also presented in Figure 5.5. The efficiency values found for this proposed method at each of these percentages in concrete are given in Table 5.5. The corresponding relationships for the overall cementing efficiencies (k_7, k_{28}, and k_{90}) at 7, 28, and 90 days for replacement levels varying from 15% to 75% were found to be

$$k_7 = 2.67\,p^2 - 3.75\,p + 1.45 \qquad (5.4)$$

$$k_{28} = 2.78\,p^2 - 3.80\,p + 1.64 \qquad (5.5)$$

$$k_{90} = 2.50\,p^2 - 3.59\,p + 1.73 \qquad (5.6)$$

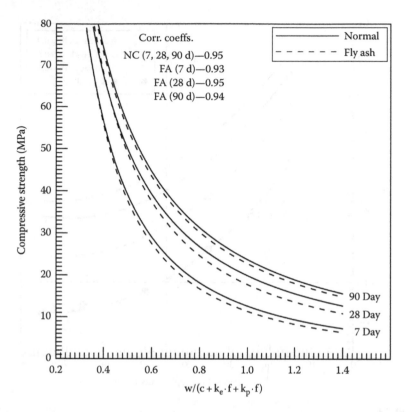

FIGURE 5.6
Effectiveness of the proposed efficiency factors. (From Rao, G.S.N., Effective utilization of fly ash in concretes for aggressive environment, PhD thesis, submitted to IIT, Madras, India, May 1996.)

The above investigations show that the efficiency of fly ash is dependent on both age $[k_e]$ and the percentage replacement $[k_p]$, unlike the single value for "k" proposed hitherto. Furthermore, the value of "k" even at 75% replacement was about 0.33 at 28 days, which was higher than the value of 0.25 reported by Smith (1967). It was observed that the overall cementing efficiency values of fly ash varied in a range of about 0.95–0.13, 1.15–0.33, and 1.25–0.43 for replacement percentages ranging from 15% to 75%, studied at 7, 28, and 90 days. The study also revealed that the efficiency values can vary slightly with the type of cement, type of aggregate, cement content, curing conditions, and so on. The efficiency values reported "k_e," "k_p," and "k" were the average values at the different percentages of replacement for ordinary Portland cement, normal type of aggregate, and normal curing conditions. Table 5.5 also proposes slightly lower values that could probably be useful in formulating codal provisions, because the recommended strengths may not be easily obtained due to variations in the quality of cement and the consistency of the other constituent materials. These were defined considering

TABLE 5.5

Summary of Fly Ash Data Used in Assessment of Efficiency

Fly Ash (%)	w/(c + f)	Comp. Strength (MPa)			Average Efficiencies and (Recommended)[a]		
		7 Day	28 Day	90 Day	7 Day	28 Day	90 Day
0	0.33–0.85	15–69	22–80	24–90	1.00	1.00	1.00
15	0.36–0.50	30–56	43–75	51–90	0.95	1.15	1.40
					(0.75)	(0.90)	(1.10)
25	0.37–0.77	17–57	22–71	27–88	0.69	0.86	1.07
					(0.55)	(0.65)	(0.75)
35	0.42–0.75	15–38	22–55	32–69	0.43	0.62	0.81
					(0.35)	(0.50)	(0.60)
50	0.36–0.56	15–40	27–54	40–79	0.25	0.44	0.55
					(0.20)	(0.35)	(0.45)
65	0.38–0.91	04–12	09–42	15–61	0.16	0.38	0.46
					—	—	(0.35)
75	0.30–0.99	02–12	8–40	10–37	0.13	0.33	0.41
					—	—	(0.25)

Source: Rao, G.S.N., Effective utilization of fly ash in concretes for aggressive environment, PhD thesis, submitted to IIT, Madras, India, May 1996.

[a] Recommended values given in brackets.

the fact that it is best to limit the percentage replacement to a maximum of 50% and even so the proposed values were only 80% of the average values obtained in the present evaluation. Also, the recommended efficiency values given in brackets are obtained by reducing the actual values obtained by about 0.20 at the later ages while they were reduced by about 0.10 for the younger-age concretes.

The methodology described in the earlier paragraphs shows the depth of investigations related to the evaluation of the efficiency of fly ash in concrete. It is only appropriate to mention at this stage that very similar investigations were performed on equally extensive and appropriate databases of concretes containing ground granulated blast furnace slag (GGBS) and limestone powder (LSP) as well as silica fume from the investigations reported in the literature (Sree Rama Kumar, 1999; Ganesh Babu, 2012; Prakash, 1996). It can be seen that these studies clearly show that the efficiency of these pozzolanic admixtures is also dependent on both age [k_e] and percentage replacement [k_p], unlike the single value for "k" proposed hitherto. Even in the case of these materials the efficiency values reported "k_e," "k_p," and "k" were only the average values at the different percentages of replacement for ordinary Portland cement, normal type of aggregate, and normal curing conditions. However, it is not proposed to show the entire evaluation procedure in arriving at the final efficiencies for each of these materials as was done for fly ash.

Thus it suffices to say that both GGBS and LSP have shown almost the same behavior as fly ash did with only the numerical values being slightly different at the various percentage replacement levels. Notable is the fact that even the LSP, which is conventionally taken as an inert powder extender by many in their efforts to produce SCCs, shows a very small amount of reactivity, though it cannot be called pozzolanic. These efficiencies have been utilized for assessing the behavior of SCCs containing the corresponding materials while evaluating their behavior in the next chapter along with that of fly ash. The corresponding equations of the efficiency calculations were also reported correspondingly at the same place. It is only appropriate to inform that silica fumes due to their extremely high fineness and significant active silica content have shown efficiencies that are severalfold higher than that of fly ash and are naturally considered to be a highly efficient pozzolanic admixture. It is only appropriate probably to look at it in its full perspective while looking at the discussions on such high-end pozzolanic materials like silica fumes in the later chapters. Notwithstanding these comments, there were experiments conducted in the laboratory on each of these materials that reaffirmed the efficiency factors as discussed and presented in the various sections. Maybe it is only appropriate to mention that the laboratory has been active on these efforts to understand the effectiveness of several of the other pozzolanic admixtures like metakaolin, zeolite, and rice husk ash (Appa Rao, 2001; Surekha, 2005; Narasimhulu, 2007) to name a few. No effort is being made presently to elaborate the behavior of SCCs as the literature available on all these materials is too scanty even to make an attempt.

5.6.3 Factors Influencing the Efficiency of Fly Ash

In an earlier review, it was reported that the cementing efficiency of fly ash will vary depending on the type of cement, fly ash, age, and so on (Munday, 1983). A later investigation showed that the efficiency of fly ash is not only dependent on the physical and chemical characteristics of fly ash, but also influenced by mix design parameters, strength range, addition level, and age (Dhir, 1985; Hassaballah, 1993). In contrast, a study by Bijen (1993) reports that the effect of the water–cement ratio on efficiency was minor even for different types of cements. It was also stated by them that while the type of fly ash has no significant influence on the efficiency factors, the type of cement had some effect. Also, the "k" value increased with age for ordinary Portland cement, and the curing period and temperature were found to have some effect on the efficiency. It was felt that the different conclusions drawn by the above investigators and even by many others earlier were all based on limited experimental investigations. It was thus decided to reevaluate the data chosen earlier to study and assess the effect of fly ash over a wide range of water–cementitious materials ratios, percentage replacements, and strengths at different ages.

Earlier investigations clearly show that the mix design procedures for fly ash in general adopted two different methodologies. In the first place, the investigators used a semi-empirical approach based on extensive experimental investigations while others preferred the efficiency concepts. The semi-empirical design procedures select the maximum replacement possible for a given strength. This obviously is due to the fact that in the case of pozzolanic replacements like fly ash (which are considerably cheaper than cement), the maximum economy can be obtained by having the maximum replacement. Also, it is crucial to note that there is a maximum replacement possible for any given strength, which was not recognized or perceived by a few as they have limited the replacement percentages to fairly low values at all strengths.

At this stage, it can be seen that the efficiency approach, being based on the cementing efficiency of the fly ash, will generally limit the maximum replacement percentage possible. A typical relation between the strength and maximum replacement percentage possible is presented in Figure 5.7 for fly ash from earlier investigations (Munday, 1983). This relation originally given for the [w/(c + f)] ratio to the fly ash cement ratio is now presented in a direct form, in terms of the compressive strength to the maximum percentage replacement possible.

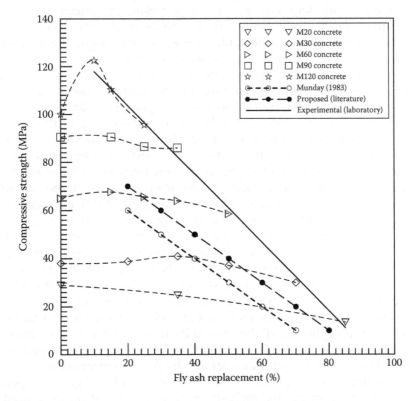

FIGURE 5.7
Maximum permissible replacements at any strength. (From Rao, G.S.N., Effective utilization of fly ash in concretes for aggressive environment, PhD thesis, submitted to IIT, Madras, India, May 1996.)

The figure also shows that it is possible to obtain strengths below the maximum possible at that particular percentage through an appropriate design taking into consideration the efficiency values. This is one of the reasons why there have been several strengths reported for the same percentage replacement or as a corollary several possible replacement levels (below a certain maximum) for a particular strength. At this stage, the maximum percentage replacements realized at any specific strength from the data available were once again reviewed. These results show that it is possible to exceed the previously defined limits by Munday (1983) by about 10% at all strength levels.

This higher value can probably be taken to be possible limit as the experimental investigations in the laboratory show that even higher levels can be incorporated as can be seen from Figure 5.6. Furthermore, if for any other reason (durability or corrosion considerations) any percentage of replacement lower than the maximum possible is preferred, the efficiency approach can easily predict the strengths possible. One should note that as described earlier it is even possible to have the specific efficiency curves defined for the different efficiencies like durability, corrosion, and so on, apart from the ones to account for variations such as type of pozzolan, age, curing conditions, temperature, addition of chemical admixtures or workability required, and so on. It is also possible to define, like in the case of strength, the maximum replacements possible for any other parameter of interest.

One of the important factors that need an absolute clarity and also a perfect definition is that the water–cement ratio to strength relationship for the normally vibrated concretes. This is needed for comparing it with the strength to water cementitious materials ratio of the pozzolanic concretes so that the efficiency values can be properly established. However, in most cases researchers feel that this should be established with the constituents that were present in the concretes with which the pozzolanic concretes were also made. Such an effort to establish these relationships for the normally vibrated concretes in each specific case to compare with the pozzolanic concretes means that the research effort doubles. Obviously, this is one of the primary bottlenecks that could be cited for establishing appropriate strength efficiency factors for any specific pozzolanic material. Having said this, it is obvious that most national bodies like ACI, BS, DIN, and even the most recent Euro norms provide specific water–cement ratio to strength relationships which could be used for this purpose. In fact, if normal concretes do not conform to these meticulously and comprehensively tested and accepted norms proposed, obviously the acceptability of the cement itself in most cases as the material of the quality that is mentioned (be it 33.5, 43.5, or 53.5 grade as in the case of Euro cements) is very much in question. The ACI recommendations, while not supporting the concept of strength grade of cement, still give out specific relationships for normal- and high-strength concretes that could be used in the present context though the relationship for high-strength concretes that is appropriate still needs to be identified among the various alternatives that are proposed by the recommendations. Incidentally, a very important observation that can be made looking at most research efforts toward the development of SCCs is that the increased powder content required for SCCs is predominantly supplied through fly ash in most research efforts. This obviously is also prudent because of the fact that the fly ash is freely available and is the cheapest material that could be used as the additional powder content. Apart from this, fly ash possesses a reasonable level of pozzolanicity and will help both the strength and performance levels of these concretes. Moreover, at lower percentages of up to about 30%, which is the level of the powder content that has been used in most investigations of SCCs, it not only helps the lubricating effect through the additional powder factor but could also aid the fluidity because of the spherical nature of the individual fly ash particles. This is probably one of the reasons that many of the investigations did not encounter the need for additional wetting water for the additional powder content and the increased surface area that needs to be wetted at these lower fly ash content levels.

Extending this discussion a little further into the effect of fly ash in concretes which was investigated extensively over the years, the strength to water–cement ratio relationships for these fly ash concretes is not very difficult to obtain. One may go even a small step further in pointing out that the ACI 211.4 recommendations for high-strength concretes specifically include the need for fly ash along with a suitable superplasticizer. Naturally, the available relationships for fly ash or GGBS concrete in conjunction with superplasticizers and for the maximum size of aggregate of 20 mm are already available for direct comparison. Moreover, the code comprehensively presents the water content requirement for each aggregate size and even an aggregate proportioning broadly to arrive at a reasonably high-strength concrete. A minor modulation of the constituents proposed for aggregates in particular probably could result in reasonably acceptable SCCs, through the use of a slightly higher superplasticizer content with the corresponding reduction in the water content and if need be a small quantity of a viscosity-modifying agent. In fact, some of the mix design methodologies discussed earlier specifically suggest that the ACI or the Euro regulations be used for ascertaining the cement content required for a particular strength, though without mentioning that it is only feasible if and only if a certain water content is presupposed to get a fix on the water–cement ratio. It may not be out of place to note that most investigations on SCCs that have always been conducted on a limited number of normally vibrated concretes are concretes which are almost conforming to the above norms of ACI 211.4, which permits the researchers to compare the resulting concrete strength characteristics at least to those that are expected. A limited check of this nature alone will be able to substantially help and clarify the present status of most of the SCC studies. Another way to look at it is that the comparison made with at least those concretes conforms to the limited spectrum of ACI 211.4 recommendations to ensure that the constituents do perform as expected. A further corollary to this particular line of discussion is that by and large the performance of concrete with a certain pozzolanic type and content level is generally dependent on the strength characteristics of the final material, though a few special characteristics due to the increased powder content and possible physical changes in the mechanical structure of concrete may be different, which will be discussed exclusively and separately. This then helps in arriving at an understanding of even the performance of the resulting concrete composite with the type of pozzolan involved from the vast literature that is available in print. Some of these pertinent aspects will be studied further to ensure that a comprehensive understanding of the SCC composites that are possible with the different pozzolanic admixtures available can be understood more comprehensively and appropriately in the later parts.

As a first step, a panoramic view of the water–cement ratio to strength relations adopted by a few national bodies is presently looked at to understand their appropriateness while keeping in mind the enormous number of variables that could influence the ratios in terms of the constituents, admixtures, and modifications that keep pace with the demands of the constructed facilities. Some of these notable advancements could be listed as those in the dry production process in cements and cements of higher grades, chemical admixtures that could result in substantially lower water requirements for the same workability, and ultimately a reinvention of the significant capabilities of the pozzolanic mineral admixtures including the newer ones apart from the advancements in fibrous materials that could augment the uniformity of the behavior of the resulting cementitious matrices.

Having observed that the present-day concretes have been customarily modified significantly in several ways in recent times, there is a basic doubt that is expressed by many, which is whether there is any wisdom at all in still attaching any significance to the water–cement ratio to strength relation that was originally postulated for normal concretes and that too for only concretes of marginal strengths. However, the fact remains that many national and international bodies still use relations very similar to the original Abram's relation in the mix design methodologies that are proposed by them (Abrams, 1920). Also, the fact remains that this sort of an approach presents the possibility of putting together composites with relative ease and also have a comparison of the resulting materials.

It is also to be noted that there is a lot of research and developmental effort that has gone into and is going into a better understanding of the performance of concrete. Probably, singularly concrete or cementitious composites in all are the most highly researched materials in the present-day structural and building materials scenario. Even so there is often a lot of divergence in the methodologies of the design, production, and resultant concrete, often simply attributed to the variations in the characteristics of the constituent materials. Maybe it is only appropriate, at least at this point, to have fresh relook at the entire scenario and reassess objectively and without inhibitions and reservations to arrive at some conscientious relationships that could pave the way for a more reliable and comfortable design of the concrete composites.

A general approach for the design of a concrete mix is to evaluate the properties of the constituent materials available—cement, aggregates, and water. Based on the above, for a required design strength and workability, the water–cement ratio and water content are determined. From these, the cement content is obtained and the total aggregate content is worked out either by the absolute volume method or by the weight method. Then an appropriate ratio of the fine aggregate to the total aggregate is calculated based on the maximum size of the coarse aggregate and the grading of the fine aggregate, as in the case of BIS (IS:10262, 1983), BS (Neville, 1987) and ACI (ACI 211.1 [1997] and 211.4 [1998]) while the grading of the total aggregate is ensured in DIN 1045 (1988) and Euro standards (Sika Concrete Handbook, 2005).

However, a designer should also be concerned about the performance of the concrete in terms of the green state and hardened state characteristics, in addition to economy, even while adopting the design mix. The durability requirements are taken care of by specifying limits on the water–cement ratio and cement content. Also, structural concretes require, apart from strength and durability, limitations on the maximum size of the aggregate and the workability requirements based on the elements size (particularly its thickness and depth), the reinforcement detailing (spacing), and the placement and the compaction procedures. However, in the present context the requirements for durability or the suitability of the concrete for any specific application has not been considered.

Keeping all these aspects in the background it appears that most fundamental to the design of a concrete mix is the strength to water–cement ratio relation. Knowing that concrete strength is significantly dependent on the age or the maturity, the present discussions concentrate only on the 28-day strength to water–cement ratio relation for normal concretes.

5.6.4 Water–Cement Ratio to the Strength Relation

The strength concrete, for the given cement and aggregates, depends primarily on the water–cement ratio, if a proper compaction is assured basically through an adequate workability. Actually the strength is influenced by the porosity, which depends upon the water–cement ratio as it indicates the excess water, not required for the hydration of cement. For a direct comparison of the water–cement ratio to strength relations of the different methods, it is necessary to take cognizance of the following factors as per the various standards: standard physical requirements of OPC; cement strength; type, size, and gradation of aggregates used; and also the type and size of the standard test specimen adopted. A comparison of these relations as adopted by the different methodologies is presented in Figure 5.8, considering that the cement adopted is about 42.5 MPa grade and also assuming that the effects of the other parameters are not very significant. This means that the resulting concretes are reasonably proportioned and are also adequately compactable

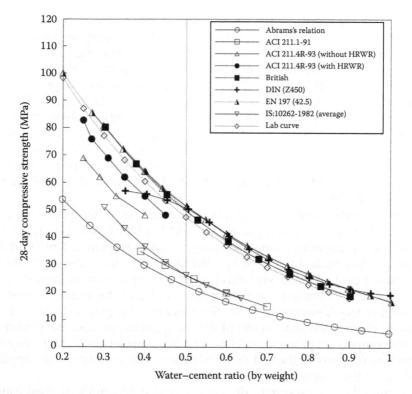

FIGURE 5.8
Comparison of the different water–cement ratio to strength relations.

so that the final resulting mass is not very much influenced by the aforementioned parameters significantly and the final resulting concrete is of acceptable homogeneity.

Figure 5.8 presents the water to cement relations as per the ACI, British, German, Euro, and Indian standards and the ACI relations in particular are presented for both the normal- and high-strength concrete alternatives. The figure also contains the original Abram's relationship as well as a curve that was developed in the laboratory. A comparison of water–cement ratios obtained for a specific strength by the four methods yields widely varying water–cement ratios, and if one looks at the design in full, the water content required is not so very different to result in such a high variation. The mix design methodology also shows that the maximum design strengths for normal concretes according to these methods also vary significantly. It should also be noted that the American standards have separate specifications for the design of concretes above 55 MPa, which in addition contain fly ash.

Though different standards stipulate that the design data should be used only as a guideline and the final design should be obtained by trial mixes utilizing the actual constituents for each project, it is to be expected that the design data yields fairly good results under standard conditions and only needs alterations due to the variations in the materials available at the site and their deviation from the standard conditions. It is also clear that the British, earlier DIN, and Euro relations are very close to each other and will also result in the highest strength for a specific water–cement ratio. Further, a laboratory relation based on the results of the studies at the IIT Madras and some normal concretes that were taken from the reported literature as a part of the investigations on concrete with mineral admixtures like fly ash, silica fume, and GGBFS was also superimposed on these relations. This also can be seen to be very close to the above upper bound relations. The information presented in the earlier paragraphs show that the expected strength for a specific water–cement ratio is much higher in the British, DIN, and Euro methods than the indian and ACI relations. The relation developed in the laboratory at the institute, based on the laboratory investigations, and also a small amount of other data from the reported literature show that the strengths much closer to the maximum proposed by these methods were possible even without a very strong control on the various parameters associated with the concrete constituents.

At this stage, it was felt that one simple method would be to look at the relationship for the 42.5 grade cement that is often used in most countries as the one for which an appropriate relation is required. The actual Euro specifications for the water–cement ratio to strength of the different cements used in structural construction are presented in Figure 5.9. Maybe it is also appropriate to inform at this stage that these were very close to the original DIN specifications, which were derived from the investigations of Walz (1971) and are also known as "Walz curves" or "Roll curves," either in the name of the investigator or from the methodology by which they are derived which is explained below. The fundamental Walz curve was derived from investigations on 830 laboratory concrete mixtures and the strength of the corresponding cements. In principle, the investigation showed that the relationship between "water–cement ratio" and the "ratio of the strength of concrete to the strength of mortar at 28 days" for any cement strength was nearly unique and the variations if any were within ±5%. It also showed that irrespective of the actual cement strength the strength of cement and the strength of concrete were nearly the same at a water–cement ratio of 0.485 (almost 0.5). The German recommendations for the water–cement ratio to strength relationships for concretes with different cement strengths were obtained from this unified relationship and hence the name Roll curves. It is also important to note that these were originally derived for the Z35, Z45, and Z55 cements of the DIN code and remained almost the same in the Euro code, with the drooping portions at lower water–cement ratios being removed, negating the effect of compacts and inadequacy. It can also be seen that the present 32.5, 42.5, and 52.5 cement grades will result in concretes of strength 42.5, 52.5, and 62.5, respectively, at a water–cement ratio of 0.5. The other regulations that are prescribed for these cements are probably not of significance at this point and are not being discussed. For the present it is enough to say

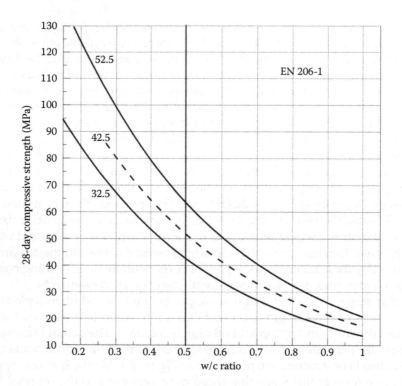

FIGURE 5.9
Water–cement ratio to strength relations of Euro norms. (From Sika Concrete Handbook, Sika Services AG Corporate Construction, CH-8048 Zürich, Switzerland, 2005.)

that the typical water–cement ratio to strength relationship for the 42.5 grade cement that is used most can be presented as

$$f_c = 155 \cdot e^{-2.2(w/c)}$$

A note of caution at this stage is to clearly recognize that the cement grade as prescribed in the German code should not be confused with the cement grades presented in the Indian standards that bear the same number, as the method of assessing the cement strengths is entirely different. The cement strength as prescribed in the Indian standards is obtained at a very low water–cement ratio related to the consistency (a much older if not outdated cement testing method) and if one wants to apply the proposed methodology the cement strength compliance with the prescribed regulations must be checked at a water–cement ratio of 0.5, and correction should be appropriately made to arrive at a similar relationship for a cement of the particular strength. Another important fact is that the relationships presented in Figure 5.9 are the 28-day concrete compressive strength of 150 mm cube.

This being so it would also be appropriate to look at the recommendations of the ACI that might probably be applicable to SCCs under consideration at present. It is well known that the SCCs utilize both pozzolanic materials like fly ash or GGBS along with the superplasticizers in the making. To be broadly representative we can also say that the maximum size of coarse aggregate used in most cases is the 20-mm aggregate. Considering these two factors, it is only appropriate to say that the water–cement ratio to strength relations that are proposed for 20-mm aggregates with high-range water-reducing admixtures (HRWR) could be most appropriate to be considered as representative relationship. It should also be recognized that these are essentially cylinder strengths (150 mm dia. × 300 mm long cylinders) and the actual relationship can be presented as

$$f_c = 155 \cdot e^{-2.6(w/c)}$$

It is best to inform at this particular stage that these relationships are marginally modified to be able to arrive at a simple relationship wherein the equations are the same with only the exponent changing from 2.2 to 2.6 in the case of the American code from the Euro standard. In all evaluations of the efficiency of a particular pozzolanic material, or any other comparison requiring the water–cement ratio to strength relationships, these two equations are used throughout the text for all formulations and assessments other than the earlier works reported on fly ash and also later silica fumes.

It is only proper at this stage to also see how the cube and cylinder compressive strengths are related in general. There have been several efforts in this particular direction but one of the most recent ones on this is the tabular form presented in the EN 206, which actually goes well beyond the 100 MPa concrete strength, which is normally not so easily available. The values presented are graphically plotted and the best relationship is established, which after minor modifications is also recorded in the figure itself (Figure 5.10).

Armed with this information it is probably not very difficult to see how the earlier two equations representing the EN 42.5 cements for cube strength and the ACI, 20 mm, high-strength concrete relationship with HRWR can be correlated (Figure 5.11). The resulting correlation of the ACI cube strength transformed to the cylinder strength of EN 42.5 shows a reasonably fair equivalence, within the limits of ±2 MPa surprisingly. This fact is probably one more fact that not only substantiates the usefulness of equations suggested but also supports the strength evaluations of the entire range of different SCC pozzolanic admixtures in the later chapters. It is also to be re-emphasized at this particular stage that for any research effort it is necessary to have a standard base for comparison which is obviously well known and cementitious composites is the basic water–cement ratio to strength relationship of the normal concrete with which all other comparisons can be made.

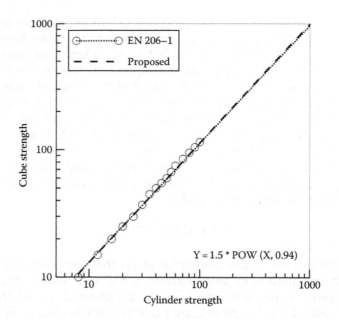

FIGURE 5.10
Relationship between cube and cylinder compressive strengths.

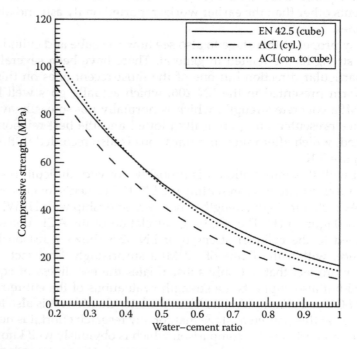

FIGURE 5.11
Correlation between the cube and cylinder strengths.

Needless to say that as a precaution almost every research effort conducts tests on a few normal concretes and these could easily be compared with the EN or ACI recommendations as the case may be so that the further assessments attempted on the modifications in the cementitious composites can be appropriately explained.

5.7 Effects of Pozzolanic Addition on Consistency

The need for ensuring adequate powder material in the making of SCCs is obviously well recognized. It is also made abundantly clear that the use of at least the low-end pozzolanic admixtures for this purpose will certainly not only provide the powder content required but also help in several ways—by providing secondary cementitious compounds to enhance the strength, reduce the permeability, reduce heat of hydration, all of which have a bearing on the performance of the self-consolidating cementitious composites. One simple fact that immediately comes to mind is that such very fine materials in the matrix will need a certain additional quantity of water for wetting, even if it were to be controlled through the use of more recent high-efficiency superplasticizers based on poly-carboxylic esters. So long as one looks at smaller additions of materials like fly ash and GGBS for low-strength concretes requiring only about 400–500 kg/m^3 of cementitious materials one can control the water requirements through a certain modulation of superplasticizers. However, if the cementitious material content is far in excess of these normal levels for reasons of higher strengths or the pozzolanic admixture that is being used in several times finer than the cement like silica fumes it becomes necessary to look at the water content more critically, and managing it with superplasticizers alone becomes unmanageable. Another aspect of self-compactability is to ensure that the granular skeleton is losing the ability to flow. This flow is aided by a coating of the fine powders including the pozzolanic admixtures. It is generally recognized that this fine coating of paste should at least be in sufficient quantity for an about two-layer thickness to even out the possible rough and uneven exterior surface of the aggregates. In fact, it is preferable to have little more than this quantity to ensure that it could also fill the other additional voids created due to the loosening and wall effects that come into play. Apart from this, there is another complex mechanism involving the hydration and the secondary cementitious materials formation as well as in more dominant filler effect, particularly with the finer pozzolanic admixtures. The further aspects in this regard are discussed in the ensuing topic and also later during the methodology proposed for ensuring high-performance self-consolidating cementitious composites.

5.8 Packing and Optimal Granular Skeleton

The previous chapter presents a comprehensive overview of the different approaches followed by the several researchers hitherto for achieving SCCs of specified characteristics. With a view to get better insights into the concretes resulting from these different methodologies, the actual mix proportions were calculated and were also graphically presented. The salient features of the resulting mixes, in terms of the major parameters that govern the behavior of SCCs, were all critically analyzed and presented in Table 4.2 earlier. The probable strengths that these mixes could have, based on the strength to water–binder ratio relationships of ACI and Euro norms, were also reported. This comparison helps in understanding the makeup of the constituents for ensuring self-compactability in concretes. In repeating and reiterating what was observed it can once again be said that each of these concentrated on the effectiveness of a specific concept, while ensuring that the other factors are also modulated to realize the ultimate goal. The primary factor accepted by all is that the fine content should be increased to ensure adequate flowability, while reducing correspondingly the coarse aggregate content. The need for an understanding of the actual gradation of aggregates and their appropriate combination to ensure the flowability with much less fines is recognized by Hu (2011), showing that well-graded aggregate could considerably reduce the yield stress and consequently the viscosity of concrete. Some have discussed the loosening effect due to the continuous gradation. The factors concerning the maximum size of aggregate and the distribution in relation to the passing ability of SCCs are of specific importance and need better understanding. The need to limit the maximum size of the aggregate to much lower values, from a value of 20–25 mm to about 10–12 mm or even lower, to ensure effective packing and good compactability also needs to be appropriately addressed. The fact that a granular skeleton of the highest packing density possible–yet having the grading characteristics that could facilitate free flow through the loosening effect (similar to the granular flow in silos) while being aided by a paste content that is sufficient to ensure a reasonable coating thickness to facilitate the sliding of the aggregates (with superplasticizers and VMA) and resulting in an optimum cement content– is the way forward for achieving an SCC. Marquardt (2002) has shown that the increased fines needed could require additional quantity of water for wetting which could increase the water–binder ratios, thus decreasing the strengths possible. These factors and their consequences are known to affect even normal concretes and require a better grip for ensuring SCCs of higher performance.

To have a clear picture of the resulting concrete from the mix design procedures available to date, a brief summary of the results was once again compiled presenting clearly the actual quantities and their range from the different methodologies presented in the earlier chapter (Table 5.6).

TABLE 5.6

Summary of the Recommendations for SCC

Reference	(p) Powder (kg/m³)	(c) Cement (kg/m³)	(F) Fly ash (kg/m³)	(w/p) wt./(vol.)	(w) Water (kg/m³)	(CA) (vol.)	Strength est./(expt.) (MPa)
Oka (1995, 2000)	660–700	460–490	200–210	(0.9–1.0)	200–210	28	71–77
JSCE (1999)	500–600	350–420	150–180	(0.85–1.15)	155–175	28–35	63–82
EFFNARC (2002)	400–600	280–420	120–180	(0.8–1.10)	<200	28–35	65–85
EFFNARC (2005)	380–600	265–420	114–180	(0.85–1.10)	150–210	27–36	65–85
Marquardt (2002)	500–600	300–380	150–180	0.26–0.28	135–145	—	50–90
Su (2002)	425–553	200–350	127–165	0.30–0.41	170–176	26–28	(15–75)
Yu (2005)	500–650	300–385	150–195	<0.41	178–228	33	25–60
ACI 237 (2007)	400–475	200–415	120–142	0.32–0.45	160–180	28–32	—
Sebaibi (2013)	250–650	225–585	(S), 25–65	(0.30–0.90)	130–170	—	(15–75)
Ferrara (2007)	420–605	330–470	126–182	0.32–0.40	150–215	—	(65–85)
Range	*250–700*	*200–585*	*114–210*	*0.30–0.45*	*150–228*	*26–36*	*(15–85)*
Literature	*300–900*	*200–800*	*F/G/L/S/M*	*0.25–0.45*	*130–300*	*28–36*	*(12–86)*
Authors	*400–700*	*200–500*	*F/G/L/S/M*	*0.20–0.70*	*160–200*	*Comb. grad*	*(20–120+)*
Ideal (F/G)	*400–650*	*200–400*	*F/G (30%–50%)*	*0.60–0.35*	*170–190*	*Comb. grad*	*30–70*
Ideal (S/M)	*450–650*	*400–550*	*S/M (10%–20%)*	*0.20–0.45*	*150–170*	*Comb. grad*	*60–120*

Notes: (F) Fly ash 10%–70%, (G) GGBS 10%–70%, (L) LSP 10%–50%, (S) silica fume 10%–20%, (M) metakaolin (10%–20%), (max. agg. size–10–40 m); (assumed 30% fly ash in all designs).

The values in italics are not specified in the recommendations and are calculated or proposed.

The table incidentally presents the quantities of cement and fly ash, the water–powder ratio, and the water content as well as the coarse aggregate content ranges from the different methodologies and the estimated or the experimentally obtained strengths. From this, the actual overall range of SCCs was also presented. Not satisfied with the information from the theoretical mix design methodologies presented, large experimental data on the concretes available in the literature was also compiled and the range of this particular dataset in terms of the constituents is also enclosed in Table 5.6. It is important to note that the dataset compiled contains information on SCCs containing different powder constituents as admixtures that include apart from fly ash, GGBS, LSP, silica fume, and metakaolin as well, to name a few.

It is only appropriate that these factors are best addressed through a comprehensive evaluation of the SCC mixtures reported in the literature so that appropriate guidelines for producing high-performance SCCs with reasonable confidence can be suggested. It is in line with this that the proposed methodology was formulated and verified first with the results of an experimental investigation conducted in the laboratory.

5.9 Proposed Methodology for High-Performance SCCs

In simple terms, high performance requires two specific aspects to be satisfied, a very low porosity to restrain the environmental ingress and a very low reactivity to the environment so that the degradation of the composite is significantly reduced. For low porosity, the physical parameters associated with grading and compaction of the ingredients including cement are the primary factor, while ensuring that the available cement and the other supplementary cementitious materials do hydrate appropriately and bind the composite to achieve the highest strength possible. The second aspect is to limit the water content to the minimum required while taking into account the self-compactability requirements by balancing the amount of water, the superplasticizer, and if required the VMA contents. This is to ensure the lowest possible cementitious materials content that is the most reactive component in the system and yet achieve the highest strength possible. In this bargain, one should not lose track of the minimum cement requirements to ensure an adequate lifespan of the structure in the given environment. These are all highly complex and mutually contradicting requirements have to be addressed to ensure the highest performance in the concrete composite designed and yet be economical.

In the earlier part, an effort was made to look at the presently available methods and the concepts used in them to arrive at SCCs. While the concepts in themselves have been enumerated and illustrated by many, it is necessary to fix some definite rules and also boundaries, preferably based on the experience to date. Presently, in an effort to understand the behavior of SCCs that were hitherto reported, the authors compiled an extensive database of the same and tried to look at them in the light of what is accepted world over for the design of conventional concrete mixtures. It was also felt that instead of mandating arbitrary algorithms to look for the relationships involved in the makeup of the constitutes of these SCCs, the authors looked for possible correlations to the existing methodologies for the design of conventionally vibrated concretes. After the several iterations needed to fix appropriately the design rules for the SCCs, the following conceptual procedure has evolved. It can be said that the authors tried to take all the parameters and their combinations form both the normal and SCC perspectives. In particular, the requirements of minimum powder content, maximum packing density for any specific strength grade of concrete were all addressed the best they could be. Due to the rigor involved through the use of such a large database it is felt appropriate to presume that this method will ensure the best design and that only needs tweaking in terms of the plasticizer and VMA contents at the time of mixing the concrete to account for the changes in moisture of the aggregates or variations in the fineness of the constituents. The method also suggests a procedure for ensuring appropriate packing density in terms of the combined aggregate gradation in line with some of the existing codal provisions.

5.9.1 Strength Assessment of the Pozzolanic Cementitious Contents

In the first place, the mix design details of over 850 SCCs reported in the literature with different pozzolans at the varying percentages adopted were compiled. However, with the variations in the type of pozzolans and their percentages of replacement, even disregarding the possible variations in the aggregate characteristics, the water, and superplasticizer contents adopted, one can never establish any specific relationships, not to mention strength to water–cementitious materials ratio relationship as in the case of conventional concretes. Thus, as a first step only about 80 concretes containing 30% fly ash (26%–35%) as the cementitious material were considered for assessment. The cementitious efficiency of the fly ash as a supplementary cementitious material was taken into account as reported earlier (Ganesh Babu, 1996). Figure 5.12 presents a picture of the strength to the water–cementitious materials ratios of these concretes along with a best fit for the same. In principle, this relationship should naturally be the same as that reported for conventional concretes and to compare, the strength to water–cement ratio relationships that correspond to the ACI and the Euro norms were both superimposed on the same. It is clear from the figure that the best-fit relationship is very close to that proposed by the ACI, while the Euro relation as expected lies well above. Even so it can also be seen that the maximum strengths reported at any specific water–cementitious materials ratio exceeded even that predicted by

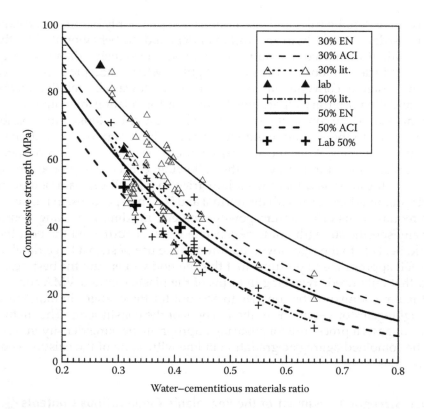

FIGURE 5.12
Strength to water–cementitious materials ratios of fly ash concretes.

Euro norms. The figure also contains a few values that were obtained in the laboratory to indicate that the predicted strengths could be achieved without much difficulty. It is to be informed that similar assessments of strength to the effective water–cementitious materials ratio relations for the other SCCs with the different replacement levels ranging from 10% to 50% fly ash in the database above have all shown a very similar trend. Thus, it looks obvious that, as already observed by Liu (2010) earlier, the cementitious efficiencies of fly ash reported by the author earlier (Ganesh Babu, 1996) were valid for SCCs also.

5.9.2 Water Content, Plasticizer, and VMA Interactions

It is now necessary to assess the water content required for the increased fine content mandatory for the SCCs, while ensuring that it is limited to a range, a minimum that will adequately support the plasticizing effect and a maximum that will require the minimum plasticizer and maybe if required a minimum VMA content. Incidentally, this water content will affect the cementitious

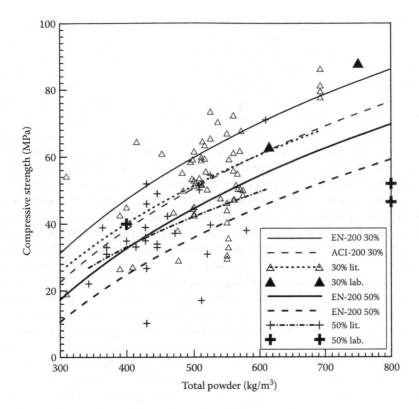

FIGURE 5.13
Total powder content corresponding to strength of fly ash concretes.

materials content (having accepted the strength to water–cement ratio relation earlier) and to understand dynamics the total powder content associated with the corresponding strengths of the concretes was looked at. This
is presented in Figure 5.13 along with the best-fit relationship to arrive at an
average from the literature data. On this, the curves that represent the total
powder content with 30% fly ash at the different strengths as per the ACI
and the Euro norms at two different chosen minimum and maximum water
content of 160 and 200 kg/m^3 were also superimposed (having seen that 180
kg/m^3 is probably the transition point from the studies of Felekoglu (2007) as
reported earlier). It is to be emphasized that in the experience of the author,
there is an optimum water content range for the mix to effectively plasticize
without the aid of very large quantities of superplasticizer, which will naturally increase the cost of the mix. In fact, it can be seen from Table 5.6 that
the water content varied over a wide range, from 130 to 300 kg/m^3 along
with the cementitious materials ranging from 300 to 900 kg/m^3 in the literature. This is because of the fact that on the one hand researchers have used
significantly lower cementitious materials content necessitating only limited

water requirements through the use of higher amounts of superplasticizer and VMA. On the other hand, they have used very large powder content to ensure the required plasticity with relative ease resulting in higher water content and even plasticizer contents. The figure also shows that the range chosen represents the upper bound values for both the ACI and Euro norms as was the case in the previous figure. The laboratory data though very limited also shows that the same were adequate keeping in mind the fact that the authors limited the plasticizer content to about 0.5%–1.5% of the cementitious material from cost considerations. This gives the designer the flexibility to modulate the water content requirement keeping in mind the powder content proposed, depending on the strength and only an interpolation or a simple calculation of the same for a different powder content, is ever necessitated. This may be necessary due to the increased minimum powder content required in the case of most of the lower-strength concretes to ensure enough paste content (presently recommended to be 400 kg/m^3) or to limit the powder content to an appropriate value to redress the higher paste and water content and the associated higher shrinkage, particularly at the higher strengths (presently recommended to be 700 kg/m^3). It is to be understood that the plasticizer and, if required, the VMA contents are to be normally limited to 1.0%–2.5% and 0.0%–0.5% (for low- and high-strength concretes respectively), essentially from cost considerations. Another fact that is often not recognized at this stage is that it is relatively easy to achieve self-contracting concretes of medium strength (approximately 40–70 MPa) as the powder content required is most apt for coating the aggregates with a thin layer to achieve fluidity without making the mixture a paste-dominant one (as in very-high-strength concretes) and getting into trouble with both segregation and shrinkage. Also, for lower-strength concretes if we fix a 30% limit on pozzolans like fly ash or GGBS, the powder content available may not be sufficient for this desired coating level on the aggregates. The minimum of 400 kg/m^3 is recommended from that consideration, which will also limit or at least reduce the need for VMA.

5.9.3 Aggregate Grading and Proportioning on the Packing and Loosening Aspects

After having understood the requirements of water and cementitious contents, it now remains to be seen how to apportion the remaining volume within the aggregate skeleton of coarse and fine aggregates required. For this, it is necessary to understand how to ensure an optimal packing with the required loosening effect to achieve the highest strength possible for the concrete composite designed. One look at the information discussed earlier presents clearly that though it is the practice in the design of concrete composites to delineate the percentages of the coarse and fine aggregates, the best way is to adopt a combined grading curve in concretes, more so in the case of SCCs. Such a continuous grading is obviously appropriate to create

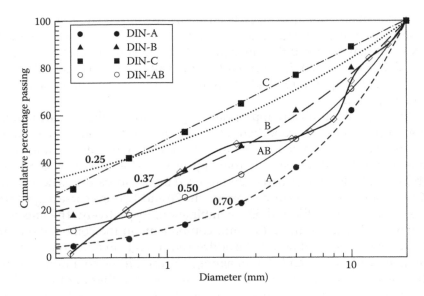

FIGURE 5.14
Appropriateness of the methodology of continuous grading (DIN).

the necessary loosening effect while ensuring the best packing. It may also be noted that the DIN specifications or presently the Euro norms adopt such an approach and the grading curves A, B, and C, as desired by these norms are shown in Figure 5.14.

It is well established that a grading that results in the highest packing of the aggregates may not be able to produce the highest-strength concrete. de Larrard (1999) explains the concept of the requirement of the water–powder mixture coating on all the packed aggregates to ensure an appropriate impervious mass of concrete after hydration. A number of reports of SCCs have already discussed the grading distribution of the aggregate matrix (Marquardt, 2002; Brouwers, 2005; Ferrara, 2007). Literature shows that the relation for the grading distribution of the aggregate in terms of the ratio of diameter to the maximum diameter could be defined by an exponent of value ranging from 0.70 to 0.25, with that for the Fuller's grading being 0.50. Incidentally these values for the DIN-A and DIN-B curves will be 0.70 and 0.37, with the average of both (say DIN-AB) being at 0.5 and the best fit for DIN-C being 0.25 though it is actually a log linear relationship. A further refinement can be done in these by also including the minimum aggregate dimension (using some value in the sand regime) as was proposed by Funk (1994) and was adopted by a few like Ferrara (2007), or by including all the fines, even cement and the other mineral admixtures, though these fines are essentially required for coating the aggregates to bind them together and not for packing the granular skeleton. It can also be noted that in an effort to simply satisfy the norm of about 40% volume of the fine aggregates in the mortar phase, attempts have been made to simply suggest combined grading curves corresponding to exponents of 0.37 (DIN-B) and even 0.25 (DIN-C). It may be noted that an exponent of 0.45 is recommended by ACI for SCCs. Brouwers (2005) observed that the value of 0.25 is probably valid if the powders required for coating are included in the computations of the granular grading.

At this stage, it is probably appropriate to point out that the design methods using specifications for proportions of both coarse and fine aggregates in the mix will not really be able to generate a combined aggregate grading curve that is continuous as can be seen from Figure 5.14. This curve was generated with the available well-graded materials used in the laboratory as desired for SCCs by some researchers by combining the 20 and 12 mm coarse aggregates and sand in a 20%, 30%, and 50% combination, respectively. The effects of the general absence of the 2–6/8 mm components in the combinations are a matter of concern and need to be addressed though an additional component along with the 20 and 12 mm aggregates used if the continuous distribution aspect has to be addressed. It can also be seen that even the DIN-A curve will be able to satisfy the loosening effect required, and incidentally the DIN code recommends a continuous grading curve line between A and B, which was used for SCCs by Marquardt (2002), without experiencing any difficulty in producing high strengths at relatively lower cement contents. In fact, in the opinion of the author (and the experience with these gradings as can be seen later in the verification), grading is precisely the reason why these SCCs exhibit such high strengths as reported by Marquardt (2002). Also, in the experience of the author with several pozzolanic admixtures like fly ash, silica fume, GGBS, and so on, using a combined aggregate grading curve DIN-AB for the coarse and fine aggregates (Fuller's curve with 0.50 as the exponent without considering the minimum aggregate size or including the fine powders involved), supported by an adequate amount of cementitious binder of about 400 kg/m^3 minimum for coating these aggregates and by adjusting the superplasticizer content (adding a very small quantity of VMA if required), resulted in a fairly compact mass of SCC mixture, which probably had the maximum compressive strength for the given cement content. Needless to say that recommendations of exponents like 0.37 or below (DIN-B to DIN-C curves) will only lead to mixes with very high sand content that may require higher fines for coating and higher water for wetting, leading to lower strengths. The need for neglecting the aggregate interlocking effects in the shear calculations and the absence of intrusion effect and across-the-grain fracture effect will also have to be investigated properly in such very high sand–containing mixtures from structural considerations to name a few, apart from possible higher shrinkage due to additional fines and higher air entrainment due to the thixotropy from plasticizers and VMAs.

Experience in the laboratory shows that the maximum size of the aggregate also has an effect in this scenario. It may also be seen that in the literature, while the American and British standards generally recommend a 20–25 mm aggregate the DIN recommendations use 16 mm for conventional structural construction. One advantage of using a lower sized aggregate is that there is a much lower chance of segregation in the mix in the case of SCCs. The maximum size of the aggregates in the investigations by Marquardt (2002) was 16 mm and that by Ferrara (2007) was only 12 mm with the concretes achieving higher strengths, of up to 80 MPa. In fact, if one attempts to achieve very-high-strength SCCs (of over 100 MPa) it is preferable to have the maximum size of aggregate limited to 16 mm or even lower at 12 or 8 mm, which could also permit the possible use of thinner sections that are less heavy. The reason for recommending a minimum cementitious materials content of 400 kg/m^3 and 8–16 mm aggregate in the author's recommendations essentially stems from these discussions.

5.9.4 Optimal Utilization of Pozzolanic Materials for High-Performance SCCs

The fact that the additional powder content required in SCCs need not necessarily be the cement or a cementitious material is reasonably acceptable. Notwithstanding the fact while there is a possibility for large quantities of LSP being available from the cement factories, it has been the practice from the very day the concept of SCCs was initiated by Okamura (1995) that fly ash has been the material of choice invariably. One can also see that barring a few exceptions most research efforts toward the production of SCCs have always been utilizing mostly fly ash for the additional powder content required in SCCs. However, the use of other powders including those that are not effective as pozzolans is not precluded in the manufacture of SCCs at any point. Also, in a broad sense, a specific observation that can even be made at this stage is that there is no understanding of how to arrive at an optimal solution with the multitude of pozzolanic materials that are available with their varying and specific physical and chemical characteristics. In a broad sense though the use of the higher-end pozzolanic materials like silica fume, metakaolin, rice husk ash (RHA) obviously appear to be the better choice for achieving SCCs of higher strength ranges. The fact that there is no such specific effort to arrive at high-strength SCCs even while studying them with silica fumes as the powder extenders was indeed surprising. This does not mean that there have been no high-strength SCCs with these high-end pozzolans, as they were indeed used in building several high-rise structures like the *Burj Khalifa*. The reason why some of them do not get reflected in the literature in their proper perspective is probably because of the fact that in making those SCC composites the designers have used these high-end pozzolans in combination with other lower-end powders making them mostly ternary composites. The formulation and application of these materials is best taken up at a later stage after we get to understand the basic fundamental principles that are associated with the strength and performance of SCCs containing a specific pozzolanic material.

This would mean that an understanding of the different powder constituents as secondary cementitious materials (specifically their pozzolanic reactivity) should be addressed exclusively. In fact, apart from a limited work by Liu (2010) from the UCL group headed by Domone (the author of the UCL method, 2009), there is very little information on the pozzolanic aspect itself so far. Even the few studies that used silica fumes as the powder extender in the SCCs designed (Sebaibi, 2013) have not really looked into the reactivity of such high-end pozzolanic materials and their contribution to the strength of concrete in any way. It then remains to be seen how to address the enormously complex domain of possibilities with several different pozzolanic admixtures with their varying reactivities. Looking at the information available in the literature and considering the economics involved in the various options, one basic fact that can be easily postulated is that for the lower-strength SCCs it is only appropriate to suggest the use of the lower-end pozzolanic additions like fly ash, GGBS, and LSP. The cementitious efficiency of these being significantly lower than 1.0, a replacement methodology utilizing the efficiency concept will naturally increase the total powder content. These will also supplement the powder volume necessary without substantially increasing the cost, and in fact reducing it through a reduction in the cement content is an appropriate replacement methodology through the efficiency concept that is followed. The lower cementitious efficiencies of these materials are thus utilized advantageously not only to increase the powder content but also to lower the environmental reactivity of the material compared to an OPC concrete of the same grade. There could also be synergistic effects of the chloride-binding capability while using them in marine environment. Even the heat of hydration can be controlled and modulated to suit the needs of the construction sequence in massive constructions.

After a close look at the literature available on these types of materials like fly ash and GGBS, with the experience generated through investigations in the laboratory, a broad outline of limits for arriving at specifically the lower-strength SCCs in the range of 30–70 MPa for structural applications is also presented in Table 5.2. It may be noted that the minimum of the cementitious materials content required for the SCCs was taken to be 400 kg/m^3 (to have the level of comfort required while allowing the minimum dosage of superplasticizers and if at all required the least amount of VMA) with cement content limited to about 200 kg/m^3. However, it is absolutely essential to ensure that the minimum cement content and the maximum water–cement ratio required for the environmental zone will still be the governing factors. The water content was also pegged at the level of 170–190 kg/m^3 to lower the superplasticizer requirements. In direct contrast to this for arriving at higher-strength SCCs in the range of 70–120 MPa for structural applications requiring very high strength and performance, Table 5.2 presents the cementitious materials content between 450 and 650 kg/m^3. It should be noted that the maximum cementitious materials content was limited to 650 kg/m^3, specifically to ensure that the shrinkage and creep characteristics of the resulting composites are kept in check, particularly at the very high stress levels to which these will be subjected to. The water content was also pegged at the level of 150–170 kg/m^3 to ensure that the lower water cementitious materials ratio is lower (even as low as 0.20, naturally accounting for the cementitious efficiency of the high-end pozzolanic material like silica fumes). Yet another important observation at this stage is the fact that both the superplasticizers and the VMAs used in the production of SCCs contain a substantial quantity of water, which should be taken into account in calculating the total water content that is projected above. One of the simplest and probably the easiest methods would be to simply add the quantities of the superplasticizer as well as the VMA (if it is used) to the water content to ensure that the strength levels for which the concrete is being generated are not affected in any way. Another observation at this stage from experience in the laboratory with several of the concretes containing pozzolanic materials at the various levels is to recognize that there is a very sharp edge in the water content required for a collapse slump from a stage where the concrete is flexible yet highly thixotropic exhibiting a slump of anywhere between 50 and 100 mm only. At a water content less than this, anymore addition of superplasticizer will not result in the collapse slump or self-compacting material that is required (and if ever it is noticed that such a thing happens it is probably because of the water in the superplasticizer rather than the superplasticizer itself that is the cause). It should also be added that the additional water content required at this stage will probably change the water–cementitious materials ratio at best in the third decimal level and is literally only a few grams and can only be judged by the engineer on site. However, once a reasonable level is established, a small additional water content is will not have any problems in ensuring the self-compactability of the concrete and the process is repeatable.

5.10 Efficacy of the Proposed Methodology

The methodology for arriving at SCCs of high performance in terms of the basic governing parameters, the efficiency of the cementitious materials content, the wetting water requirements for the increased fine constituents in particular, and the appropriateness of the combined grading of the coarse and fine aggregates, has all been looked at through the laboratory data available on fly ash concretes (Rao, 1996) conforming to the required self-compactability as presented in Table 5.7. The table clearly shows that fly ash–based SCCs ranging from about 25 to 110 MPa could be produced by effectively replacing 70%–25% of the cement in well-designed concrete mixes by considering the

TABLE 5.7

Results of the Laboratory Investigations on SCCs

Concrete	(c) Cement (kg/m³)	(f) Fly Ash (%)	(w/p) Ratio	SP (%)	Slump Flow (cm)	Strength 28 Day (MPa)	Strength 90 Day (MPa)
D625	375	25	0.37	0.5	63	59.6	75.4
C425	400	25	0.31	1.25	69	67.0	76.0
C625	600	25	0.25	1.5	63	98.0	110.0
F235	167	35	0.72	0.0	56	24.8	32.3
C435	400	35	0.31	1.5	70	63.0	72.0
C635	600	35	0.27	2.0	76	88.0	105.0
C250	200	50	0.41	1.25	58	40.0	52.0
C450	400	50	0.31	1.75	72	46.5	60.0
C450R	400	50	0.31	1.13	69	51.9	54.5
C270	200	70	0.42	1.5	74	20.2	27.2
C270R	200	70	0.41	0.75	52	21.3	27.9
C6P10[a]	333	10(SF)	0.40	1.0	51	86.3	95.3
C2T12[a]	581	10(SF)	0.24	3.5	67	120	125.4

Sources: Prakash, P.V.S., Development and behavioural characteristics of silica fume concretes for aggressive environment, PhD thesis, submitted to IIT, Madras, India, May 1996; Rao, G.S.N., Effective utilization of fly ash in concretes for aggressive environment, PhD thesis, submitted to IIT, Madras, India, May 1996.

[a] Concretes containing silica fume.

efficiency factors proposed earlier for fly ash (Ganesh Babu, 1996). A combined aggregate of 20 mm maximum size conforming to DIN-A grading was adopted in all concretes. It can also be seen that these concretes containing 25%–70% can be modulated at several levels of cementitious material content and concretes of different consistencies could be produced with the different water and superplasticizer combinations. Obviously it is a complex interaction and there is only a very thin line between the changes in concrete from a cohesive thixotropic mix to a completely flowing mass, and the water as well as superplasticizer content together will influence this change. It can also be seen that while most investigators have been presenting SCCs with similar cementitious materials to be resulting in strengths in the range of only 30 to 60 MPa for replacements at 30%–50%, the table shows that is possible to achieve much higher strengths (strengths in the range of 50–110 MPa), by addressing the cementitious materials, the water and superplasticizer contents, and the grading characteristics appropriately. It may be noted that even at 70% replacement concretes in the range of 20–30 MPa could be produced, though the setting and early strength gain characteristics could be adversely affected. The table also contains results of procedures on concrete mixtures that were repeated just to prove the repeatability of the results of the concrete produced.

Also, for the other pozzolans like silica fumes, understandably, the efficiency relationships have to be appropriately chosen from the available literature (Ganesh Babu, 1995, 1996, 2000, 2012). Table 5.2 presents a couple of SCCs containing silica fumes at 10% from an earlier investigation (Prakash, 1996) similar to the investigation reported by Sebaibi (2013) earlier. The mix proportions of the first of these are very similar to that adopted by Sebaibi, a concrete designed for 50 MPa, yet resulting in 70 MPa. The concrete reported in Table 5.2 resulted in a strength of 85 MPa at 28 days and reached 95 MPa at 90 days. These results indicate that concretes of much higher strengths, ranging up to 125 MPa, can be produced by modulating the cementitious materials content appropriately along with the water and superplasticizer contents as indicated in the proposed methodology. This shows that for producing economical and high-performance SCCs, by ensuring a minimum powder content in the mix, one can use any of the low-end pozzolans like fly ash, GGBS, volcanic tufts, or even LSP with due recognition of their actual cementitious efficiencies. Alternatively, by appropriately utilizing the higher-efficiency pozzolans like silica fumes, metakaolin, rice husk ash, and so on, it is possible to limit the total cementitious materials content to significantly lower values, ensuring very-high-strength, high-performance self-contracting concretes with reasonable confidence.

However, not satisfied with the strength performance alone, the durability performance of these concretes was also presented to ensure that they satisfy all the requirements for structural applications. An assessment of the performance characteristics of the fly ash concretes indicates that there is a significant reduction in the 6-hour charge (Rapid Chloride Permeability Test values (RCPT) values), the charge decreasing from greater than 3000 coulombs in conventional concretes to values below 400, 300, 200, and 100 coulombs for the concretes with 25%, 35%, 50%, and 70% replacement of cement through the efficiency concept. This may be due to both the pore filling and pozzolanic effects of the fly ash apart from its chloride-binding effect at these percentages. Other durability characteristics like the acid attack, corrosion rate, and time to cracking characteristics have also shown a significant reduction due to the decreased cement contents with effective replacement. However, there is a very small increase in the water absorption characteristics of high-volume fly ash concretes, which was still well below the prescribed limit for good concretes as per CEB-FIP and did not in any way affect the performance characteristics of these concretes. It can also be seen that in Table 5.6 the ranges for the various constituents recommended were broader to accommodate both the lower-efficiency pozzolans for low- and normal-strength concretes at the lower limit, while the higher limit is suitable for the much finer (including nano) pozzolans like silica fumes wherein the increased wetting water requirements have to be compensated particularly at the significantly lower water content required to limit the cementitious contents in very-high-strength concretes. Naturally, the stability or cohesion required is to be adjusted with the addition of the VMA, also taking into account the available paste content and its cohesion due to the plasticizer.

References

Abrams, D.A., Design of concrete mixtures, Bulletin No.1, Structural Materials Research Laboratory, Lewis Institute, Chicago, IL, 1920, 20pp.

ACI 211.1-91, *Standard Practice for Selecting Proportions for Normal Heavyweight, and Mass Concrete*, ACI Manual of Concrete Practice, American Concrete Institute, Farmington Hills, MI, 1997, 38pp.

ACI 211.4R-93, *Guide for Selecting Proportions for High-Strength Concrete with Portland Cement and Fly Ash*, ACI Manual of Concrete Practice, American Concrete Institute, Farmington Hills, MI, 1998, 13pp.

ACI 237, *Self-Consolidating Concrete*, American Concrete Institute, Farmington Hills, MI, 2007, 30pp.

Appa Rao, Ch.V., Behaviour of concretes with metakaoline, MS thesis, submitted to Indian Institute of Technology, Madras, India, 2001.

Bijen, J., and Van Selst, R., Cement equivalence factors for fly ash, *Cement and Concrete Research*, 23, 1993, 1029–1039.

BRE (Building Research Establishment), *Design of Normal Concrete Mixes*, Building Research Establishment, Watford, U.K., 1988, 38p.

Brouwers, H.J.H. and Radix, H.J., Self-compacting concrete: Theoretical and experimental study, *Cement and Concrete Research*, 35(11), 2005, 2116–2136.

BS 6588, *Specification for Portland Pulverized-Fuel Ash Cement*, British Standards Institution, London, U.K., 1985, p. 18.

BS 6610, *Specification for Pozzolanic Cement with Pulverized-Fuel Ash as Pozzolana*, British Standards Institution, London, U.K., 1985, p. 20.

CEB-FIP, *Durable Concrete Structures: Design Guide*, Thomas Thelford, London, U.K., 1992, p. 128.

CEB-FIP Model Code, Design code, Thomas Thelford, London, U.K., 1994, p. 128.

de Larrard, F., *Concrete Mixture-Proportioning—A Scientific Approach*, Modern Concrete Technology Series, No. 9, E & FN Spon, London, 1999, 421pp.

Dhir, R.K., Ho, N.Y., and Munday, J.G.L., Pulverised fuel ash in structural precast concrete, *Concrete*, 19(6), June 1985, 32–35.

DIN 1045, *Beton und Stahlbeton*, Beton Verlag GMBH, Koln, Germany, 1988.

Domone, P., Proportioning of self-compacting concrete—the UCL method, 2009.

EFNARC, *Specification and Guidelines for Self-Compacting Concrete*, EFNARC, Surrey, U.K., 2002, p. 32.

EFNARC, *The European Guidelines for Self-Compacting Concrete*, EFNARC, Surrey, U.K., 2005, p. 68.

Felekoglu, B., Turkel, S., and Baradan, B., Effect of water/cement ratio on the fresh and hardened properties of self-compacting concrete, *Build Environment*, 42(4), 2007, 1795–1802.

Ferrara, L., Park, Y., and Shah, S.P., A method for mix-design of fiber-reinforced self-compacting concrete, *Cement and Concrete Research*, 37(6), 2007, 957–971.

Funk, J.E. and Dinger, D.R., *Predictive Control of Crowded Particulate Suspension Applied to Ceramic Manufacturing*, Kluwer Academic Press, Dordrecht, the Netherlands, 1994.

Ganesh Babu, K., High performance concrete, in *International Symposium on Innovative World of Concrete*, ICI-IWC-93, Bangalore, India, Vol. II, September 3, 1993, pp. 169–180.

Ganesh Babu, K. and Chandra Sekhar, B., Efficiency of limestone powder in SCC, *MRS Proceedings*, 1488, 2012.

Ganesh Babu, K. and Prakash, P.V.S., Efficiency of silica fume in concretes, *Cement and Concrete Research Journal*, 25(6), 1995, 1273–1283.

Ganesh Babu, K., Raja, G.L.V., and Srinivasa Rao, P., Concrete mix design methodology—An appraisal of the current codal provisions, *Indian Concrete Journal*, 66, 1992, 87–95.

Ganesh Babu, K. and Rao, G.S.N., Efficiency of fly ash in concrete with age, *Cement and Concrete Research Journal*, 26(3), 1996, 465–474.

Ganesh Babu, K., Rao, G.S.N., and Prakash, P.V.S., Efficiency of pozzolans in cement composites, in *Concrete 2000*, Dundee, U.K., Vol. 1, 1993, pp. 497–509.

Ganesh Babu, K. and Sree Rama Kumar, V., Efficiency of GGBS in concrete, *Cement and Concrete Research Journal*, 30(7), 2000, 1031–1036.

Grube, H., Kerkhoff, B., and DIN EN 1045-2, The new German concrete standards DIN EN 206-1 and DIN EN 1045-2 as basis for the design of durable constructions, *Beton*, 51(3), 2001, 19–28.

Hansen, T.C. and Hedegaard, S.E., Modified rule of constant water content for constant consistency of fresh fly ash concrete mixes, *Materials and Structures*, 25(6), 1992, 347–354.

Hassaballah, A. and Wanzel, T.H., Defining the water to cementitious ratio in Fly ash concrete, in *Concrete 2000*, Vol. 1, 1993, pp. 497–509.

Hu, J. and Wang, K., Effect of coarse aggregate characteristics on concrete rheology, *Construction and Building Materials*, 25(3), 2011, 1196–1204.

IS:10262-1982, *Recommended Guidelines for Concrete Mix Design*, Bureau of Indian Standards, New Delhi, India, 1983.

JSCE, *Recommendation for Self-Compacting Concrete*, Guidelines for Concrete, No. 31, JSCE, Tokyo, Japan, 1999.

Liu, M., Self-compacting concrete with different levels of pulverized fuel ash, *Construction and Building Materials*, 24, 2010, 1245–1252.

Marquardt, I., Determination of the composition of self-compacting concretes on the basis of the water requirements of the constituent materials—Presentation of a new mix concept. *Betonwerk + Fertigteiltechnik BFT*, 11, 2002, 22–30.

Munday, J.G.L., Ong, L.T., and Dhir, R.K., Mix proportioning of concrete with PFA : Critical review, Fly Ash, Silica Fume, Slag and Other Mineral Bi-products in concrete, ACI Publication SP-79, Vol. 1, 1983, pp. 267–288.

Narasimhulu, K., Natural and calcined zeolites in concrete, PhD thesis, submitted to Indian Institute of Technology, Madras, India, 2007.

Neville, A.M., *Properties of Concrete*, Longman Scientific & Technical, Harlow, U.K., 1987.

Okamura, H. and Ouchi, M., Self-compacting concrete, *Japan Concrete Institute, Journal of Advanced Concrete Technology*, 1(1), 2003, 5–15.

Okamura, H. and Ozawa, K., Mix-design for self-compacting concrete, *Concrete Library, JSCE*, 25, 1995, 107–120.

Popovics, S., What do we know about the contribution of fly ash to the strength of concrete? *Second International Conference on Fly Ash, Silica Fume, Slag and Natural Pozzolanas in Concrete*, ACI SP-91, 1, 1986, pp. 313–331.

Prakash, P.V.S., Development and behavioural characteristics of silica fume concretes for aggressive environment, PhD thesis, submitted to IIT Madras, India, May 1996.

Rao, G.S.N., Effective utilization of fly ash in concretes for aggressive environment, PhD thesis, submitted to IIT, Madras, India, May 1996.

Schiessl, P. and Hardtl, R., Efficiency of fly ash in concrete—Evaluation of "ibac" test results, Technical Report of Institute fur Bauforschung, RWTH, Aachen, 1991, pp. 1–31.

Sebaibi, N., Benzerzour, M., Sebaibi, Y., and Abriak, N., Composition of self-compacting concrete using the compressible packing model, the Chinese method and the European standard, *Construction and Building Materials*, 43, 2013, pp. 382–388.

Sika Services AG Corporate Construction, *Sika Concrete Handbook*, Sika Services AG Corporate Construction, Zürich, Switzerland, 2005.

Smith, I.A., The design of fly ash concretes, *Proceedings of the Institution of Civil Engineers*, London, U.K., 36, 1967, 769–790.

Sree Rama Kumar, V., Behaviour of GGBS in concrete Composites, MS thesis, submitted to Indian Institute of Technology, Madras, 1999.

Su, N., Hsu, K.C., and Chai, H.W., A simple mix design method for self-compacting concrete, *Cement and Concrete Research*, 31, 2001, pp. 1799–1807.

Surekha, S., Performance of rice husk ash concretes, MS thesis, submitted to Indian Institute of Technology, Madras, India, 2005.

Walz, K., *Herstellung von beton nach DIN 1045, Betontechnologische arbeitsunterlagen*, Beton-Verlag GmbH, Dusseldorf, Germany, 1971, p. 89.

Yu, Z., Pan, Z., and Liu, X., Optimal mixture design of high performance self-compacting concrete, in *First International Symposium on Design, Performance and Use of Self-Consolidating Concrete*, 2005, pp. 181–190.

Zia, P., Leming, M.L., and Ahmed, S.H., High performance concretes: A state-of-the-Art report. SHRP/FR-91-103, North Carolina State University, 1991.

6

SCCs Based on Powder Extenders
and Low-End Pozzolans

6.1 Introduction

The concept of supplementing cement with additional fines from available resources to self-consolidate concrete has been explained in several ways by several authors starting from the initial efforts of Okamura (1995). The use of low-end pozzolans like fly ash, which are available in plenty and also have physical characteristics mostly similar to that of cement, was an obvious choice in most of the research efforts reported. The one specific limitation in most of these research efforts was that, while the advantages of supplementary cementitious materials was well understood the need for an actual quantification of the same was recognized only by a handful of researchers. Some of these aspects and how to address these particular limitations have already been discussed earlier. At this point the effort in the next two chapters is to examine the results of the various researchers and their contributions to the behavior of self-compacting concretes (SCCs) to arrive at some broad outlines of how each of these materials contributes to the performance of SCCs. Incidentally, as is the case with most compilations of these particular types of materials or concepts, the simplest option is to review them using relevant parameters suggested earlier by various authors. However, while this would be certainly a good exposition of the various efforts and their relevance in advancing the study of formulations and characteristics of the SCCs, it would not be directly useful to the professionals involved in the application of this material in the field. It is with this view that a database was compiled for each group of the SCCs prepared using different cementitious extenders to look for a broad, theoretically sound, and acceptable order to ensure that they would be better understood so that they could be better formulated.

6.2 Concept of Powder Extenders

The need for increasing the performance of concretes even in the most aggressive environments requires no specific emphasis. Several possibilities toward achieving this objective have been proposed and these can be broadly understood through the two groups of materials that actually make the concrete what it is, namely, the cementitious binder component and the aggregate filler component. In the earlier parts, we looked at the methodologies of extending the performance of concrete individually through modulations of these two groups of materials. In fact, it is well understood that the cementitious components should adequately coat the aggregate fillers to ensure a cohesive mass, with the interstitial spaces being filled by the hydration products.

To ensure that the total void space is minimized to the extent possible, several efforts have been made to arrive at an appropriate packing of the aggregates, even while fully appreciating the fact that the closest packing of aggregates will not result in the highest strength of concrete. In a way, the idea of filling the space available from the largest aggregate size fraction with the next size and the remaining from the next is often advocated, which then is moderated to have a minimum loosening effect as was discussed earlier. The need to account for a minimum coating thickness of the cementitious materials which bind the mass together is also to be appropriately considered at this stage. A typical representation of the improvements in packing density and the corresponding reduction in void space as well as the effect of paste coating on the effective aggregate sizes and packing was presented by Kwan (2001). A typical representation of the concept of how the paste layer smoothens even an irregular aggregate is shown in Figure 6.1.

The next aspect that needs a better understanding is the fact that for a free flow of the granular aggregates it is but obvious that the packing should

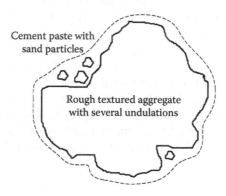

FIGURE 6.1
Cement paste layer over an uneven aggregate surface. (After Kwan, A.K.H. and Mora, C.F., *Mag. Concrete Res.*, 53, 91, 2001.)

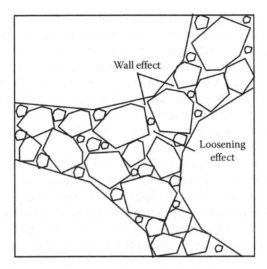

FIGURE 6.2
Wall effect and loosening effect. (After de Larrard, F., *Concrete Mixture Proportioning: A Scientific Approach*, E&FN Spon, London, U.K., 1999.)

be loose enough to ensure free flow without the arching effect in particular. The simple fact that the aggregate gradation chosen to achieve the highest packing density (like the Fullers packing for spheres) is not to be attempted and availability of a small fraction of intermediate sizes will go a long way in providing the loosening effect required (Figure 6.2) was demonstrated by de Larrard (1999). The effect of the phenomena in an aggregate mixture of multiple sizes was also recognized and discussed by de Larrard et al. (2002). Also, the effect of such a loosening effect near the surface of a member (termed the wall effect) was also explained as can be seen in Figure 6.2. Another factor that needs to be addressed at this stage is the fact that the aggregates are not perfect spheres as depicted even if they were to be river gravel, and certainly not so if they were to be obtained by crushing.

It is also known that aggregate grading plays a similar role of ensuring an effective packing and realizing a coherent mass of concrete with the least amount of cementitious materials to achieve economy. As already stated there is a need to understand how the solid packing is affected by the thin layer of paste and mortar, which in fact defines the consistency. Laboratory observations will indeed show that particularly when the water–cement ratios are too small, even with the presence of a superplasticizer, the coarser particles get fully covered with a thick coating of cement paste or mortar but apparently lie rotating on top without participating in the mixing process, which is a sort of a dry segregation of the larger particles alone. Considering that in SCC the aggregates have to slide freely in the mortar matrix and yet should not segregate, it is obvious that the undulations in the aggregate surface, and maybe even the roughness in the case of crushed aggregates, have to be covered and compensated for ensuring an effective self-compactable mix. A very similar observation was also made earlier by Kwan (2001). The concept of how the different sizes of coarse aggregates coated with the cement appear to pack in the concrete and the requirement for the additional mortar and paste to fill the voids in the concrete matrix is the theme of the presentation in Figure 6.3. One can see that the presentation is basically in two dimensions for simplicity and understanding while the actual packing phenomenon has to be understood in its three-dimensional format.

The need for powder extenders being obvious, it is only appropriate to look for the possible candidate materials that will do the needful. The most important characteristic of such a material that could perform the role of a powder extender is, first, that it should be in the same particle size range and distribution as that of cement so that it is physically compatible with the needs of normal concrete. Another way to look at it is to look for a slightly higher fineness so that it can act as the filler and produce a densifying effect, which may marginally improve the strength of the cementitious paste in the system. Alternatively, one could choose a very similar material in the particle size range of cement, but which can also exhibit pozzolanic or supplementary cementitious material characteristics, imparting both the filler effect and the pozzolanic effect to the system and thus improving its performance.

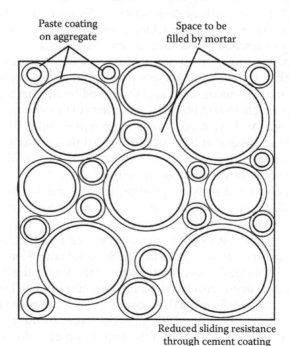

FIGURE 6.3
Packing of different sizes of paste-covered aggregates.

Fly ash or pulverized fuel ash, a material that is abundantly available from the electrostatic precipitators of thermal power stations, is probably one material that is most suitable for acting both as a powder extender as well as the pozzolanic supplementary cementitious material. It is also a fact that for reasons of economy and to ensure that the use of the pyro-processed cements that produce an equivalent amount of CO_2 for every tonne produced is limited to the extent possible, research on fly ash has been going on for several decades now. It is thus obvious that there is a wealth of information available about the suitability of the different fly ashes, methodology of designing cementitious composites, and their performance. The fact that, though SCCs are marginally different in the green state, they will not chemically differ in their performance characteristics in any way from the normally vibrated concretes or flowing concretes is only obvious. Thus, probably it is only appropriate to first discuss the behavior of fly ash in concrete and the characteristics of fly ash–based SCCs, before one looks at even the simple powder extenders. Also, it is a fact that one of the most commonly available powder extenders to the cement industry is obviously limestone powder (LSP), which in fact is a part of the raw meal in the production of cement. However, in recent years it was seen that in its fine form after grinding, LSP could also exhibit a small amount of pozzolanic activity by itself. If one has to understand some of these factors, it may be necessary to first fully comprehend what exactly are the physical and chemical dynamics of fly ash in the production of SCC so that it could also be extended for systems containing LSP-like materials.

6.3 SCCs Incorporating Fly Ash

An effective solution for requirement of powder extenders in SCCs is to adopt the ubiquitous pozzolanic industrial by-products like fly ash or slag. Most research efforts in SCC to date have used fly ash content of around 30% without accounting for the efficiency. The method of incorporating pozzolanic admixtures at different percentages of replacement was discussed earlier. The initial developments of SCCs in Japan in the late 1980s mainly for the highly congested reinforced structures in seismic regions mostly used fly ash (Ozawa, 1989). Also, the spherical nature of fly ash particles improves rheological properties apart from reducing concrete cracking due to the heat of hydration (Kurita, 1998). Significant improvement in the rheological properties of concretes containing fly ash was reported by Miura (1993). The need for viscosity-modifying admixtures to ensure cohesive and non-segregating SCC mixtures during handling operations and to enhance stability beyond that offered by sand and paste was advocated by Khayat (1997). However, studies to date have been limited to SCCs containing fly ash at about 30%. The development of SCCs with fly ash is a positive contribution to the sustainability of concrete technology. The fact that the use of fly ash reduces the demand for cement, fine fillers, and sand was recognized by Khurana (2001). It was also seen that the defects due to improper or inadequate vibration for compacting the concrete, which often reduce the durability of the structure, are completely avoided (Okamura, 2000). It was reported that concretes containing high volumes of fly ash of about 60% could result in excellent mechanical and durability properties (Bilodeau, 2000).

As a part of the several extensive investigations on concretes of various strength grades and their modifications with most available chemical and mineral admixtures, a very large database was generated and compared with the information available in the international domain. These investigations always tried to look for design practices that could ensure cementitious composites of the highest strength and performance. They were invariably studied through several test methods to understand the durability and corrosion aspects as well as their performance in different environments. It is enough to say that several of the pozzolanic materials available today like fly ash, silica fumes, ground granulated blast furnace slag (GGBS), metakaolin, rice husk ash, and so on have all been studied in the laboratory for their effectiveness as structural concrete composites. The investigations were also extended to include study of fibers of various types and their effects in terms of the strength and fracture characteristics. One of the largest experimental investigations on fly ash was done by Rao (1996) and the results of this study culminated in the formulation and substantiation of the two-part efficiency concept of general efficiency and percentage efficiency factors that were described in the previous chapter (Ganesh Babu and Rao, 1993; Ganesh Babu et al. 1993). In continuation of these efforts, investigations on fly ash–based SCCs were also undertaken and a broad outline of the results was reported earlier (Ganesh Babu, 2005a,b,c). A simple review of the status of fly ash in SCCs along with the results of a few different experimental investigations and those in the laboratory was also a part of this.

Also, more appropriately, it is necessary that the range of the cementitious materials content in conjunction with the other parameters like water requirement and strength characteristics should be critically analyzed afresh through an assessment of the available data of the experimental investigations on SCCs. A critical understanding of the broad database obtained from the earlier research efforts that resulted in the preparation of SCCs of acceptable characteristics is the right place to start any such correlations and assessment. As already stated, a very large experimental database of a fairly large number of investigations reported in the literature was compiled, and this database has been used in the present study. The references that are related to this particular effort alone were listed separately under each of the different powder extenders, so there is a clear picture of the spectrum that was chosen (References for Evaluations on Fly Ash). The present list is limited to those that presented the complete details of the mixtures and their strength characteristics. Some of the references containing only one or two mixes were also included as they were seen to be representative of the larger spectrum. It is always possible to include the several other investigations that were published before or after this particular database was generated, but it is felt that it will not change the broad conclusions of the assessment. SCCs from a few individual papers in a couple of international conferences were also compiled. For the sake of brevity, these references were simply listed as conference volumes (RILEM, 1999 and 2003; ACBM, 2002) as only a few mixtures were chosen from them, mostly single mixes investigated.

In the first place, all the concretes selected were sorted out into the groups that are relevant, specifically in relation to the pozzolanic admixture used, such as fly ash, LSP, GGBS, silica fumes, metakaolin, and others, to understand the effect of each of these admixtures appropriately. A total of over a couple of thousand mixes were looked at from a very large number of references and after rejecting several of them for either large or very large size aggregates, significantly lower or higher plasticizer concentrations apart from insufficient data available, finally, about 1000 mixes were chosen, which were then subdivided into the above pozzolanic admixture groups for evaluation. At this point, it was also felt that each group should be analyzed in its own rightful perspective so that there was a possibility of understanding the individual groups first and if possible generate broader guidelines which could be used further. This helps in not only understanding each group separately, but clearly presents a methodology by which the groups can be analyzed in terms of their efficiencies at their respective percentages of replacement. An effort was made to define the complete picture of these self-consolidating concretes as can be seen later. It is also noted that as powder admixtures go, the simplest and probably the supposedly inert powder that is freely available to the user and one that has been investigated by a larger fraction of the researchers is LSP. However, knowing that an alternative in terms of fly ash was the original powder used at the start of the efforts toward SCC development, the first discussions were all on the fly ash–based SCCs.

One of the first objectives of the preparation of SCCs is to get a broad fix on the quantity of cementitious materials needed to ensure a thixotropic paste required both for coating the aggregates adequately (assumed to be broadly for the minimum double layer of thickness apart from filling in of the undulations in the aggregates for an adequate separation of the aggregate to aid their free movement and flow) and to fill the void spaces with the aid of the mortar formed in combination with the fine aggregate. In this picture, the aggregate skeleton thus gets loosely packed and tends to flow freely with the aid of the cement paste and mortar around it. While there are several theoretical and experimental assessments by a large number of investigators over a period of time, it appears as if most of these opinions vary over a wide range, essentially because the cementitious paste is generally made up of cement and a pozzolanic admixture with two different specific gravities. This makes the volume definitions go into a disarray and even the weight definitions are not easy to arrive at. The proposed range for the quantity of cementitious material was observed to be 380–600 kg/m³ as suggested by the EFNARC's Euro specifications, which appears to be the general range if fly ash was to be the powder extender. Most investigations concentrated on fly ash as the extender and were also primarily interested only in the development of the general-purpose concretes in the lower range of 20–40 MPa, even though sometimes, inadvertently, about 60 MPa was achieved.

The two important aspects that define the strength characteristics of concrete are essentially the water–cement ratio and the water content required, which actually is the part that decides the cementitious binder component in concrete with the remaining being the inert filler. As many of the investigators have utilized only limited quantities of fly ash, and also because of the fact that the role of fly ash was understood reasonably well over the years, first the fundamental relation between the water–cement ratio and strength of these SCCs was looked at. Figure 6.4 presents the picture that ensues from such an effort. As can be seen from the figure at an approximate water–powder ratio of about 0.45 it is possible to have concretes of strengths ranging anywhere between 25 and 60 MPa, while the same at 0.3 could range from 40 to almost 80 MPa indicating a very wide scatter. Just to have a better perspective, the water–cement ratios to strengthen relationships of both the ACI and Euro norms are superimposed in the figure.

A further factor that is relevant in terms of strength is the water content at which it has been realized. Figure 6.5 presents the variations of the water

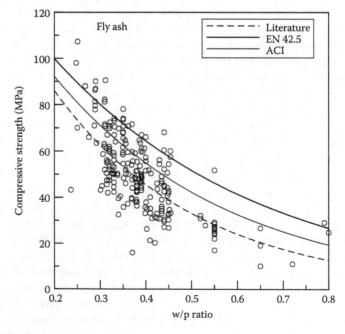

FIGURE 6.4
Strength variation with water–powder ratio.

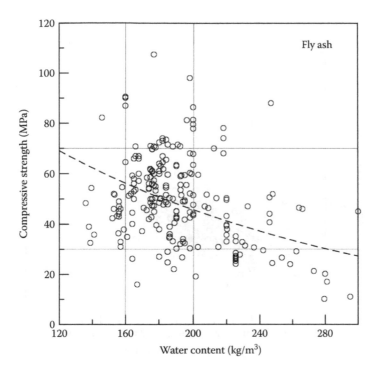

FIGURE 6.5
Effect of water content on compressive strength.

content with compressive strength, indicating an even higher scattering. Just to limit the water content, yet ensuring an appropriate minimum to regulate the more expensive superplasticizers needed at lower levels, Figure 6.5 shows that a substantially large number of the concretes fall within the limits of 160–200 kg/m³, allowing the possibility of making 30–70 MPa fly ash concretes.

The second aspect that is also equally important could be the water requirements for the increased cementitious materials content in these concretes. It can be seen from Figure 6.6 that by restricting the water content to between 160 and 200 kg/m³ the powder content could be generally limited to 400–650 kg/m³, which is indeed the near-optimal range of powder content required for SCCs. The suggested limitations and the approximate strength values that were shown to be possible to achieve with these limitations can be seen to be agreeing closely with the recommendations of both the EFNARC Euro norms as well as the ACI recommendations discussed earlier.

Even though the above limitations on the parameters present a broad outline, it is obvious that these are not enough for an effective design, which is only possible through a more critical look at the data and in specific parts that define the behavior more clearly. In view of this, the data was once again reassessed with only the concretes containing 30% fly ash to start with.

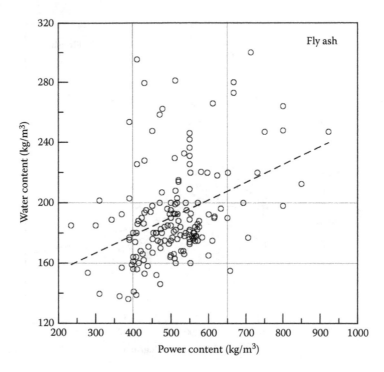

FIGURE 6.6
Water requirements for total powder content.

However, this time an attempt was made to fit the complete data of all the concretes available into a single mosaic depicting graphically the broad steps involved in the design of a concrete mix. However, it is to be remembered that as was already discussed and presented earlier, the strength of these pozzolanic concretes is related to the effective cementitious materials ratio that considers the efficiency of the particular pozzolan at the appropriate level of replacement. In view of this first aspect that is presented in Figure 6.7 is the strength to water–powder ratio plot (bottom right corner). The figure also contains the ACI and the EN relationships (for the C42.5 cement) that were discussed earlier which were, however, corrected for the efficiency of the fly ash at 30% that is presently under consideration. The evaluation of the efficiency of fly ash at the various percentages in all these presentations was done using the relationships presented in the previous chapter. Figure 6.7 also contains an average relationship in the same format for the results that are available from the literature along with a few points that depict the experimental results from the laboratory. This part of the presentation was earlier discussed thoroughly in the previous chapter while presenting how the strength to water–cementitious materials ratios is dependent on the efficiency of the pozzolanic admixture at that level of replacement. However,

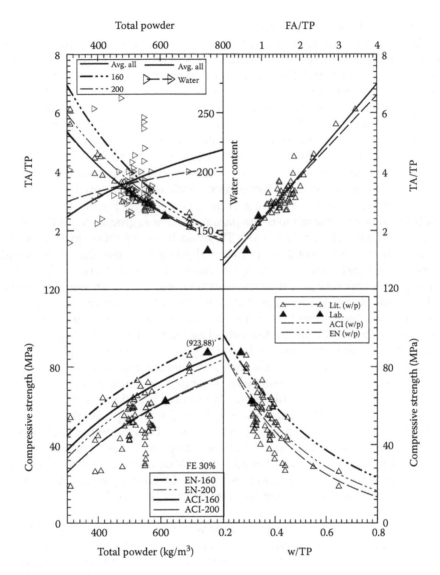

FIGURE 6.7
Correlations between the constituents in SCCs with 30% fly ash.

the relationship for the 28-day efficiency (k_{28}) at the various percentages of replacement (p′) is presented below once again for ready reference.

$$k_{28} = 2.78(p')^2 - 3.80p' + 1.64$$

From this point onward, the steps are quite simple and follow the circular clockwise pattern with the next quadrant presenting the total powder content for that compressive strength in these concretes (bottom left corner). Incidentally, this quadrant also presents the expected total powder content for both the water–cement ratio relationships in ACI and EN at two different water contents of 160 and 200 kg/m^3, which were obtained out of the previous discussions. The top left quadrant presents the total aggregate to total powder ratio (TA/TP) in relation to the total powder also showing the theoretical TA/TP variations for both the water contents of 160 and 200 kg/m^3 considered earlier. It may be noted that the water content associated with the total powder content was also presented in the same quadrant for the sake of completion, though it is seen to have congested the presentation. As already stated, the variation of the water content with the total powder content is largely dependent on the dynamics of the paste system. Finally, the top right quadrant shows the variation of the parameter fine aggregate to total powder ratio (FA/TP) in relation to the TA/TP. In each one of these quadrants, the probable average value of the experimental points was also shown. It should be recognized that some of them do not present a specific trend as the choice of the powder content and thus the resulting water content required varied highly in the various studies reported. Some of these aberrations are not really relevant to the discussions that are on hand as we can see that the results present a broad correlation of the theoretical predictions with the experimental results from the literature and the laboratory as well. These correlations at the 30% level alone not being satisfactory, the relevant mosaics at the different concretes at the varying percentage levels of replacements, namely, 10, 20, 30, 40, 50, and also >50% available, were also plotted individually. These figures show that at all these levels also the results exhibit an appropriate correlation with the theory that was proposed, which could be used with confidence to design the concretes of any specific grade at any specific percentage.

Before it is taken for granted that these figures do represent the relationships between the various parameters at the various levels of fly ash in SCCs, a few specific observations and recommendations may be in order.

- The first aspect is that the strength to water–cement ratio relationships presented for each percentage are as per the ACI and Euro norms (EN 42.5 cements). These indeed represent the cylinder and cube strengths of the concretes and also they are nearly the same (±2 to 5 MPa) if the relevant corrections are made as has already been seen.

- These equations are corrected for the efficiency of the fly ash at the relevant percentage to represent the water–powder ratio to strength relation at that percentage through the fly ash efficiency relation (Equation 5.5 presenting the 28-day efficiency, "k_{28}" for fly ash).

- It should be noted that the method can be used for the design of a specific strength of concrete at a particular age using the other equations available (Equations 5.4 through 5.6 that present the 7-, 28- and 90-day efficiencies, k_7, k_{28}, and k_{90} for fly ash).

- We can also see that the strength values reached the EN specifications at the relevant water–cement ratios showing that it is possible to achieve the particular strength if we could control the other parameters appropriately.

- The second aspect is the choice of the water content, presently taken to be varying from 160 to 200 kg/m³. This depends primarily on the interrelationships between the cement, the characteristics of the pozzolanic admixture, the water content, and finally the type and compatibility of the superplasticizer itself. It is certainly possible to play around a little with the values (reducing them to almost as low as 140 kg/m³) to reduce the cementitious materials content of high-volume pozzolanic concretes with low-end pozzolanic admixtures (fly ash) or for realizing very-high-strength concretes (+100 MPa) while utilizing the high-end pozzolanic admixtures like silica fumes to keep both the water to effective cementitious materials ratio and the cement content in check (by considering the cementitious efficiency of these pozzolans appropriately). Suggestions that much lower water content could be achieved appear to show that it is the water content in the superplasticizer that was probably primarily responsible for this effect and not the significantly higher superplasticizer content itself. The use of water content in excess of 200 kg/m³ in general is not really essential and in case there is a need a very small quantity of additional superplasticizer may always be able to do the needful.

- The third point of interest is that the relationship between the total powder and the TA/TP that was presented in the top left quadrant is nearly the same for all replacement percentages of a particular pozzolanic admixture. In the case of fly ash, this total aggregate content was seen to be defined by the relation:

$$(TA/TP) = 10.98 \cdot e^{-0.0024 \cdot TP}$$

- The water content that is presented in the same quadrant as already stated could be always modulated to regulate the paste content through the water superplasticizer combination modifications and controls.

- The next aspect is to get a fix on the FA/TP to TA/TP relation (presented in the top right quadrant) which appears to be coherent with the laboratory experimental results. Even this relationship was seen to be nearly the same for all replacement percentages of a particular pozzolanic admixture. Presently, for fly ash, this fine aggregate content was defined by the relation:

$$(TA/TP) = 1.55(FA/TP) + 0.79$$

- The last and final point that one needs to have a perfect under-standing of is the fact that though the relationships between the total powder and TA/TP and FA/TP are indeed the same for all fly ash percentages in the cementitious materials of the SCCs and are defined by the relationships given above, it is recommended that instead of these fine aggregate and coarse aggregate proportions a combined grading of the aggregate confirming to the average of the DIN-A and DIN-B (DIN-AB) is the most appropriate. This combined grading specifically ensures the availability of the frac-tions between 6 and 2 mm that are essential for the loosening effect in the aggregate mass for free flow aided by the paste system in the matrix.

- One highly pertinent point that could easily be seen through a close look at these graphs is that in most cases almost all concretes that did not achieve the expected strength levels appear to be deficient in terms of the cementitious materials content utilized. While it is possible to modulate the cementitious materials content through a combined modulation of the water and superplasticizer content, one should keep in mind that this value is related to the water to effective cementitious materials ratio (as it is related to the water–cement ratio in the case of normal concretes). Choosing to utilize extremely lower cementitious materials content will certainly involve a compromise in terms of the powder requirements in the mix resulting in a porous matrix that will have a lower strength as can be seen. The nomograms that are being presented below fol-lowing this observation show how to assess the minimum powder content required at the two water contents of 160 and 200 kg/m^3 but will also help in calculating the powder content required at other water content.

To be more proactive in lending a helping hand to the practitioner on site, these evaluations of the data collected on a large number of SCCs from the literature at the different percentages ranging from 10% to 60% in the form of the nomograms that are being discussed above have all been compiled and placed in Appendix FA-6 at the end of this chapter. This is to ensure that all these large nomograms can be consolidated at one location so that the explaining text can have the continuity in presentation.

Incidentally, some of the studies on fly ash–based SCCs have also reported concretes with 0% fly ash (only with cement as the powder) and it was felt that evaluating these as was done for the SCCs containing fly ash would give us the confidence that the earlier accepted relations of water–cement ratio to strength suggested for the normally vibrated concretes by ACI and EN are indeed valid and we also get a clear idea of how the other constituents

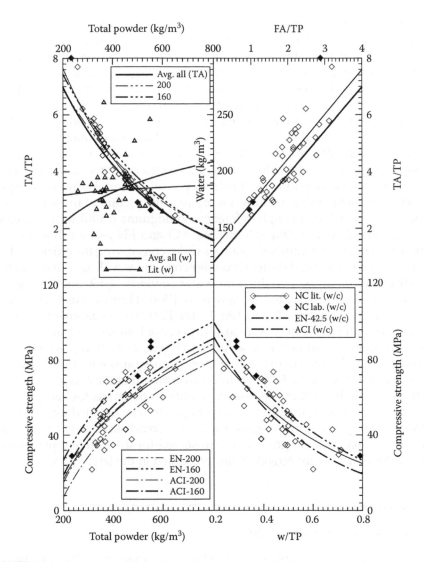

FIGURE 6.8
Correlations between the constituents in SCCs without fly ash.

are distributed in the SCCs with only cement. Figure 6.8 presents the same showing clearly the validity and acceptability of the presently existing codal specifications are certainly valid for the SCCs also. To be more specific, most of the strengths of the normal concrete have exceeded the predicted values from the water–cement ratio strength relationships by the ACI and Euro norms. Also the cement content appears to be very nearly in the range that it needs to be from theory and probably this is the primary reason that most of the strength values have exceeded the strengths predicted by the ACI and Euro norms. Following this the various relationships between the different groups containing fly ash at the different percentages of 10%–60% are presented in Appendix FA-6 (Figures FA6.1 through FA6.6). Figure FA6.7 presents the complete results for all the concretes and the possible relationships arising out of these to show the reader the complexity in addressing this complete data in one go for arriving at a reasonable methodology prediction.

Once we were able to establish the efficacy of the various interrelations between the SCCs at the different fly ash concentrations in the cementitious constituents, an effort was made to put all these together once again to understand and assimilate how a more realistic picture of the same could be obtained to arrive at simple reference nomograms useful to the practicing engineers. Stated in simple terms the ACI and EN relationships for the strength to water–cementitious materials ratio (considering the efficiency factors presented earlier) and the total powder contents as well as both the aggregate distributions at the two different water contents of 160 and 200 kg/m^3 were replotted and presented in Appendix FA-6 (Figures FA6.8 and FA6.9 for ACI and Figures FA6.10 and FA6.11 for EN). The experimental results available from the laboratory were also superposed, showing clearly that the proposed strength can easily be achieved (Figure FA6.10, representing the cube strength). The idea of presenting these nomograms with the two different water contents, 160 and 200 kg/m^3, is to ensure that the executive in field could decide for himself the optimal cementitious materials for his operation and choose the water content that is appropriate. The detailed explanations in the preceding paragraphs regarding fly ash could serve as guidelines for most discussions that are going to be presented for the different pozzolanic admixtures that were investigated and presented in the later parts.

6.4 SCCs Incorporating LSP

The effect of LSP that cannot be classified as a pozzolan, industrial or natural, on concrete has never been fully understood. Initially, it was assumed that this material is inert and performs only the pore-filling effect in the matrix apart from being a powder extender in SCCs. The double peak in the heat of hydration resulting from experimental calorimeter studies suggests that apart from accelerating the hydration reactions LSP is not just an inert additive but may be an active constituent in the hydration process as reported earlier (Ramachandran, 1986; Poppe, 2005). Instead of going into the chemical aspects of the reaction chemistry, it is necessary to get a fix on the capability of the LSP to improve the strength of the SCCs.

The strength characteristics of SCCs containing LSP as a mineral admixture were assessed from the results of over 250 SCC mixtures reported in the literature (References for Evaluations on LSP). In principle, all the data collected was categorized into concretes containing different percentages of LSP, 10%–50%, along with the details of the constituents and their strength characteristics. An assessment of the 28-day average cementitious strength efficiency (k_{28}) of the LSP at the various percentages in these concretes resulted in the following relation:

$$k_{28} = 1.01 - 0.025p' + 0.00024(p')^2$$

where (p′) is the percentage of LSP in the concrete. The methodology of assessing the cementitious efficiency of powder material in concrete was already explained earlier with reference to fly ash, which is essentially the same for LSP also. The strength to water–cement ratio relationship of the C 42.5 grade Euro cement was used for establishing these efficiency factors. A more detailed account of the complete methodology used for the evaluation of the efficiency of LSP was presented earlier by Ganesh Babu (2001). These evaluations clearly show that LSP participates in the cement hydration process imparting a definite improvement in strength.

However, a detailed evaluation of the behavior of these concretes is not possible just by looking into the water–cementitious materials ratio to strength relations alone. The actual cementitious materials content in these concretes was also looked at, as the cement content is the one that defines the economics of the concrete of a given strength ultimately. This will also give an opportunity to design these concretes with some confidence knowing the water–cement ratio and cementitious material required for a specific strength at the different percentages of LSP.

As in the case of fly ash, the relation between the water–cement ratio and strength for these SCCs was looked at (Figure 6.8). The water–cement ratio to strength relationships proposed by superimposing ACI and Euro norms show that it is possible to achieve strengths that are in line with the proposed norms with the cementitious systems containing LSP also. It is obvious from the figure that at a water–powder ratio of 0.25 the strengths of the concretes varied from 40 to 90 MPa while at a water–powder ratio of 0.55 the strengths varied from 10 to 40 MPa, indicating the large scattering possible. With a proper understanding of the effect of water–cementitious materials ratio through the efficiency factors at the different replacement percentages, it is proposed to see if this scattering could be addressed as was done earlier in the case of fly ash.

A look at the amount of water that is required at the different compressive strengths of the above system, presented in Figure 6.9, shows that most of the concretes have water content between 160 and 200 kg/m³, resulting in concretes of strength varying from 30 to 80 MPa.

A further look into the dynamics of the total cementitious materials content and their wetting water requirements for producing SCC of acceptable quality, shown in Figure 6.10, indicates that with powder

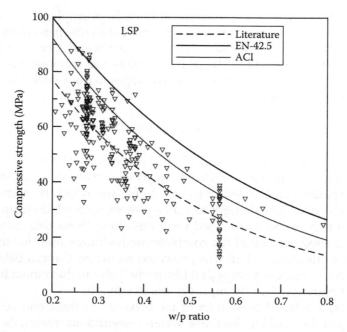

FIGURE 6.9
Water–powder ratio strength relation for SCCs with LSP.

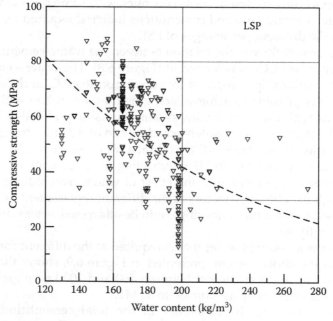

FIGURE 6.10
Variation of water content with strength in SCCs with LSP.

content varying from 400 to 650 kg/m³ most of the concretes had the same requirements of 160 to 200 kg/m³ as in the case of fly ash. This was indeed expected as fly ash, the cement, and even LSP generally have very similar particle sizes and also distributions (but for the fact that fly ash particles coming out of the flue gas systems of thermal power stations are spherical in nature).

To have a specific understanding of the interrelations between the various constituents in these concretes containing LSP the data of all the concretes containing 40% LSP in the system was looked at critically. The methodology and the distribution of the various interrelations between the parameters that are being discussed are very similar to those presented for the fly ash–containing SCCs discussed earlier and will not be repeated. The corresponding interrelations between these parameters can be seen in Figure 6.11. A critical look at the figure indicates that SCCs with strengths of almost 70 MPa or more can be achieved at a replacement of 40% LSP in a total powder content of approximately 600 kg/m³ though there is a large scattering as can be expected with the different water and superplasticizer interactions with a total powder (Figure 6.12). Also, one can see that the TA/TP and FA/TP values are also following a definite trend as was the case in fly ash too.

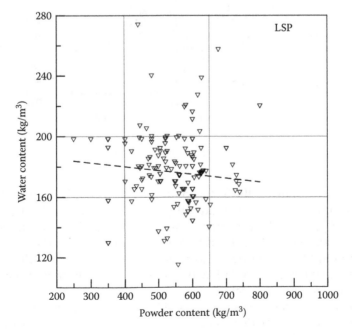

FIGURE 6.11
Water requirements of SCCs with LSP.

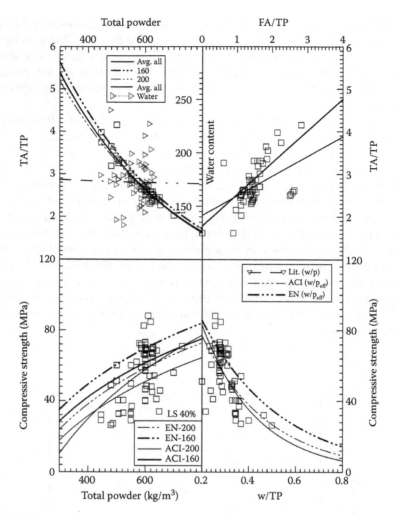

FIGURE 6.12
Correlations between the constituents in SCCs with 40% LSP.

Similar correlations for the SCCs containing 10%–50% LSP were also made and the resultant nomograms are all presented in Appendix LS-6 similar to fly ash. Appendix LS-6 contains the relationships between the different groups containing LSP at the different percentages of 10%–50% (Figures LS6.1 through LS6.5). Figure LS6.6 presents the complete results for all the concretes and the possible relationships arising out of these to show the complexity in addressing this data.

After arriving at these relationships of SCCs with different concentrations of LSP and effort was made to simplify the system and put them all together considering the ACI and EN relationships at the water content of 160 and 200 kg/m³ as was presented for fly ash (Figures LS6.7 and LS6.8 for ACI and

Figures LS6.9 and LS6.10 for EN). These figures can be directly used to arrive at the compositions for SCCs of either cylinder or cube strength at the various percentages of LSP.

6.5 SCCs Incorporating GGBS

The manufacture of slag cements through inter-grinding of GGBS with cement clinker has been in vogue for a long time. The effectiveness of granulated slag in such cements even at very high proportions is also well recognized and accepted. The granulation process as well as the slag has undergone several changes to ensure an increase in its reactivity. The use of slag as a supplementary cementitious material has also been recognized for a long time, and there have been several investigations regarding what is termed as "lag activity index" to understand its effectiveness. ASTM C989 (1994) defines slag activity index as the percentage ratio of the average compressive strength of the combined slag—cement (50%–50%) mortar cubes to the average compressive strength of reference cement mortar cubes at a designated age. The properties of GGBS influencing its reactivity were observed to be the glass content, chemical composition, mineralogical composition, fineness of grinding, and type of activation provided (Hooton, 1983). It was suggested that to attain strengths comparable to normal concretes the total cementitious materials containing GGBS are to be increased to 20% and 65% for replacements of 10% and 50% cement (Swamy, 1990). In an effort to bring about clarity regarding the efficiency of GGBS, Ganesh Babu (2000) analyzed the large volume of data available and concluded that the compressive strength of GGBS concretes depends both on the age and the percentage replacement level similar to that seen in the case of fly ash and silica fume concretes studied earlier in this laboratory (Ganesh Babu, 1993, 1995, 2000). These evaluations and further studies in the laboratory by Sree Rama Kumar (1999) resulted in the following relation for the 28-day efficiency (k_{28}) of GGBS.

$$k_{28} = 1.25 - 0.017\,p' + 0.000095(p')^2$$

In line with the methodology adopted for other pozzolans, the water–powder ratio to strength relations of SCCs containing GGBS was studied through the data available in the literature (References for Evaluations on LSP). Apart from this, the water and powder contents at which these strengths were obtained were also critically analyzed. Without going into the details of how each of these relationships looks like, as was done previously, it is enough to say that SCCs of strengths ranging from 30 to 90 MPa could be produced at the water content of 160 to 200 kg/m³ with powder

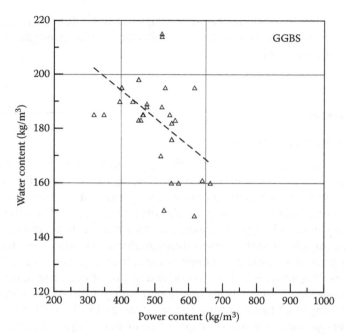

FIGURE 6.13
Water requirements of SCCs with GGBS.

content of 400 to 650 kg/m³, very similar to the SCCs with fly ash and LSP discussed earlier (Figure 6.13).

Having observed these facts, all the data collected was categorized into SCCs containing the different percentages of GGBS, 10%–50% available, along with the details of the constituents and their strength characteristics. The typical relations between the different constituents in the SCCs containing a 40% replacement are presented in Figure 6.14. It can be clearly seen that the concretes of about 60–80 MPa could be achieved through the proposed methodology using the efficiency of GGBS and the relevant percentage. This, of course, is in line with what has been observed so far for concretes with other pozzolans like fly ash and LSP.

Similar appraisals of the available SCCs with GGBS at the different percentages of 10%–50% have been presented in Appendix GG-6 at the end of this chapter (Figures LS6.1 through LS6.5). Figure GG6.6 presents the complete results for all the concretes and the possible relationships arising out of these. For arriving at the distribution of the various constituents of SCCs at various percentages of GGBS directly, considering the ACI and EN strength relationships and the water contents of 160 and 200 kg/m³ nomograms were presented (Figures GG6.7 and GG6.8 for ACI and Figures GG6.9 and GG6.10 for EN). These figures can be used by the practicing engineer to look at the options for the compositions of SCCs keeping in mind that the ACI presents the cylinder strength and the EN Euro regulations present the cube strength for C42.5 cements.

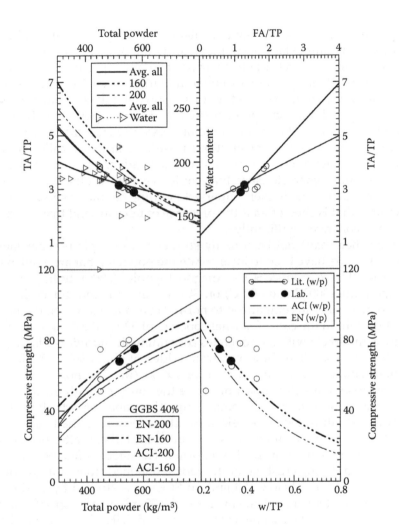

FIGURE 6.14
Correlations between the constituents in SCCs with 40% GGBS.

6.6 SCCs through Other Inert Powder Extenders

The use of locally available materials has understandably always taken precedence over other factors particularly in the development of the materials of construction from time immemorial. Researchers have also looked for alternatives for powder extenders for SCCs either from considerations of economy or for the development of a superior or even a different product that may have other applications. It is only appropriate to say that these materials are categorized as other powder extenders specifically because they probably do not show any pozzolanic reactivity but will still be able to perform this simple task of supplementing the powder content and may even have a small effect due to their poor filling capacity in some cases. Some of these inert powder extender materials used by researchers in producing SCCs could be listed as glass powder (glass), quartz powder (quartz), recycled materials powder (RC), and rubber powder (RB), just to name a few. However it is important to understand that some of these materials apart from not being cementitious (resulting in the so-called dilution effect in the cementitious composites) can also adversely impact even the original strength characteristics that could have been otherwise possible with cement alone due to their imperfect bonding characteristics. One classic example that is cited often is the presence of mica in sand that affects the strength of concrete significantly.

Though there have not been many studies with simple powder materials, investigators have tried to incorporate the powders that are available on hand in an effort to produce SCCs. Anagnostopoulos (2009) studied the effect of temperature (300°C and 600°C) on SCCs using different fillers like slag, LSP, and glass filler in comparison to normal concrete. However, the limited cement content in the case of normal concrete of 330 and 375 kg may not be directly comparable with the concretes containing the additional fillers, particularly slag. The fillers, however, were added at about 30%–40% in these variations containing a water content of 190 kg/m³. The normal concretes were obviously not self-compacting while the concretes with the additional powders were seen to be self-compacting. The paper reports the residual strengths and ultrasonic pulse velocities after exposure to the temperatures studied. The variations in the stress–strain characteristics after exposure to the temperature regimes were also reported. The effect of fine powder fractions obtained during crushing of the rubble obtained from building demolition (rubble powder below 150 µm) on SCCs was studied by Corinaldesi (2011). The powder had a Blaine fineness of 0.99 m²/g and a specific gravity of 2.15 and was studied along with fly ash and LSP. The concretes had a cement content of 440 kg/m³ along with an additional powder content of about 100 kg/m³. The water content was 200 kg/m³ resulting in a water–cement ratio of 0.45. The rubble powder was also studied with the recycled coarse aggregate of 15 and 6 mm apart from the natural limestone aggregate of 15 mm. It was demonstrated that concretes with a slump flow of 700–800 mm can be successfully produced with these materials, all of which achieved the strength of about 40 MPa at 28 days. Recycled powders produced by heating the crushed recycled aggregates to 300°C to loosen the cementitious materials and their removal by attrition were used as an additive (160 to 190 kg/m³) in SCCs prepared with 366 kg/m³ of cement with the water content of 183 kg/m³ studied by Quan (2011). The concretes have all shown that they are self-compacting at a slump flow higher than 600 mm and have resulted in achieving strengths of more than 60 MPa. This was also the case with the LSP concrete used for comparison though the strengths of slag concretes reached almost 80 MPa. The compressive to tensile strength and modulus of concrete relationships were also studied. The drying shrinkage characteristics were all found to be related to the weight loss reasonably well.

Presently, as was the case in understanding the previous materials, the water–powder ratio to strength relations of SCCs containing these powders were studied through the data available to see if they could also be positioned appropriately in the total scheme of powder extenders for SCC. However, it was clearly seen that these materials did not make any significant contribution to the strength characteristics of the SCCs, and hence the water–cement ratio to strength relationships were plotted instead of the water–powder ratio to strength relationships as can be seen in Figure 6.13. However, as the wetting water requirements are still influenced by the other materials that are in the system and the volume occupied by these powders has to be accounted for, the remaining part of the presentation in Figure 6.13 remains similar to that of the earlier presentation considering the powder content in the system. Looking at Figure 6.15, it can be seen that most of the materials resulted in strengths that correspond to the cement content alone. The slight increase that was seen for recycled powders is probably due to the very fine dust that is available as a part of the system due to crushing that could have participated as a pore filler. Evaluations of specific contributions from such materials are not possible with the limited amount of information that is available but can be attempted if more specific data is generated. This shows that it may be possible to use these materials as inert powder extenders without considering them in the cementitious contribution to the concrete matrix. Even so it is advised that their use be restricted to levels below 15% or 20% to ensure that the dilution effects that include the additional water requirement will not cause any serious problems to the strength and performance of these SCCs.

6.7 Practical Limitations on Powder Fillers

It is abundantly clear from the earlier discussions that powder extenders could either be simple inert fillers or pozzolanic materials. The concept of using inert fillers should only be attempted in circumstances where there is no possibility to get even the simplest of them like fly ash. Even then it would be preferable to have them as the further extenders beyond the pozzolanic additives that are available. The second aspect is that any powder extender beyond approximately 15%–20% would certainly require an additional quantity of wetting water for its participation in the concrete composites along with the aggregates. Some of these requirements could be modulated through the use of a small quantity of the superplasticizer in most cases. However, at the higher levels of replacements and larger cementitious materials content,

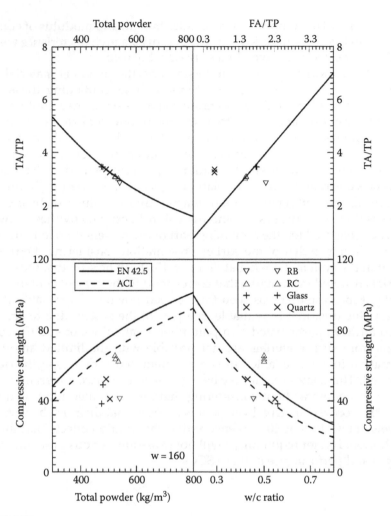

FIGURE 6.15
Correlations between the constituents in SCCs with simple powders.

the water content required needs to be certainly addressed as discussed. In view of this scenario where the powders can impart certain advantages in terms of fluidity but will require additional wetting water, it is suggested that the total powder content be retained between 400 and 650 kg/m³. At values above or below these, the need for both expert interaction and experimental evaluation is essential. One specific obvious disadvantage is that the unburnt carbon should be kept within acceptable limits (preferably below 1%) to have a good control on the superplasticizer required. Finally, the compatibility of the superplasticizers, particularly for the supplementary cementitious materials like fly ash and GGBS, may need to be established appropriately before they are approved for any specific project.

Appendix FA-6

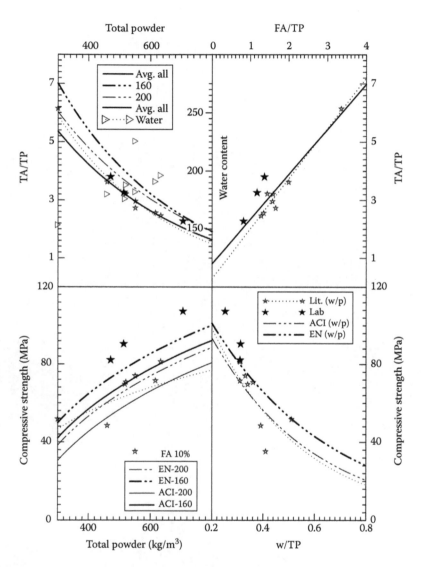

FIGURE FA6.1
Correlations between the constituents in SCCs with 10% fly ash.

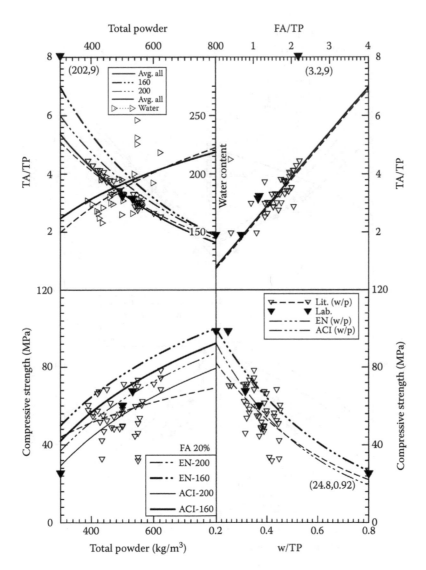

FIGURE FA6.2
Correlations between the constituents in SCCs with 20% fly ash.

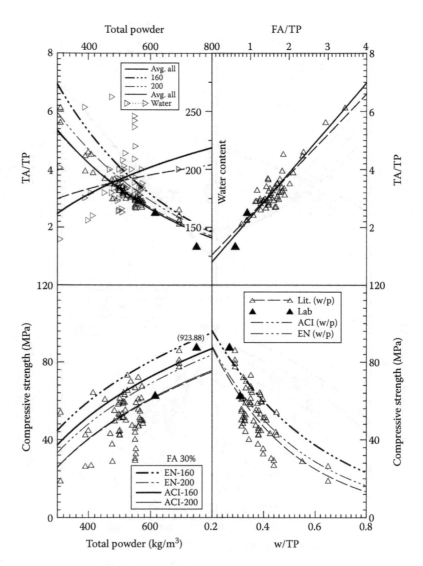

FIGURE FA6.3
Correlations between the constituents in SCCs with 30% fly ash.

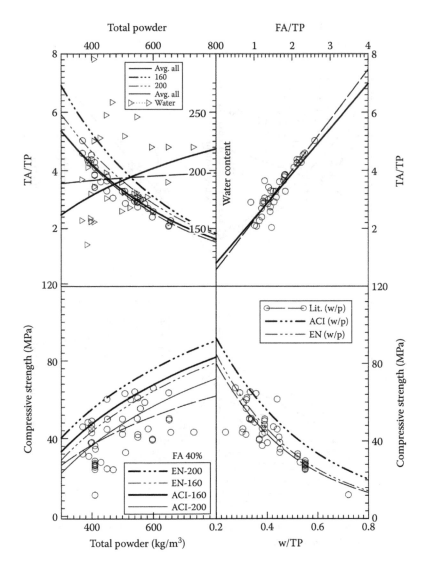

FIGURE FA6.4
Correlations between the constituents in SCCs with 40% fly ash.

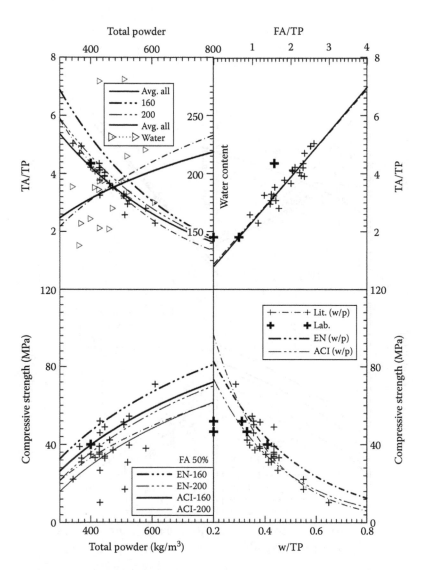

FIGURE FA6.5
Correlations between the constituents in SCCs with 50% fly ash.

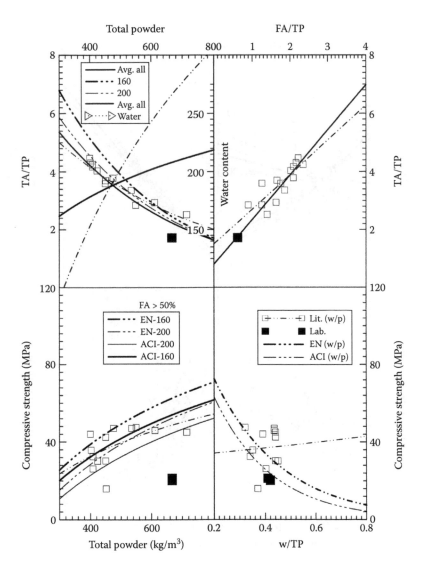

FIGURE FA6.6
Correlations between the constituents in SCCs with 60% fly ash.

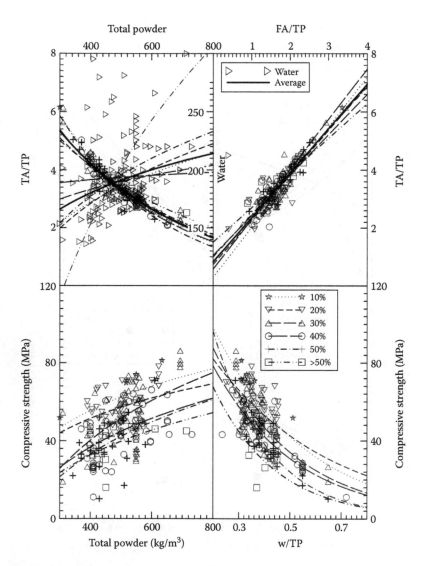

FIGURE FA6.7
Distribution of constituents in SCCs containing fly ash.

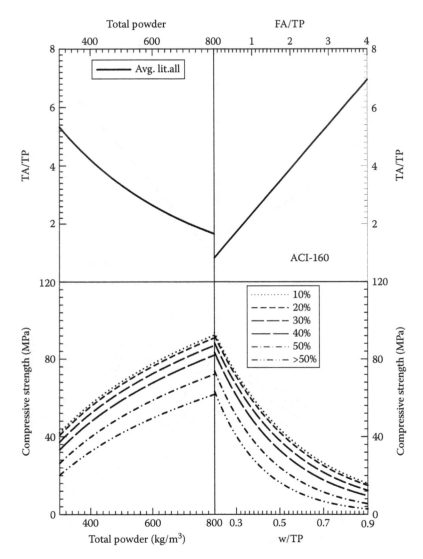

FIGURE FA6.8
Nomogram for design SCCs with ACI at 160 kg/m³ water (cyl.).

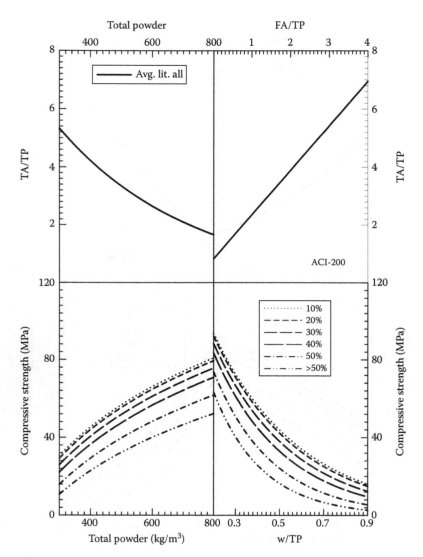

FIGURE FA6.9
Nomogram for design SCCs with ACI at 200 kg/m³ water(cyl.).

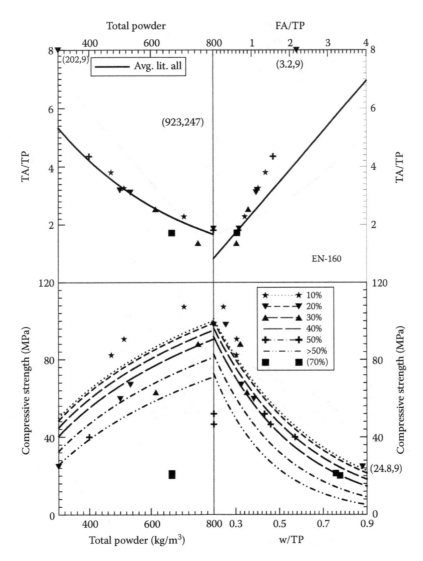

FIGURE FA6.10
Nomogram for design SCCs with EN 42.5 at 160 kg/m³ water (cube).

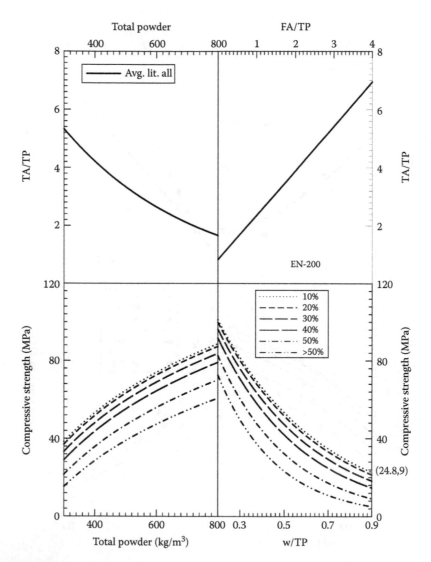

FIGURE FA6.11
Nomogram for design SCCs with EN 42.5 at 200 kg/m³ water (cube).

Appendix LS-6

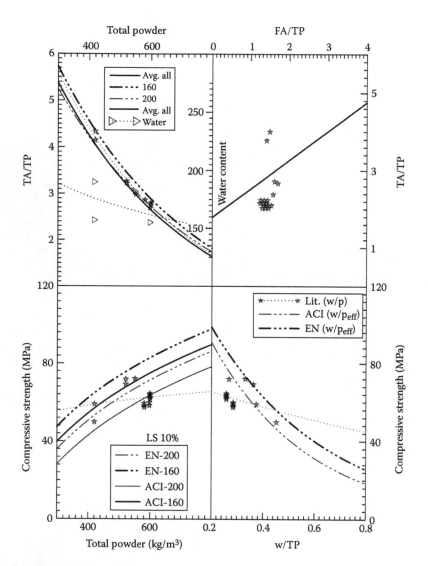

FIGURE LS6.1
Correlations between the constituents in SCCs with 10% LSP.

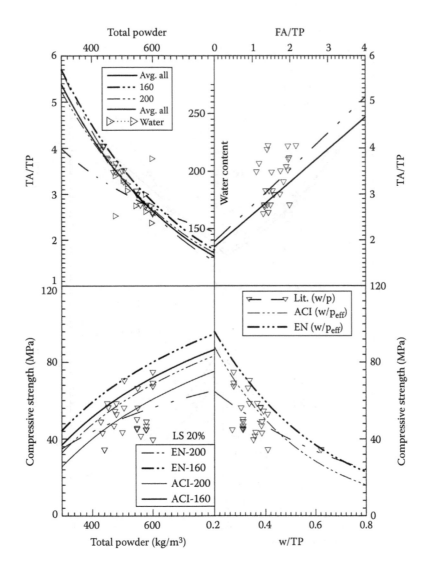

FIGURE LS6.2
Correlations between the constituents in SCCs with 20% LSP.

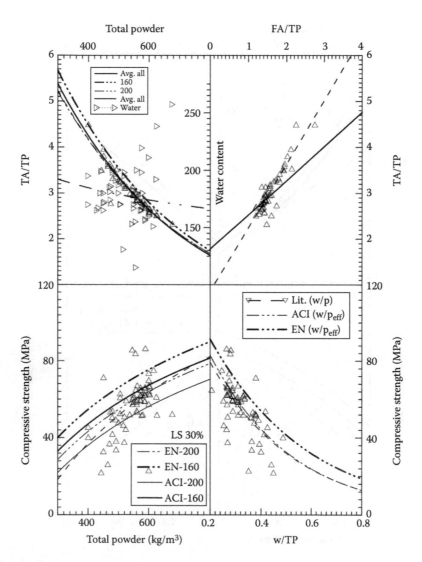

FIGURE LS6.3
Correlations between the constituents in SCCs with 30% LSP.

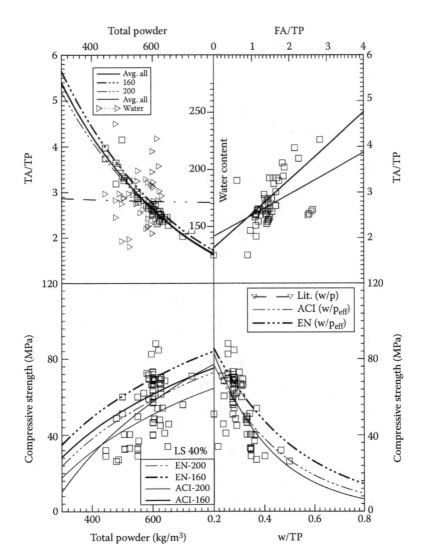

FIGURE LS6.4
Correlations between the constituents in SCCs with 40% LSP.

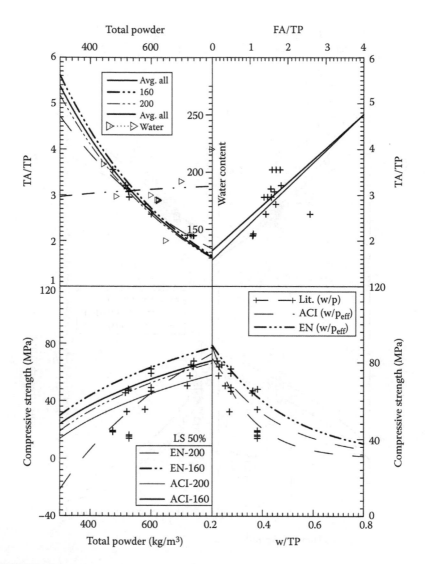

FIGURE LS6.5
Correlations between the constituents in SCCs with 50% LSP.

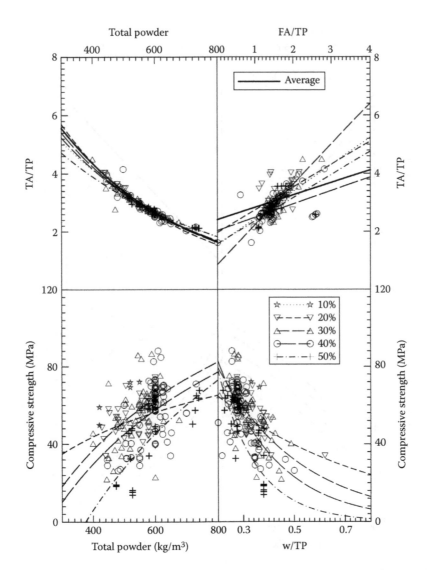

FIGURE LS6.6
Distribution of constituents in SCCs containing fly ash.

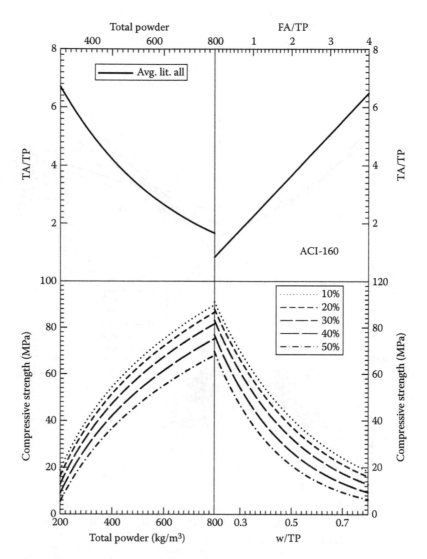

FIGURE LS6.7
Nomogram for design SCCs with ACI at 160 kg/m³ water (cyl.).

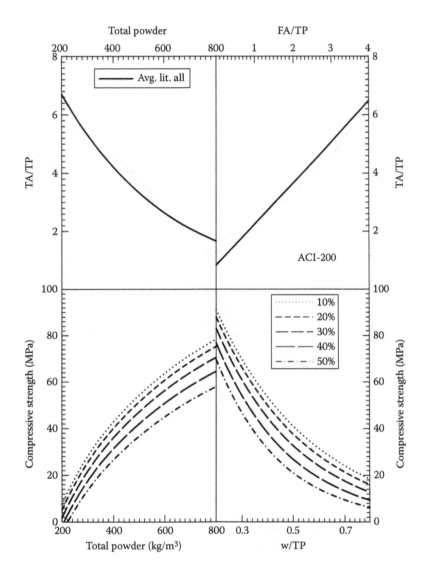

FIGURE LS6.8
Nomogram for design SCCs with ACI at 200 kg/m³ water (cyl.).

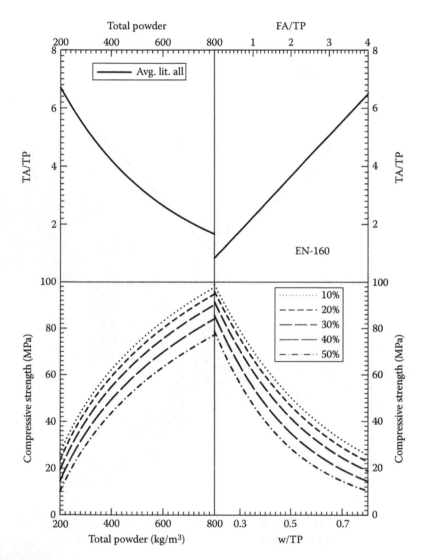

FIGURE LS6.9
Nomogram for design SCCs with EN 42.5 at 160 kg/m³ water (cube).

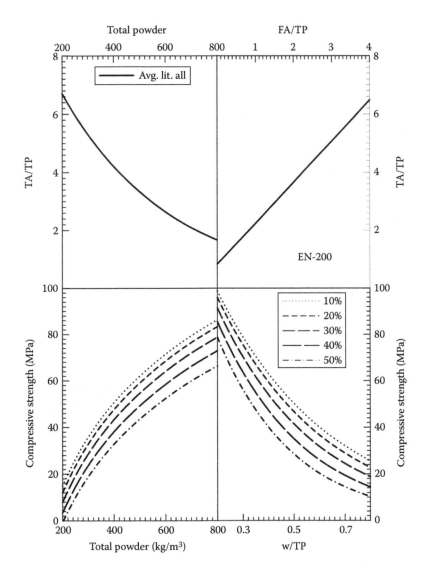

FIGURE LS6.10
Nomogram for design SCCs with EN 42.5 at 200 kg/m³ water (cube).

Appendix GG-6

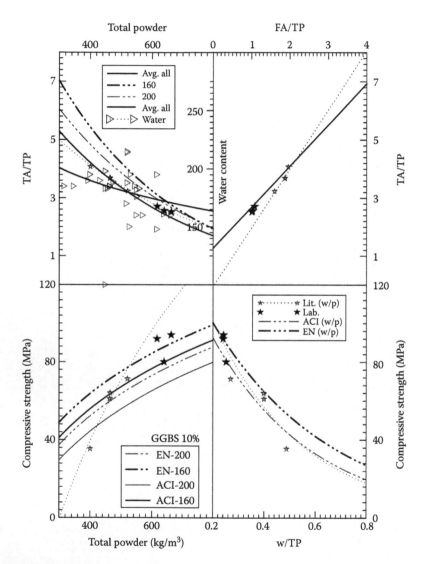

FIGURE GG6.1
Correlations between the constituents in SCCs with 10% GGBS.

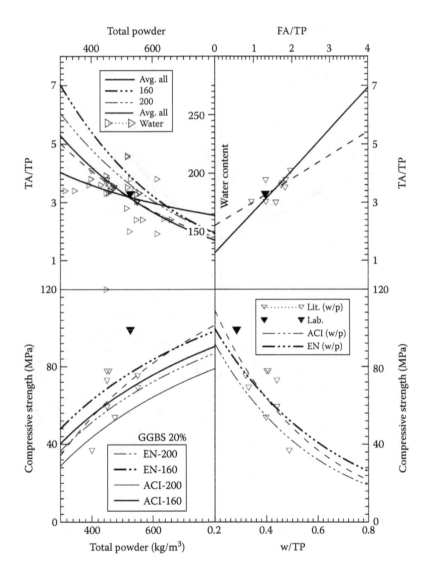

FIGURE GG6.2
Correlations between the constituents in SCCs with 20% GGBS.

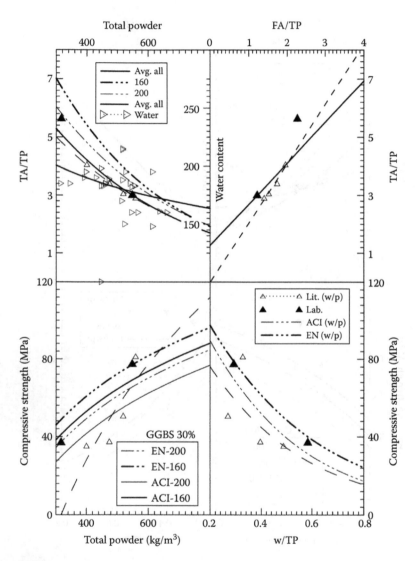

FIGURE GG6.3
Correlations between the constituents in SCCs with 30% GGBS.

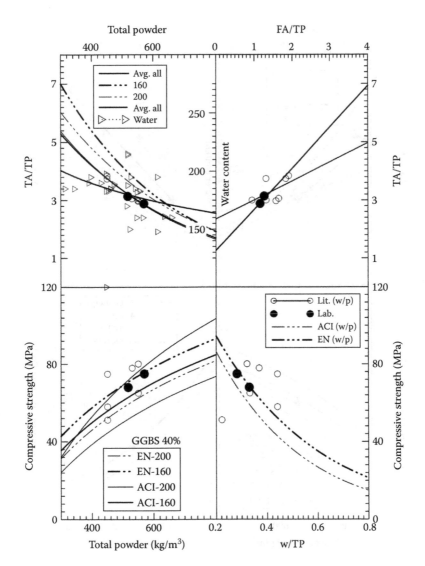

FIGURE GG6.4
Correlations between the constituents in SCCs with 40% GGBS.

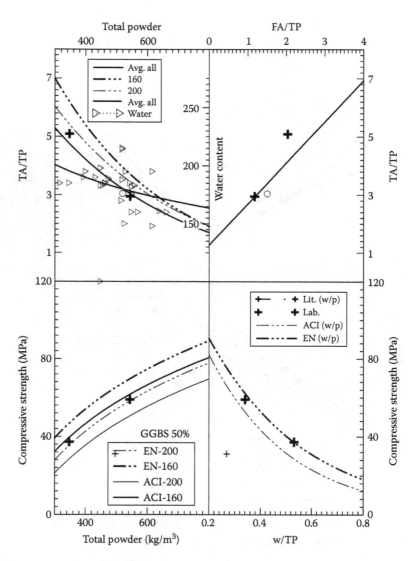

FIGURE GG6.5
Correlations between the constituents in SCCs with 50% GGBS.

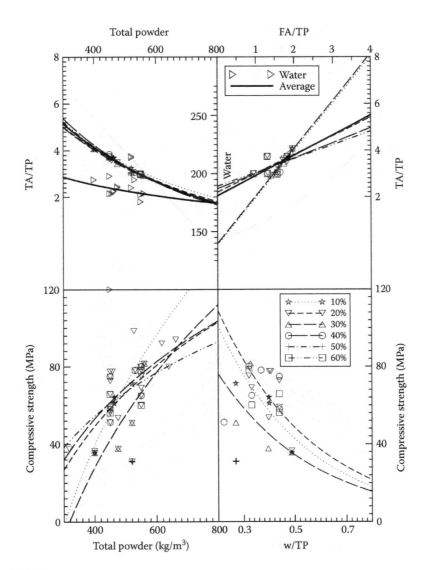

FIGURE GG6.6
Distribution of constituents in SCCs containing fly ash.

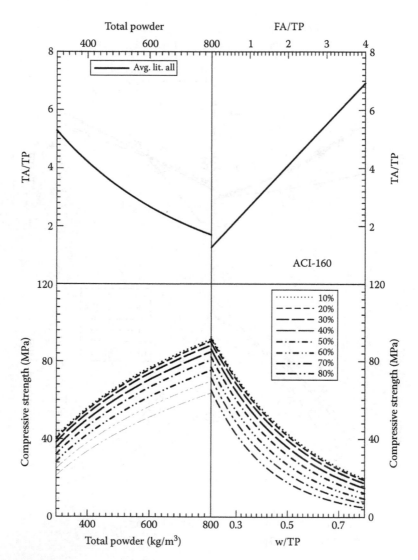

FIGURE GG6.7

Nomogram for design SCCs with ACI at 160 kg/m³ water (cyl.).

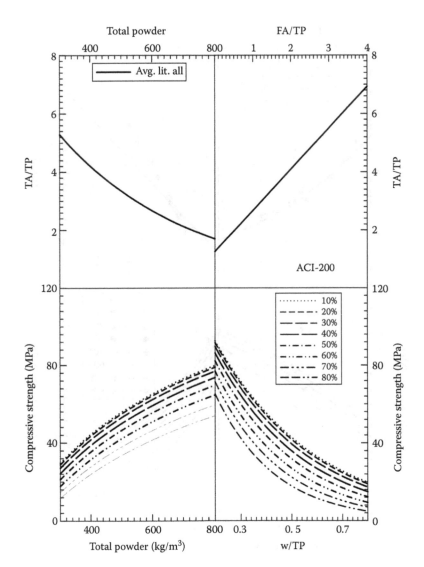

FIGURE GG6.8
Nomogram for design SCCs with ACI at 200 kg/m³ water (cyl.).

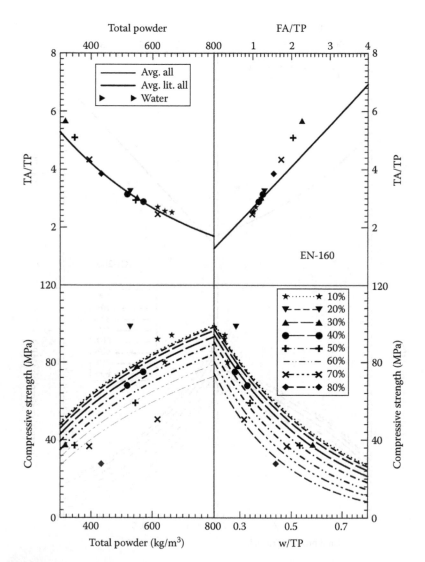

FIGURE GG6.9
Nomogram for design SCCs with EN 42.5 at 160 kg/m³ water (cube).

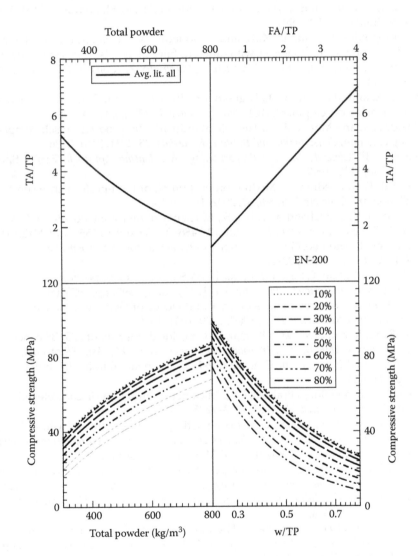

FIGURE GG6.10
Nomogram for design SCCs with EN 42.5 at 200 kg/m³ water (cube).

References

ASTM C989, Standard specification for ground granulated blast-furnace slag for use in concrete and mortars, Annual Book of ASTM Standards, ASTM, West Conshohocken, PA, 1994.

Anagnostopoulos, N., Sideris, K.K., and Georgiadis, A., Mechanical characteristics of self-compacting concretes with different filler materials, exposed to elevated temperatures, *Materials and Structures/Materiaux et Constructions*, 42, 2009, 1393–1405.

Bilodeau, A. and Malhotra, V.M., High-volume fly ash system: Concrete solution for sustainable development, *ACI Materials Journal.*, 97(1), 2000, 41–48.

Corinaldesi, V. and Moriconi, G., The role of industrial by-products in self-compacting concrete, *Construction and Building Materials*, 25, 2011, 3181–3186.

de Larrard, F., *Concrete Mixture Proportioning: A Scientific Approach*, E&FN Spon, London, U.K., 1999.

de Larrard, F. and Sedran, T., Mixture-proportioning of high-performance concrete, *Cement and Concrete Research*, 32, 2002, 1699–1704.

Ganesh Babu, K. and Chandra Sekhar, B., Efficiency of limestone powder in SCC, in *MRS Proceedings*, p. 1488, 2012. imrc12-1488-7b-039. doi:10.1557/opl.2012.1542.

Ganesh Babu, K. and Rao, G.S.N., Efficiency of fly ash in concrete, *Cement and Concrete Composites*, 15, 1993, 223–229.

Ganesh Babu, K., Rao, G.S.N. and Prakash, P.V.S., Efficiency of pozzolans in cement composites, in *Concrete 2000*, Dundee, U.K., Vol. 1, 1993, pp. 497–509.

Ganesh Babu, K. and Sree Rama Kumar, V., Efficiency of GGBS in concrete, *Cement and Concrete Research Journal*, 30(7), 2000, 1031–1036.

Ganesh Babu, K. and Sree Rama Kumar, V., Chloride diffusivity of GGFBS concretes, in *Seventh CANMET/ACI International Conference on Fly Ash, Silica Fume, Slag and Natural Pozzolans in Concrete (SP-199)*, Chennai, Tamilnadu, India, 2001, pp. 611–622.

Ganesh Babu, K. and Surya Prakash, P.V., Efficiency of silica fume in concretes, *Cement and Concrete Research*, 25, 1995, 1273–1283.

Ganesh Babu, K. et al., Design of self compacting concretes with fly ash, in *Second North American Conference on the Design and Use of Self-Consolidating Concrete*, Center for Advanced Cement-Based Materials, Chicago, IL, 2005a.

Ganesh Babu, K. et al., Self compacting concrete with fly ash, in *Proceedings of the Second International Symposium on Concrete Technology for Sustainable Development, With Emphasis on Infrastructure*, American Concrete Institute, Hyderabad, India, 2005b, pp. 605–614.

Ganesh Babu, K. et al., Fly ash based High Performance Cementitious Composites, in *Material Science of Concrete*, Special Volume. Indo-US Workshop on High-Performance Cement-Based Concrete Composites, American Ceramic Society, Chennai, Tamilnadu, India, 2005c, pp. 171–182.

Hooton, R.D. and Emery, J.J., Glass content determination and strength development predictions for vitrified blast furnace slag, *ACI SP 79*, Detroit, MI, 1983, pp. 943–962.

Khayat, K.H. and Guizani, Z., Use of viscosity-modifying admixture to enhance stability of fluid concrete, *ACI Materials Journal*, 94(4), 1997, 332–341.

Khurana, R. and Saccone, R., Fly ash in self-compacting concrete, fly ash, silica fume, slag and natural pozzolans in concrete, in *ACI SP-199*, American Concrete Institute, Chennai, Tamilnadu, India, 2001, pp. 259–274.

Kurita, M. and Nomura, T., Highly-flowable steel fiber-reinforced concrete containing fly ash in *Sixth CANMET/ACI/JCI International Conference on Fly Ash, Silica Fume, Slag and Natural Pozzolans in Concrete*, Vol. II, Malhotra, V.M. (ed.), Tokushima, Japan, in *SP 178*, 1998, pp. 159–175.

Kwan, A.K.H. and Mora, C.F., Effects of various shape parameters on packing of aggregate particles, *Magazine of Concrete Research*, 53(2), 2001, 91–100.

Miura, N., Takeda, N., Chikamatsu, R., and Sogo, S., Application of super workable concrete to reinforced concrete structures with difficult construction conditions, in *High Performance Concrete in Severe Environments*, Zia, P. (ed.), ACI SP 140, Minneapolis, Minnesota, in *ACI SP 140*, 1993, pp. 163–186.

Okamura, H. and Ozawa, K., Mix-design for self-compacting concrete, *Concrete Library, JSCE*, 25, 1995, 107–120.

Okamura, H., Ozawa, K., and Ouchi, M., Self-compacting concrete, *Structural Concrete*, 1(1), 2000, 3–17.

Ozawa, K., Maekawa, K., Kunishima, M., and Okamura, H., Performance of concrete based on the durability design of concrete structures, in *Second East Asia-Pacific Conference on Structural Engineering and Construction*, Chiang Mai, Thailand, 1989.

Poppe, A. and De Schutter, G., Heat of hydration in the presence of high filler contents, *Cement and Concrete Research*, 2005, 35, 2290–2299.

Quan, H., Study on properties of self-compacting concrete with recycled powder, *Advanced Materials Research*, 250–253, 2011, 866–869.

Ramachandran, V.S. and Zhang, C.M., Influence of Ca CO_3 on the hydration and microstructural characteristics of tricalcium silicate, *Il Cemento*, 3, 1986, 129–152.

Rao, G.S.N., Effective utilisation of fly ash in concrete for aggressive environments, PhD thesis, submitted to Indian Institute of Technology Madras, Chennai, India, 1996.

Sree Rama Kumar, V., Behaviour of GGBS in concrete composites, MS thesis, submitted to Indian Institute of Technology Madras, Chennai, India, 1999.

Swamy, R.N. and Bouikni, A., Some engineering properties of slag concrete as influenced by mix proportioning and curing, *ACI Materials Journal*, 87, 1990, 210–220.

References for Evaluations on Fly Ash

ACBM, *First North American Conference on the Design and Use of Self-Consolidating Concrete ACBM*, North Western University, Chicago, IL, 2002.

Askari, A., Sohrabi, M.R., and Rahmani, Y., An investigation into mechanical properties of self compacting concrete incorporating fly ash and silica fume at different ages of curing, *Advanced Materials Research*, 261–263, 2011, 3–7.

Atan, M.N. and Awang, H., The mechanical properties of self-compacting concrete incorporating raw rice husk ash, *European Journal of Scientific Research*, 60(1), 2011, 166–176.

Barbhuiya, S., Effects of fly ash and dolomite powder on the properties of self compacting concrete, *Construction and Building Materials*, 25(8), 2011, 3301–3305.

Bouzoubaâ, N. and Lachemi, M., Self-compacting concrete incorporating high volumes of class F fly ash: Preliminary results, *Cement and Concrete Research*, 31(3), 2001, 413–420.

Corinaldesi, V. and Moriconi, G., The role of industrial by-products in self-compacting concrete, *Construction and Building Materials*, 25(8), 2011, 3181–3186.

Gesoğlu, M., Güneyisi, E., and Özbay, E., Properties of self-compacting concretes made with binary, ternary, and quaternary cementitious blends of fly ash, blast furnace slag, and silica fume, *Construction and Building Materials*, 23(5), 2009, 1847–1854.

Gesoğlu, M. and Özbay, E., Effects of mineral admixtures on fresh and hardened properties of self-compacting concretes: Binary, ternary and quaternary systems, *Materials and Structures/Materiaux et Constructions*, 40(9), 2007, 923–937.

Gheorghe, M., Saca, N., Ghecef, C., Pintoi, R., and Radu, L., SELF compacted concrete with fly ash addition, *Romanian Journal of Materials*, 41(3), 2011, 201–210.

Gorzelańczyk, T., Moisture influence on the failure of self-compacting concrete under compression, *Archives of Civil and Mechanical Engineering*, 11(1), 2011, 45–60.

Güneyisi, E., Gesolu, M., and Özbay, E., Strength and drying shrinkage properties of self-compacting concretes incorporating multi-system blended mineral admixtures, *Construction and Building Materials*, 24(10), 2010, 1878–1887.

Li, B., Guan, A., and Zhou, M., Preparation and performances of self-compacting concrete used in the joint section between steel and concrete box girders of Edong Yangtze River Highway Bridge, *Advanced Materials Research*, 168–170, 2011, 334–340.

Ling, T.-C., Poon, C.-S., and Kou, S.-C., Influence of recycled glass content and curing conditions on the properties of self-compacting concrete after exposure to elevated temperatures, *Cement and Concrete Composites*, 34(2), 2012, 265–272.

Liu, M., Self-compacting concrete with different levels of pulverized fuel ash, *Construction and Building Materials*, 24(7), 2010, 1245–1252.

Mňahončáková, E., Pavlíková, M., Grzeszczyk, S., Rovnaníková, P., and Černý, R., Hydric thermal and mechanical properties of self-compacting concrete containing different fillers, *Construction and Building Materials*, 22(7), 2008, 1594–1600.

Poon, C.S. and Ho, D.W.S., A feasibility study on the utilization of r-FA in SCC, *Cement and Concrete Research*, 34(12), 2004, 2337–2339.

Quan, H., Study on properties of self-compacting concrete with recycled powder, *Advanced Materials Research*, 250–253, 2011, 866–869.

RILEM, *Proceedings of the First International RILEM Symposium on SCC*, Skarendahl, A. and Petersson, O. (ed.), 1999, 804p.

RILEM, *Proceedings of the Third International RILEM Symposium on SCC*, Wallevik, O. and Nielsson, I. (ed.), Reykjavik, Iceland, 2003, 1056p.

Siad, H., Mesbah, H.A., Bernard, S.K., Khelafi, H., and Mouli, M., Influence of natural pozzolan on the behavior of self-compacting concrete under sulphuric and hydrochloric acid attacks, comparative study, *Arabian Journal for Science and Engineering*, 35(1), 2010, 183–195.

Siddique, R., Aggarwal, P., and Aggarwal, Y., Influence of water/powder ratio on strength properties of self-compacting concrete containing coal fly ash and bottom ash, *Construction and Building Materials*, 29(6), 2012, 73–81.

Sonebi, M. and Cevik, A., Genetic programming based formulation for fresh and hardened properties of self-compacting concrete containing pulverised fuel ash, *Construction and Building Materials*, 23(7), 2009, 2614–2622.

Sukumar, B., Nagamani, K., and Srinivasa, R., Evaluation of strength at early ages of self-compacting concrete with high volume fly ash, *Construction and Building Materials*, 22(7), 2008, 1394–1401.

Tao, J., Yuan, Y., and Taerwe, L., Compressive strength of self-compacting concrete during high-temperature exposure, *Journal of Materials in Civil Engineering*, 22(10), 2010, Art. No. 004010QMT, 1005–1011.

Turk, K. and Karatas, M., Abrasion resistance and mechanical properties of self-compacting concrete with different dosages of fly ash/silica fume, *Indian Journal of Engineering and Materials Sciences*, 18(1), 2011, 49–60.

Ulucan, Z.Ç., Türk, K., and Karataş, M., Effect of mineral admixtures on the correlation between ultrasonic velocity and compressive strength for self-compacting concrete, *Russian Journal of Nondestructive Testing*, 44(5), 2008, 367–374.

Uysal, M., Yilmaz, K., and Ipek, M., Properties and behavior of self-compacting concrete produced with GBFS and FA additives subjected to high temperatures, *Construction and Building Materials*, 28(1), 2012, 321–326.

Vejmelková, E., Keppert, M., Grzeszczyk, S., Skaliński, B., and Černý, R., Properties of self-compacting concrete mixtures containing metakaolin and blast furnace slag, *Construction and Building Materials*, 25(3), 2011, 1325–1331.

Xie, Y., Liu, B., Yin, J., and Zhou, S., Optimum mix parameters of high-strength self-compacting concrete with ultra pulverized fly ash, *Cement and Concrete Research*, 32(3), 2002, 477–480.

Zhu, W., Gibbs, J.C., and Bartos, P.J.M., Uniformity of in situ properties of self-compacting concrete in full-scale structural elements, *Cement and Concrete Composites*, 23(1), 2001, 57–64.

References for Evaluations on LSP

Anagnostopoulos, N., Sideris, K.K., and Georgiadis, A., Mechanical characteristics of self-compacting concretes with different filler materials, exposed to elevated temperatures, *Materials and Structures/Materiaux et Constructions*, 42, 2009, 1393–1405.

Assie, S., Escadeillas, G., and Waller, V., Estimates of self-compacting concrete "potential" durability, *Construction and Building Materials*, 21, 2007, 1909–1917.

Bakhtiyari, S., Allahverdi, A., and Rais-Ghasemi, M., The influence of permanent expanded polystyrene formwork on fire resistance of self-compacting and normal vibrated concretes, *Asian Journal of Civil Engineering*, 12(3), 2011a, 353–374.

Bakhtiyari, S., Allahverdi, A., Rais-Ghasemi, M., Zarrabi, B.A., and Parhizkar, T., Self-compacting concrete containing different powders at elevated temperatures—Mechanical properties and changes in the phase composition of the paste, *Thermochimica Acta*, 514, 2011b, 74–81.

Barbhuiya, S., Effects of fly ash and dolomite powder on the properties of self-compacting concrete, *Construction and Building Materials*, 25, 2011, 3301–3305.

Benkechkache, G. and Houari, H., Deferred behaviour of the self compacting concretes based local materials, *Asian Journal of Civil Engineering*, 12(2), 2011, 219–232.

Bosiljkov, V.B., SCC mixes with poorly graded aggregate and high volume of limestone filler, *Cement and Concrete Research*, 33, 2003, 1279–1286.

Boulekbache, B., Hamrat, M., Chemrouk, M., and Amziane, S., Influence of yield stress and compressive strength on direct shear behavior of steel fibre-reinforced concrete, *Construction and Building Materials*, 27, 2012, 6–14.

Brouwers, H.J.H. and Radix, H.J., Self-compacting concrete: Theoretical and experimental study, *Cement and Concrete Research*, 35, 2005, 2116–2136.

Corinaldesi, V. and Moriconi, G., The role of industrial by-products in self-compacting concrete, *Construction and Building Materials*, 25, 2011, 3181–3186.

De Almeida Filho, F.M., El Debs, M.K., and El Debs, A.L.H.C., Bond-slip behavior of self-compacting concrete and vibrated concrete using pull-out and beam tests, *Materials and Structures/Materiaux et Constructions*, 41, 2008, 1073–1089.

Desnerck, P., De Schutter, G., and Taerwe, L., Bond behaviour of reinforcing bars in self-compacting concrete: Experimental determination by using beam tests, *Materials and Structures/Materiauxet Constructions*, 43(SUPPL. 1), 2010, 53–62.

Fares, H., Remond, S., Noumowe, A., and Cousture, A., High temperature behaviour of self-consolidating concrete. Microstructure and physicochemical properties, *Cement and Concrete Research*, 40, 2010, 488–496.

Felekoglu, B., A comparative study on the performance of sands rich and poor in fines in self-compacting concrete, *Construction and Building Materials*, 22, 2008, 646–654.

Filho, F.M.A., Barragan, B.E., Casas, J.R., and El Debs, A.L.H.C., Hardened properties of self-compacting concrete—A statistical approach, *Construction and Building Materials*, 24, 2010, 1608–1615.

Georgiadis, A.S., Sideris, K.K., and Anagnostopoulos, N.S., Properties of SCC produced with limestone filler or viscosity modifying admixture, *Journal of Materials in Civil Engineering*, 22, 2010, 352–360.

Gesoğlu, M. and Özbay, E., Effects of mineral admixtures on fresh and hardened properties of self-compacting concretes: Binary, ternary and quaternary systems, *Materials and Structures/Materiaux et Constructions*, 40(9), 2007, 923–937.

Gheorghe, M., Saca, N., Ghecef, C., Pintoi, R., and Radu, L., SELF compacted concrete with fly ash addition [Beton Autocompactant Cu CenuÅŸ̧ĊŽ ZburÇ̇Žtoare], *Revista Romana de Materiale/Romanian Journal of Materials*, 41, 2011, 201–210.

Guneyisi, E., Gesoglu, M., and Ozbay, E., Strength and drying shrinkage properties of self-compacting concretes incorporating multi-system blended mineral admixtures, *Construction and Building Materials*, 24, 2010, 1878–1887.

Ioani, A., Domsa, J., Mircea, C., and Szilagyi, H., Durability requirements in self-compacting concrete mix design, *Concrete Repair, Rehabilitation and Retrofitting II, Proceedings of the second international conference on Rehabilitation and Retrofitting II*, Cape Town, South Africa, 2008, pp. 159–165.

Mňahončáková, E., Pavlíková, M., Grzeszczyk, S., Rovnaníková, P., and Černý, R., Hydric thermal and mechanical properties of self-compacting concrete containing different fillers, *Construction and Building Materials*, 22(7), 2008, 1594–1600.

Nunes, S., Figueiras, H., Milheiro, O.P., Coutinho, J.S., and Figueiras, J., A methodology to assess robustness of SCC mixtures, *Cement and Concrete Research*, 36, 2006, 2115–2122.

Parra, C., Valcuende, M., and Gomez, F., Splitting tensile strength and modulus of elasticity of self-compacting concrete, *Construction and Building Materials*, 25, 2011, 201–207.

Quan, H., Study on properties of self-compacting concrete with recycled powder, *Advanced Materials Research*, 250–253, 2011, 866–869.

Siad, H., Mesbah, H.A., Bernard, S.K., Khelafi, H., and Mouli, M., Influence of natural pozzolan on the behavior of self-compacting concrete under sulphuric and hydrochloric acid attacks, comparative study, *The Arabian Journal for Science and Engineering*, 35, 2010, 183–195.

Tao, J., Yuan, Y., and Taerwe, L., Compressive strength of self-compacting concrete during high-temperature exposure, *Journal of Materials in Civil Engineering*, 2010, 1005–1011.

Uysal, M., Self-compacting concrete incorporating filler additives: Performance at high temperatures, *Construction and Building Materials*, 26, 2012, 701–706.

Yurtdas, I., Burlion, N., Shao, J.-F., and Li, A., Evolution of the mechanical behaviour of a high performance self-compacting concrete under drying, *Cement and Concrete Composites*, 33, 2011, 380–388.

References for Evaluations on GGBS

Anagnostopoulos, N., Sideris, K.K., and Georgiadis, A., Mechanical characteristics of self-compacting concretes with different filler materials, exposed to elevated temperatures, *Materials and Structures/Materiaux et Constructions*, 42, 2009, 1393–1405.

Boukendakdji, O., Kenai, S., Kadri, E.H., and Rouis, F., Effect of slag on the rheology of fresh self-compacted concrete, *Construction and Building Materials*, 23, 2009, 2593–2598.

Gesoglu, M., Guneyisi, E., and Ozbay, E., Properties of self-compacting concretes made with binary, ternary, and quaternary cementitious blends of fly ash, blast furnace slag, and silica fume, *Construction and Building Materials*, 23, 2009, 1847–1854.

Guneyisi, E., Gesoglu, M., and Ozbay, E., Strength and drying shrinkage properties of self-compacting concretes incorporating multi-system blended mineral admixtures, *Construction and Building Materials*, 24, 2010, 1878–1887.

Quan, H., Study on properties of self-compacting concrete with recycled powder, *Advanced Materials Research*, 250–253, 2011, 866–869.

Siad, H., Mesbah, H.A., Bernard, S.K., Khelafi, H., and Mouli, M., Influence of natural pozzolan on the behavior of self-compacting concrete under sulphuric and hydrochloric acid attacks, comparative study, *The Arabian Journal for Science and Engineering*, 35, 2010, 183–195.

Sonebi, M. and Bartos, P.J.M., Filling ability and plastic settlement of self-compacting concrete, *Materials and Structures/Materiaux et Constructions*, 35, 2002, 462–469.

Turkmen, I., Oz, A., and Aydin, A.C., Characteristics of workability, strength, and ultrasonic pulse velocity of SCC containing zeolite and slag, *Scientific Research and Essays*, 5(15), 2010, 2055–2064.

Uysal, M., Self-compacting concrete incorporating filler additives: Performance at high temperatures, *Construction and Building Materials*, 26, 2012, 701–706.

7

SCCs Based on High Efficiency and Nano Pozzolans

7.1 Introduction

The usefulness of simple or dummy materials as powder extenders for ensuring the required fines instead of the more expensive and pyro-processed cements in the making of self-consolidating concretes (SCCs) is well accepted. The efficacy of the proposal not only lies in the fact that it is more economical but was also necessitated to alleviate the problems associated with the shrinkage and temperature effects of the higher cement contents required for high-strength concretes. With this broad perspective of ensuring self-compactability through additional fines, several research efforts tried to look at the various alternative materials available for such use. The euphoria of not having to compact the material and the additional advantages of not having to deal with the continuous throb of vibrators essentially propelled the research efforts in this direction to new heights supplemented by several local, national, and international seminars and conferences on this topic. In the commotion of such a scenario, there have been sporadic attempts to use alternatives to the limestone powder and the more freely available fly ash by a few. Even the use of ground granulated blast furnace slag (GGBS) was to a certain extent limited, as the availability of GGBS in the market is limited to steel-production regions. The availability of silica fume marketed as a superior pozzolanic cementitious admixture in concrete prompted a few to look at this also as an alternative. In a way, such an effort was limited because it is several times more expensive than even cement. Looking at the broad development of SCCs over the past couple of decades, the thrust was to find only conventional levels of concrete strength and not of any special high-strength or high-performance requirements, though exclusive cementitious composites of this particular type were certainly adopted for high-rise and bridge structures. In fact, it is difficult to recall offhand a specific reference of an investigation directed toward SCCs of very high strengths. At this stage, it is good to use silica fume as material because its extreme fineness imparts a level of cohesiveness or thixotropy that is not shown by any other material in general. The second aspect is that due to its highly amorphous silica content, it offers the ideal solution to limit both cementitious materials and correspondingly water contents to achieve w/c ratio nearing the practical minimum of 0.20 that is possible to be produced with the presently available commercial construction equipment and practices. Some of these aspects and probably extensions to ensure modifications that are required to be more viable are discussed in the section on SCCs.

7.2 Concepts of High Strength and High Performance

The concepts of performance were discussed broadly in some of the earlier chapters. In fact, the perception that the use of a pozzolanic mineral admixture, fly ash, associated with the self-compactability of the material realizing a possible defect-free structure was indeed considered sufficient grounds for calling SCC a high-performance material. This is probably not appropriate. The term "high performance" was also applied to concretes of higher strength, which could, in most cases, represent a better performance due to their discontinuous pore structure below a w/c ratio of 0.45. However, the increased reactivity and the thermal effects of the higher cementitious content should be recognized and addressed specifically to ensure higher performance. In addition, the primary objective in developing these concretes was mostly to have adequate resistance to aggressive environments. In this regard, "high-performance concrete" was defined as that meeting the following requirements—the first one regarding the strength and the second one enforcing the durability criteria (Zia, 1991). The high-performance concrete defined earlier by Zia is probably still the most inclusive of all conceptually, with the dual requirement of strength and durability, namely, a minimum strength of either 3000 psi (21 MPa) within 4 hours (very early strength) or 5000 psi (34 MPa) within 24 hours (high early strength), or 10,000 psi (69 MPa) within 28 days (very high strength) with a maximum w/c ratio of 0.35, and a minimum durability factor of 80%, as determined by ASTM C 666, method. One can modify these limits for strength and the test or tests for the durability criteria depending on the environment in which the structure is expected to perform, but the essential concept still has to be the same. It is understood that the method for the assessment of durability may need to be redefined particularly, keeping in view the performance requirements of the concrete for the specific application and the environment in which it is expected to perform.

There are no specific and well-defined limits for the different strength categories of concrete, and these are changing with time and advancements in concrete technology. In general, from a broad appraisal of the terms used for concretes, it can be suggested that these could be normal strength concretes

(20–60 MPa), high-strength concretes (60–90 MPa), very high-strength concretes (90–150 MPa), and finally ultra-high-strength concretes (>150 MPa) in the present scenario. In principle, high-strength concretes can be achieved quite comfortably through the use of superplasticizers and most of the low-end pozzolanic materials as discussed earlier. However, for producing high-strength and ultra-high-strength concrete composites, it is necessary to have significantly higher-efficiency pozzolans like silica fume, metakaolin, or RHA, which could have efficiencies higher than the cement itself due to their extreme fineness and amorphous silica content that can be leveraged in the presence of superplasticizers.

7.3 SCCs Incorporating Silica Fume and Nanosilica

The use of natural pozzolans in construction has been an accepted practice from time immemorial. The past hundred years has seen a spurt in the use of artificial pozzolans, mainly obtained as waste by-products from industries and assumed to have a greater significance. The primary aim of such a use was to conserve energy and to reduce the huge stockpiles of waste generated in core-sector industries. Artificial pozzolans include fly ash, blast furnace slag, silica fume, RHA, calcined clays, and shales. Of these, fly ash is produced in very large quantities as a by-product in the process of thermal power generation. However, the pozzolanic reactivity of fly ash is rather very low and for applications requiring higher strength and related properties materials like silica fume are preferred. Silica fume is deemed one of the "new generation" construction materials. Silica fume is a by-product of the production of silicon or silicon alloys by reducing quartz in an electric arc furnace. Some SiO lost as gas is oxidized by the air and condensed, giving a very fine particulate solid, which can be collected only by bag filters. The main characteristics of interest generally are SiO_2 contents of 85%–98%, a mean particle size in the range 0.1–0.2 μm, spherical shape with a number of primary agglomerates, and finally a highly amorphous structure. The pozzolanic reactivity of condensed silica fume (CSF) was found to depend more on the chemical composition and nature of impurities than on the fineness or SiO_2 content alone. A surface layer of carbon, if present, greatly decreases the reactivity. The French regulations show that silica fume with as low as even 75% SiO_2 content is also acceptable as pozzolanic materials while most other national practices advocate silica fume with SiO_2 content of 85%–98%. Experiments in the laboratory with silica fume containing only 75% SiO_2 and even a higher range of particle size distribution prove that it is possible to achieve strengths of 120 MPa or even higher (Prakash, 1996). This is primarily because of the fact that even the limited amorphous silica content will never be fully or even partially utilized in the these cementitious material systems.

The fact that the cement capacity is not fully utilized in such concrete has been known all along.

Silica fume can be used as a simple additive in concrete or as a partial replacement for cement (from 5% to 15% by weight of cement generally). Silica fume produces two different effects in concrete as recognized by Cohen (1990), namely:

1. *Water reducing effect*—associated with the reduction in water–cementitious material ratio obtained when silica fume is added in combination with a superplasticizer while maintaining equal slump to that without silica fume

2. *Inherent effect*—associated with an increase in strength in the silica fume concrete over a similar water–cementitious material ratio concrete without silica fume but with equal slump

In a way, both these effects talk indirectly about the higher cementitious efficiency of silica fume. Silica fume can be proportioned in concrete mixtures based on an efficiency factor "k." The efficiency factor "k" can be defined as the number of parts of a Portland cement that one part of silica fume can replace in concrete without changing a specific property of the concrete. Literature shows that the "k" value varies from 2 to 5 in terms of cementitious strength efficiency. The criticism on this approach has been that the efficiency factors for the different durability criteria (carbonation, chloride permeability, etc.) are not as high as that for strength. In other words, the efficiency factor is different for different properties of the concrete. In fact, silica fume exhibits a high reactivity while being an effective filler. Both these properties combine to explain the effects of silica fume on the properties of cement-based products.

As a pozzolan, silica fume reacts with calcium hydroxide (CH) liberated during the hydration of Portland cement to produce secondary cementitious materials. The reaction of silica fume with CH can improve the quality of concrete by reducing the amount of CH. But the influence of pozzolanic reaction on durability is not so straightforward, and other complications may arise. As a filler, silica fume transforms the large concrete pores into finer ones, both in the matrix and the aggregate–cement paste interface, interfacial transition zone (ITZ), which improves the durability properties of concrete. Silica fume is also a very highly reactive pozzolan due to the higher amorphous silica and the large exposed surface area from its fineness. When used in cement systems, it produces a calcium silicate hydrate (CSH) gel with a lower C/S than the cement hydration, and consequently, it has a high capacity to incorporate foreign ions, particularly alkalis. The nature of the hydration products of silica fume and its influence on cement hydration are not completely understood even at present. Silica fume also has a definite filler effect that is believed to distribute the hydration products in a more homogeneous fashion in the space available. These two factors have the combined effect refining the pore structure when silica fume is added to cement-based mixtures.

The refinement of pore structure leads to reduced permeability and is considered the main factor responsible for the influence silica fume has on mechanical and durability properties of concrete. However, the mechanisms by which these properties are improved by addition of silica fume are not yet fully established or understood.

It was observed that in a concrete with a cement content of more than 250 kg/m^3, the water demand increased when adding silica fume when no water-reducing agents are used (Sellevold, 1987). The increased water demand per kilogram of silica fume added was in the order of 1 liter. However, water-reducing agents had a much more pronounced effect on silica fume concrete. Increased cohesiveness is the most obvious difference between the silica fume concrete and normal concrete. Because of the high cohesiveness of concrete, at equal slumps, concrete containing silica fume required more energy input for a given flow. In practice, this problem is overcome by maintaining higher slump for silica fume concrete. The major effects of silica fume on the workability of concrete are increasing its cohesiveness and stability. Consequently, bleeding is greatly reduced. The lack of bleeding in silica fume concrete makes it more vulnerable to plastic shrinkage and cracking than ordinary concrete. Plastic shrinkage and cracking take place when evaporation from a fresh concrete surface exceeds the rate of bleeding water from the concrete. The fact that bleeding can be practically eliminated in silica fume concrete makes it vulnerable with respect to cracking. Applying a proper curing procedure to the concrete surface can eliminate this problem.

The general relationship between compressive strength and w/c ratio at various percentage additions of silica fume was discussed by a few earlier researchers. In line with general practical experience, it was found that the "k"-value was higher for lower (8%) silica fume content than for higher (16%) by weight of cement. The "k" factors ranged from 2 to 5, increasing for richer mixers and decreasing with higher silica fume contents. The silica fume concrete requires protection at early ages to realize its strength potential. Silica fume makes it possible to produce very high strength (over 100 MPa) on a routine basis. Tensile and flexural strengths of silica fume concrete are related to compressive strength in a manner similar to that of normal concrete. However, if silica fume concrete is exposed to drying after only one day of curing, the tensile and flexural strengths are reduced more than those for control concrete. The brittleness of conventional concrete increases with increasing strength level, and the silica fume concrete appears to follow the same pattern as the conventional concrete in this aspect. The available data was reviewed, and it indicated that silica fume can be used to improve the bond of concrete to aggregate, reinforcing steel, various fibers, or old concrete. Cement paste and mortar containing silica fume appear to have a larger shrinkage potential than the control. This increased potential has not been found to reflect in increased concrete shrinkage. Silica fume concrete has the same thermal conductivity as normal concrete, but a lower permeability to water vapor. Fire exposure tests indicate that silica fume concrete is more vulnerable to spalling than normal concrete. High-strength silica fume concrete has superior abrasion, erosion, and wear resistance properties compared to conventional concrete.

The available data indicates that silica fume in concrete reduces the permeability more than it improves the compressive strength, that is, the efficiency factor is greater with respect to permeability than with respect to compressive strength. This finding appears to be particularly evident for low content levels of silica fume and at low concrete strength levels. A comparison of cement paste and concrete results indicates that it is particularly the aggregate interface that is improved by silica fume. Many an investigation as well as practical experience has indicated that a major potential advantage of silica fume in concrete is to improve its chemical resistance. Sulfate resistance and protection against alkali–aggregate reactions are two areas that show particular promise. Recent reports indicate that the same is true for a variety of chemically aggressive substances. The reasons for the general good performance of silica fume concrete in chemically aggressive environments include the following:

- Refined pore structure and therefore reduced transfer rates of harmful ions
- Reduced content of CH
- Low c/s of reaction products, which increases the capacity to incorporate foreign ions such as aluminum or alkalis in the lattice

The individual factors controlling the corrosion of steel in concrete are known. However, in practical situations, it is a combination of these factors that governs the risk of corrosion, and information on individual factors is insufficient to allow a direct prediction of the corrosion protection offered by different concretes. The present evidence suggests that the use of silica fume as an addition to improve concrete durability will also improve its ability to protect embedded steel from corrosion. On the other hand, the use of the silica fume to maximize cement savings in low- to medium-strength concretes may result in a shortened initiation stage, thereby increasing the risk of corrosion.

7.3.1 SCCs Incorporating Silica Fume

After having understood the effectiveness of silica fume as a pozzolanic material in cementitious composites, it was essential to have an idea of the research efforts reporting its use in SCCs. The basic advantage of silica fume and many such superfine materials is that they impart a level of thixotropy that is probably not possible with any other powder material, not even with cement. In fact, it should be recognized that these SCCs produced with silica fume as the powder extender, even with the limited quantities of around 10%–15%, could impart such a thickening effect that it would require additional superplasticizer to ensure the level of fluidity to be similar to that of other types of admixtures or with cement. The other fact that should be realized is that the additional cementitious strength efficiency of silica fume provides an opportunity for producing concretes of a much higher strength without any substantial increase in the total cementitious materials and the corresponding requirements for wetting water for self-consolidation.

It is obvious that there are not many investigations on the effectiveness of silica fume as a superior pozzolanic material in SCCs. However, there have been a few who considered it as yet another option to argument the fines in concrete design for ensuring self-compaction. In his study on the compressible packing model similar to that proposed by Su (2001) for the design of SCCs, Sebaibi (2013) used silica fume at 10% as a material for partial replacement of cement. The design adopted the w/c ratio and strength relation of the Euro norms. The concrete studied containing 400 kg/m³ with 10% silica fume resulted in a strength of about 72 MPa for the concrete design only for 50 MPa, indicating the significant strength contribution of silica fume in cementitious composites.

Extensive investigations in the laboratory on concrete containing silica fume at various percentage levels of the cementitious materials content were able to present several advantages in terms of its performance in concrete (Prakash, 1996). As a first step, details of the compositions of several silica fume concretes reported in the literature were analyzed for their cementitious strength efficiency specifically (Ganesh Babu, 1995). A similar investigation for fly ash was earlier presented revealing that the cementitious efficiency is not just one single factor for silica fume and can be understood in terms of the general efficiency and percentage efficiency factors.

7.3.2 Evaluation of Efficiency of Silica Fume

Since simple addition or replacement methods were not found to be suitable for a general understanding of the behavior of concretes with pozzolans, rational methods were developed to take into account the characteristics of the pozzolan that are known to influence the fresh and hardened state properties of the concrete. The evaluation of the efficiencies of these concretes was attempted based on the "Δw concept" discussed earlier. Figure 7.1 presents a conceptual diagram of the relationship between compressive strength and water–cementitious materials ratio for the control and silica fume concretes at any specific replacement percentage. Point "A" represents the water–cementitious materials ratio of the SF concrete at some percentage of replacement for a typical strength, while point "N" represents the water–cementitious material ratio of the control concrete at the same strength. The method attempts to bring the [w/(c + s)] ratio of the SF concrete close to that of the control concrete by applying the cementitious efficiency of silica fume "k." Now Figure 7.1 is replotted to see if a unique value of "k" can bring "A" to "N" or, in other words, the correction (Δw) required can be achieved through a unique (overall) cementitious efficiency factor "k" at all

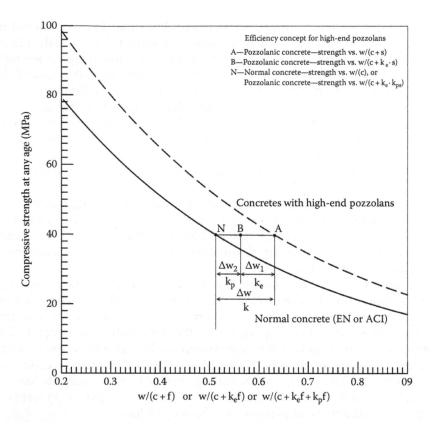

FIGURE 7.1
Conceptual diagram for evaluating the efficiency. (From Prakash, P.V.S., Development and behavioural characteristics of silica fume concretes for aggressive environment, PhD thesis submitted to Indian Institute of Technology Madras, Chennai, India, 1996.)

percentages of silica fume replacement as has been done by many researchers hitherto. It was observed that this was not possible, as the range of percentage replacements considered were far too wide. To start with, a unique value of "k" was chosen so that it brings the concretes at all percentages of replacement as close to the normal as possible, with the lowest percentages being still only slightly higher and the highest percentages being slightly lower. This value of "k," which is generally applicable for all the replacement percentages, is defined as the "general efficiency factor" (k_e). This means that point "A" now shifts to its revised location "B" due to the application of the general efficiency factor (k_e) with the axis as [w/(c + k_e·s)].

Thus, point "A" has shifted to point "B" by a distance of "Δw_1." A correction still required for bringing "B" to "N" is ($\Delta w - \Delta w_1 = \Delta w_2$) and this factor was assumed to be the effect of the different percentages of replacement. Now an additional "percentage efficiency factor" (k_p) has been evaluated as a multiplication factor to "k_e" for each of the percentages of replacement in a way similar to that adopted earlier. These two corrections together will now bring point "A" to "N" so that the w/c ratio of the control concrete and the "water to the effective cementitious materials ratio" of the silica fume concrete $[w/(c + k_e \cdot k_p \cdot s)]$ are the same for any particular strength at all percentages of replacement. This methodology finally resulted in the evaluation of the "overall efficiency factor" ($k = k_e \cdot k_p$) for any particular concrete at any particular replacement. The basic concepts of this development and its use in the design of concretes with pozzolans were presented earlier by Ganesh Babu et al. (1993). The concept is very similar to the one that was explained for fly ash earlier.

As already stated, the results of the data available were analyzed through the methodology discussed earlier. At first, for calculating the efficiencies, the w/c ratio to compressive strength relationship for control concretes is required. This was obtained by considering all the 61 values from the database provided earlier related to 0% silica fume in Figure 7.2. Based on the w/c ratio to strength curve of these control concretes, the efficiencies of SF

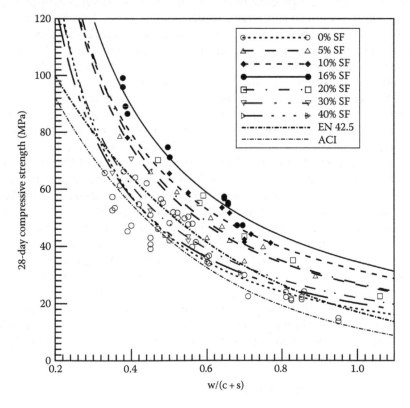

FIGURE 7.2
Compressive strength variation with $[w/(c + s)]$. (From Prakash, P.V.S., Development and behavioural characteristics of silica fume concretes for aggressive environment, PhD thesis submitted to Indian Institute of Technology Madras, Chennai, India, 1996.)

concrete were evaluated. The variations of the 28-day compressive strength with the parameter [w/(c + s)] for all concretes considered with the different percentage of replacement were also presented in Figure 6.2. The curves drawn were the best fits for each of the different percentages of replacement. It can be clearly observed that all the concretes with replacements even up to 40% show strengths higher than that of the control concrete particularly for such high-efficiency pozzolanic admixtures like silica fume. However, the improvements in strength at the different percentages of replacement at any w/c ratio varied over a wide range. In attempting these corrections, the present Euro and ACI code recommendations for normal concretes ware also kept in mind to see if the comparisons with the normal concretes made are valid with the cements available today.

Using the methodology discussed earlier, the effect of the variation of "k_e" on the strength–water–cementitious materials ratio at different percentages of SF replacement was measured. The values of "k_e" adopted in these studies were ranging between 2.0 and 5.0; however, at a "k_e" value of 3.0, the concretes containing replacements up to 16% showed strengths higher than the control, while those containing replacements above 16% showed lower strengths than the control (Figure 7.3). A value of 3.0 can be assumed as the general efficiency factor (k_e) representing the correction "Δw_1." This means that instead of using an overall efficiency, the general efficiency factor was kept constant for all the percentages of replacement.

The differences between the water–cementitious materials ratio of SF concretes, including the effect of general efficiency factor and that of the control concretes, were now computed (Δw_2) for the individual mixes. The effect of percentage replacement was then calculated by calculating an additional percentage replacement factor (k_p). Considering the average of the "k_p" values for each of the different percentages of SF replacement, the variation of compressive strength with the parameter [w/(c + $k_p \cdot k_e \cdot$ s)] is presented in Figure 7.4. It can be clearly seen that this has resulted in a reasonably close agreement with the control concrete strengths at all the percentage of replacements.

The variation of this average "k_p" with the percentage of replacement ("p_r," the percentage of SF in the total cementitious materials), at a constant general efficiency factor (k_e) of 3.0, is presented in Figure 7.5. The variation of the overall efficiency factor (k = $k_e \cdot k_p$) is also presented in Figure 7.5.

Not going into the other aspects regarding the effectiveness of silica fume in concrete at the different percentages as was given earlier, it is enough to say that these investigations presented the combined 28-day strength efficiency of silica fume defined by the relation:

$$k_{28} = 0.0044(p_r)^2 - 0.37 \cdot p_r + 8.60$$

where (p_r) is the percentage of silica fume in the concrete. This would essentially mean that the efficiency k_{28} will vary from a value of almost 7.0 at

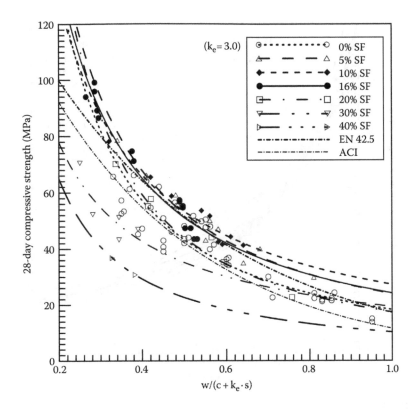

FIGURE 7.3
Compressive strength variation with [w/(c + k_es)]. (From Prakash, P.V.S., *Development and behavioural characteristics of silica fume concretes for aggressive environment*, PhD thesis submitted to Indian Institute of Technology Madras, Chennai, India, 1996.)

about 5% replacement to a value of around 1.0 at a replacement of around 40%, which will probably be just around what was reported by research workers earlier. A detailed account of the complete methodology used in the evaluation of the efficiency of silica fume in concrete was presented earlier by Ganesh Babu (1995). It can also be observed that after incorporating the effect of the efficiencies the compressive strengths of most concretes evaluated (barring a few at the very high strength region) fall within the space between the ACI and EN 42.5 relationships that were proposed earlier, as can be seen in Figure 7.6.

Similar evaluations were done for some other pozzolanic admixtures—particularly, metakaolin, RHA, and zeolite in the laboratory (Appa Rao, 2001; Surekha, 2005; Narasimhulu, 2007), which all show themselves to be reasonably higher-end pozzolanic admixtures, though certainly not at the level of silica fume. Even so many of them will be able to control the amount of cement in the mix because of their higher efficiency than the cement,

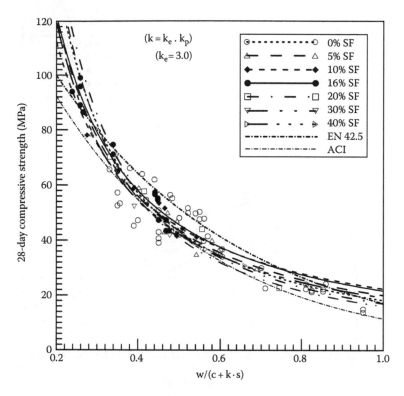

FIGURE 7.4
Compressive strength variation with $[w/(c + k_e f + k_p f)]$. (From Prakash, P.V.S., Development and behavioural characteristics of silica fume concretes for aggressive environment, PhD thesis submitted to Indian Institute of Technology Madras, Chennai, India, 1996.)

which enables the making of very high-strength concretes, such as concretes of strength beyond 100 MPa. A brief review of some of these aspects have already been published earlier giving details of the expected efficiency factors at 28 days (Ganesh Babu, 2003a, b, 2007).

It is also relevant to point out that investigations in the laboratory with different pozzolans for understanding the cementitious efficiency of these materials and their different percentages show that particularly at the very low water–cementitious material ratios required for high-strength concretes, both the water and the plasticizer contents have to be modulated to achieve an adequate workability of concrete. With these high cementitious materials contents, there is a very thin line between a highly thixotropic mass of concrete (almost rubbery), which will need a considerable compaction energy to its becoming a concrete with collapse slump, and free flow without segregation (due to the high paste content). This transition can only be realized during mixing as the various cementitious components contribute in different ways to the plasticizing effect of the water and superplasticizer combination,

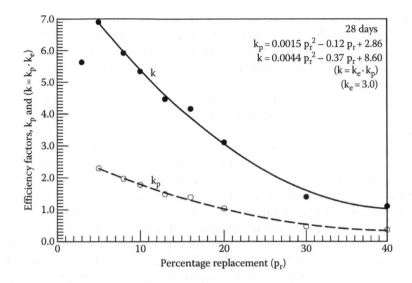

28 days

$k_p = 0.0015\, p_r^2 - 0.12\, p_r + 2.86$

$k = 0.0044\, p_r^2 - 0.37\, p_r + 8.60$

$(k = k_e \cdot k_p)$

$(k_e = 3.0)$

FIGURE 7.5
Variation of efficiency with silica fume percentage. (From Prakash, P.V.S., Development and behavioural characteristics of silica fume concretes for aggressive environment, PhD thesis submitted to Indian Institute of Technology Madras, Chennai, India, 1996.)

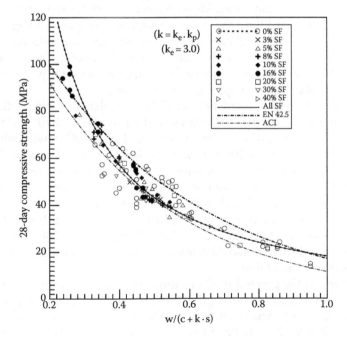

FIGURE 7.6
Efficacy of the predictions in relation to ACI and EN relations.

particularly the ones with higher fineness like silica fume where even small quantities of the finer or nanoparticles can produce significant effects.

As already stated, the investigations in the laboratory were not only directed toward the efficacy of evaluating the strength efficiency of silica fume concretes established earlier but were also extended to investigate extensively other characteristics in terms of the durability and corrosion characteristics (Ganesh Babu, 1994, 1995, 1996, 1998). As a matter of experience, it should be stated that it was possible to arrive at very high-strength concretes of the order of 120 MPa that the efficiency concept proposed. A couple of very important observations made is that silica fume concretes are in general more cohesive and thixotropic up to a certain point; a small quantity of additional water at that stage could change them from a highly cohesive matrix to a totally flowing collapse slump concrete. It is extremely important to reiterate at this stage that the experience with several concrete studies in the laboratory shows that most concretes that are cohesive and have a slump of around 100 mm (given the conditions that there is sufficient quantity of cementitious materials, around 500 kg/m^3, and a continuous grading between the DIN-A and DIN-B, without missing the fractions between 8 and 2 mm) can always be brought to a collapse slump without any difficulty. One more important factor is that unless the concrete is made to achieve the consistency problems associated with previously thixotropic materials, the strength of the concrete will be seriously reduced as the air entrapped cannot be driven out even with significant vibration effort. Most concretes for that reason were actually made to flow, and for filling the cube molds, a simple plastic mug was used instead of the normal trowel or scoop that is used for conventional concrete. In fact, it was learned, maybe the hard way with several trials, that unless these concretes were made to have a flowing consistency it will not be possible to realize the high strengths as was reported.

To understand the characteristics of SCCs with silica fume in particular, an effort was made to compile the available data to assess their performance. As is obvious from our discussion earlier, the data on SCCs with silica fume is regrettably small, but with a view to at least see if it fits into the mold of what was presented for the previous pozzolanic admixtures in concrete, the available data was analyzed in a similar fashion. Some of the important references that are related to this particular effort are listed separately as "References for evaluations on silica fume" (Guneyisi, 2010b; Ioani, 2008; Jalal 2012; Rahmani, 2011; Turk, 2011; Turk, 2010; Ulucan, 2008). It is always possible to include additional data into this framework as and when there is some information available.

For arriving at the general limits of the water–cementitious material combinations and the resultant strengths, these were specifically looked at to see if indeed such limits are possible with the limited experimental data that is available. It was then realized that the investigations in SCCs have used silica fume at 5%, 10%, and 15% only. Similar to the approach that was followed earlier for fly ash and other pozzolanic admixtures, the fundamental

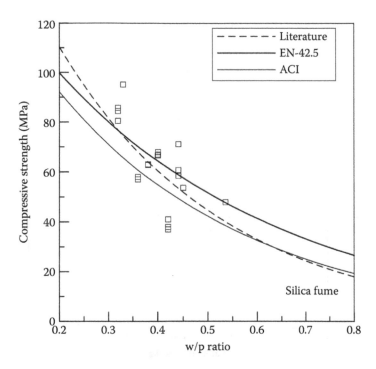

FIGURE 7.7
Strength variation with w/p ratio for SCCs with silica fume.

relation between the water–cementitious materials ratio and strength for the SCCs containing silica fume at its various percentages was looked at. Figure 7.7 presents the picture that shows that the strength values obtained, from various concretes reported were generally in excess of what was predicted through the ACI and Euro relationships. However, as can be observed, concretes of 35–70 MPa can be produced at a w/p ratio of approximately 0.45, and the strengths at the w/c ratio of 0.3 varied from 80 to 100 MPa. It is obvious that these have to be looked at through the prism of efficiency if one has to get some clarity about such wide variation.

The variation of water content with compressive strength presented in Figure 7.8 for SCCs indicates that even in the case of a significantly different material like silica fume, in terms of both physical and chemical characteristics, it still required the same water content of 160–200 kg/m³ approximately, resulting in concretes of strength ranging from 50 to 90 MPa.

Finally, the dynamics of the water content required for such a fine material at the various percentages in the total cementitious materials was looked at (Figure 7.9). Surprisingly at the approximate water content of 160–200 kg/m³, the powder content still remains to be 400–650 kg/m³ as in the case of the earlier discussed pozzolanic admixtures, which however are not anywhere near this fineness. The fact remains that even with this water content and the

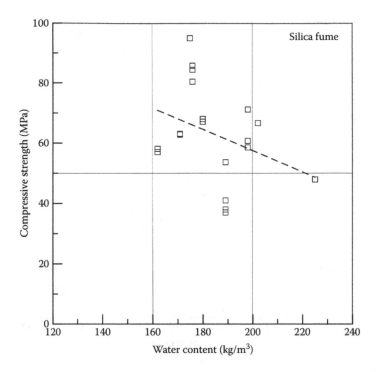

FIGURE 7.8
Variation of w/p ratio and water content with compressive strength.

cementitious materials limitation, it was still possible to arrive at strengths anywhere between 50 and 100 MPa, as can be seen. This in fact is the most important aspect of silica fume in concrete, be it self-compacting or otherwise, and it helps in achieving very high strengths at very low replacement levels of about 10%–15%.

To understand the effectiveness of silica fume at its various percentages of replacement in the cementitious materials, it was felt necessary to have the SCCs appropriately looked at in terms of the strength to water–cementitious materials ratios along with the other constituents at each of the specific percentage replacement levels. The relationships between the various constituents of the SCCs containing 15% silica fume have been presented in Figure 7.10. It is known from the previous literature that silica fume exhibits the highest efficiency at around this percentage, and thus it was felt that an understanding at this level will be the most appropriate. Figure 7.10 clearly shows that only a few concretes are reported from the literature. Secondly, these concretes have marginally fallen short of the strengths expected when compared with the relationships of the ACI and Euro regulations corrected for the strength efficiency of silica fume at this particular percentage. One critical observation could be that the cementitious materials content is a little lower than what is necessary, as can be seen from one of the concretes that

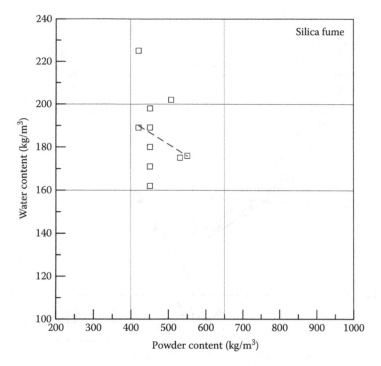

FIGURE 7.9
Variation of w/p ratio and water content with compressive strength.

exhibited a significantly lower strength. One can also say that the strength efficiency of the silica fume assumed is rather on the higher side, and a marginal reduction by about 10% (from a k value of around 4.0 to about 3.5) would have shown that these concretes other than the one that has significant powder content would all have registered the expected strength. To have a specific understanding of these relationships at the various percentages appropriately, the results of laboratory investigations discussed earlier (Surya Prakash, 1996) in which the concretes that have conformed to the self-compacting requirements were also superimposed on this particular graph. It is obvious now that the strength efficiency of silica fume concretes that was established earlier is indeed applicable for SCCs also.

To ensure that this fact can be verified appropriately at the various percentage levels possible, the results of the investigations available from the literature as well as the laboratory were analyzed through similar diagrams at 5%, 10%, 15%, and 20%, which are presented in Appendix SF-7 at the end of this chapter. It is to be noted that the laboratory investigations had concretes that confirm to SCCs at 25%, 30%, and 35%, which also show that the predictions by the proposed method are in conformity with the results. However, these are not included in the Appendix, as it is obvious that for the making of SCCs at the strengths discussed, these percentages are indeed totally uneconomical

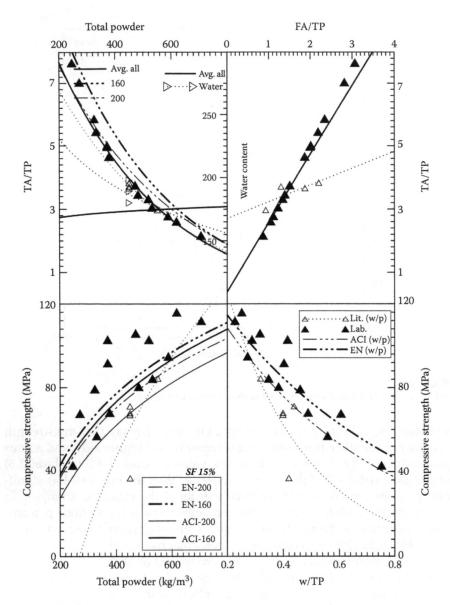

FIGURE 7.10
Correlations between the constituents in SCCs with 15% silica fume.

and superfluous. The figures containing silica fume at levels ranging from 5% to 20% are presented in Figures SF7.1 through SF7.4 in Appendix SF-7. It can be seen that in most cases the concrete composites exhibited strengths in excess of what is expected from the code. Just to show how difficult it is to recognize these facts from the raw data, the complete information of all the concretes at all levels of replacement is given in Figure SF7.5. After ensuring that these concretes can be defined through their self-compacting characteristics at the various percentages through the efficiency concept, the nomograms presenting the complete relationships between strength, cementitious materials, and aggregates corresponding to the ACI and EN relationships at the two proposed water contents of 160 and 200 kg/m³ are presented in Figures SF7.6 through SF7.9. The corresponding laboratory data was only plotted on the EN relationships at 160 kg/m³ water content, which depicts the cube strength, and the smallest w/c ratio that is presented would be the one of interest to actual concretes practitioners.

7.3.3 SCCs with Nanosilica

The use of nanomaterials to improve the performance of cementitious composites has taken root in the research industry, and several alternatives have been tried to date. In fact, microsilica or silica fume contains a small quantity of nanoparticles, and in a few cases, concretes made with silica fume were even reported to be nano cementitious composites. Nanosilica produced synthetically as an emulsion of ultrafine amorphous colloidal silica is considered as superior grade by Collepardi (2004), because of its higher fineness and amorphous silica content compared to that of silica fume (a fineness of 1–50 nm compared to 0.01–1 µm and an amorphous silica content of >99% compared to that about 90%). Incidentally, these values for fly ash appear to be in the region of 0.1–40 µm and 35%–40%. Their studies compared the strengths and collaboration characteristics of SCCs made with silica fume, nanosilica, and fly ash using concretes containing about 450 kg/m³ of cementitious materials with a water content of 174 kg/m³ and a water–cementitious materials ratio 0.44. The fly ash and silica fume were used at about 15% while the nanosilica was used at about 5% and the remaining quantity supplemented by fly ash. The investigations have shown that all these concretes resulted in strengths ranging from about 60 to 70 MPa. A simple question that one could ask is the need for silica fume if the fly ash could result in the same strength, not to mention the need for nanosilica. While it is obvious that the effectiveness of these materials could be reflected differently in terms of their specific durability characteristics, in terms of strength they all appear to perform at the same level. It is also a question why the silica fume concrete did not show a significant improvement in strength compared to that with fly ash, even though it is known to have cementitious efficiency several times that of fly ash.

There are a few other investigators also who studied the characteristics of concretes containing nanosilica (Maghsoudi, 2010; Biricik, 2014; Montgomery, 2016, etc.), but the concrete strengths reported by them were also around the same as that reported by Collepardi, who is in fact one of the preeminent experts on the use of admixtures in concrete. Montgomery (2016) studied the effect of nanosilica at 1%, 3%, and 5% on the strength characteristics in cement pastes as well as hydration process through an electron microscopy and infrared spectroscopy. The studies indicate that cement replacements of just 1% appear to yield the maximum strength, ultimately reaching a value of about 85–100 MPa. The improvement in the strength characteristics is well supported through SEM and infrared spectroscopy results. The microstructural characteristics of cement mortars containing nanosilica, silica fume, and fly ash—and 5% and 10% replacement with the w/c ratio 0.5—were studied by Biricik (2014) along with the strength characteristics at 7 and 28 days. The investigations showed an improvement in the strength characteristics of mortars containing nanosilica, improving from 30 to 40 MPa in the case of 5% and 30 to 60 MPa in the case of 10%, which was significantly higher than that for fly ash and silica fume. It was observed that nanosilica indeed acted as the nucleating agent for the formation of C–S–H gel from C_3S and C_2S. The pozzolanic reactivities observed through thermos-gravimetric analysis also support these observations.

Maghsoudi (2010) studied the effect of nanosilica on the flow characteristics of SCC along with the compressive strength, modulus, and shrinkage characteristics. The concretes contain cementitious materials of 400 kg/m^3 with 10% microsilica in one case and an additional 2.4 L of nanosilica apart from an additional 300 kg/m^3 of limestone powder with 210 kg/m^3 of water. The compressive strength of SCCs consisting of nanosilica is higher than those of specimens without, increasing from around 40 MPa to about 60 MPa. In addition, the durability was expected to be better because of the pore refinement. Quercia (2014) studied two different types of nanosilica in SCC, both having similar particle size distributions but produced through two different processes: fumed powder silica and precipitated silica in colloidal suspension. The concretes investigated contained 340 kg/m^3 of cement along with almost an equal quantity of powder materials supplemented through limestone powder and microsand along with 13 kg/m^3 of the aforementioned nanosilica in two of the concretes, at a constant water–binder ratio of 0.45. The fact that the maximum strength achieved for both the control and the nanosilica concretes ranged between 85 and 90 MPa shows that the effort is not toward the development of very high-strength concretes with the water–cementitious materials ratio of 0.45. The porosity and size distribution valuations from all the concretes including the control appear to be very similar. However, the Norwegian rapid chloride permeability, chloride diffusion, and freeze–thaw tests show a significant improvement in the performance of the concretes containing nanosilica. It can be seen that most investigations have not concentrated on extracting the complete benefit of utilizing high-performance amorphous silica compounds in ensuring high-strength concretes of over 100 MPa. Such a low proportion of additions will only be able to modulate the internal pore structure by filling the pores and slight hydration effects locally improving the performance characteristics, but will not be able to significantly participate in improving the strength characteristics through the pozzolanic reactivity that is needed for their strength development. It is understood that silica fume exhibits the highest beneficial effects in terms of both strength and performance characteristics at approximately 15% minimum, and in the case of nanosilica, such an understanding is not yet reached with the limited investigations available in literature.

The fact that the strengths of the concretes investigated even with the utilization of nanosilica have not led to the development of super high-strength concretes (beyond may be 120 MPa), higher strength than that possible with silica fume, makes it difficult to justify the very efficacy of attempting to utilize such a sophisticated and expensive material for the making of concretes though they are certainly of academic interest. A layman's explanation may be that the reactive components could not utilize the higher amorphous silica content that was present, either within the timeframe or because of the nonavailability of pore space for the hydrated products.

7.4 SCCs Incorporating Calcined Clays

There have also been several other powder materials known to exhibit a significant pozzolanic effect in the literature, though none of them are in the league of silica fume that contains purely amorphous silica in most cases. Calcined clays like metakaolin and zeolite come under this category. Investigations on such materials in the context of SCC are very few and probably may not be sufficient to get a clear picture of the contribution. However, for the sake of completion and to project the possibility of having them as candidate materials for powder extension in SCCs, these few investigations and results are also looked at critically. Vejmelkova (2011) studied two concretes containing 40% metakaolin and a blast furnace slag cement with 56% slag to establish the fresh and hardened state characteristics of the concretes. It is surprising that he felt that the slag concretes show zero yield stress. The observation that the strength gain rate of metakaolin concretes is far higher than that of the other is understandable from the fact that the efficiency of metakaolin is high. The characteristics like specific heat capacity and thermal conductivity were also reported for these two concretes. It was also observed that metakaolin concretes performed better in terms of the freeze–thaw resistance because of the finer pore structure compared to that of slag concrete, with a mass loss of 0.396 kg/m^2 after 56 cycles (considered to be in the good category), while the slack concrete exhibited mass loss in the satisfactory category as prescribed by Polish standards. Cassagnabere (2010) suggested the use of metakaolin as a clinker replacement material of 12.5%–25% in cement. His studies in a precast factory suggest that for slip-forming applications and self-compacting applications, the optimum percentages were 25% and 17.5%, respectively, with the compressive strength and porosity not being affected. This obviously is not appropriate, as it is possible to achieve much higher strengths and performance through appropriate utilization of metakaolin as a cement replacement material in the production of concrete rather than mixing it at the clinker stage. In any case, the strength characteristics of the cement produced might have altered significantly, and these would be responsible for the changes that take place in the green state or the hardened state performance of concrete. It is easier to account for the strength and performance improvements accrued from the replacements effected through the earlier suggested efficiency approach.

Guneyisi (2010) studied 65 concrete mixtures at two water–binder ratios, 0.32 and 0.44, with the first 43 concretes at a total cementitious materials content of 550 and the rest at 450 kg/m^3. The mixes contain both binary and ternary mixes containing fly ash and GGBS with replacement ratios at 20%, 40%, and 60% and silica fume and metakaolin replacements at 5%, 10%, and 15% by weight of the total binder in various combinations. Studies on SCCs containing natural zeolite and slag and their combinations at 10%, 20%, and 30% in 400 kg/m^3 of cementitious materials at a water–cementitious materials ratio of 0.40 (water content of 195 kg/m^3) were reported by Türkmen (2010). Workability studies indicate that concretes with zeolite levels beyond 10% could not result in flowing concretes and the compressive strengths reduced from about 50 to 30 MPa at the maximum 30% addition of natural zeolite in these concretes. An important observation at this stage is that with the fineness of 6350 cm^2/g, the zeolites are much finer and being clays may have a tendency to adsorb water on their surface while the fineness itself requires a much larger water content. Apart from this, zeolites have to be calcined to ensure the activation of the silica to have an effect of pozzolanic reaction in cementitious composites without which the performance of natural zeolite would be similar to that of any other inert powders affecting only the pore-filling characteristics because of the fineness associated. As the number of concretes containing these materials is very few and in most cases complete information is not available, the performance of these SCCs is discussed along with other pozzolanic materials that are not clays being reported in the next part. In fact, there were two extensive investigations on the behavior of concretes containing metakaolin and zeolite. Studies on metakaolin (Appa Rao, 2001; Ganesh Babu, 2003a, b, 2007) looked at the efficiency of metakaolin in particular from the information available in the literature. Concretes designed according to the concept of efficiency resulted in strengths of around 90 MPa at 5%–10% replacements and cements. A separate investigation on natural and calcined zeolites was also undertaken in the laboratory (Narasimhulu, 2007; Ganesh Babu, 2008), which looked at the characteristics of zeolites and the effects of calcination apart from the possible strength contribution through pozzolanic reactivity.

7.5 SCCs Incorporating Rice Husk Ash

Apart from the calcined clays discussed earlier, a few other materials are also known to be capable of imparting strength through their pozzolanic reaction as they contain a fair quantity of amorphous silica. One of the best sources of almost pure amorphous silica is probably RHA, obtained from the controlled incineration of rice husk between 400°C and 800°C, as reported by Mehta (1994) earlier. While this is true, the fact remains that it has to be finely ground before it can be used as an effective pozzolanic admixture in cementitious composites. The importance of the grinding effort is obvious from the fact that when used in the raw form without grinding, their effects are substantially different because of their highly porous microstructure and the flaky mass that actually results after the burning process. Studies of Nor Atan (2011) on six SCCs containing ternary blends of Portland cement, raw RHA, and fine limestone powder, pulverized fuel ash, or silica fume combinations with each one at 15% are an ample testimony to this. The concretes had a cementitious materials content of 475 kg/m^3. The water content was varied from 185 kg/m^3 for normal concrete to 255 kg/m^3 for concrete containing silica fume and raw RHA combination. With the observation that the strength of these concretes reduced from a range of around 45 MPa for the normal concrete to about half for the concrete containing 15% silica fume shows that raw RHA even with the 92% SiO$_2$ content is not effective. This is probably because raw RHA is by itself a highly porous material and will result in a concrete containing several weak planes that lower the strength of concretes significantly. In contrast, Sua-Iam (2012) studied concretes containing RHA at 10%, 20%, and 40% of the two w/c ratio 0.28 and 0.33 with the cementitious materials content of 550 kg/m^3. The studies included slump flow, compressive strength, and ultrasonic pulse velocity, and the concretes show strengths varying from about 35 to 60 MPa. The maximum strength was obtained at the 20% replacement level. Ahmadi (2007) looked at the behavior of normal and SCCs containing 0%, 10%, and 20% RHA at a constant water–binder ratio of 0.4 and 0.35,with the water content of 184 and 161 kg/m^3 resulting in the cementitious material content of 460 kg/m^3. The ordinary concretes containing RHA resulted in strengths of 60–75 MPa, while the SCCs appear to have achieved 80–95 MPa at the end of the 180-day period. The significant change in compressive strength observed is hard to explain for the SCCs. Ahmadi also studied the flexural strengths and modulus for these concretes. The strength efficiency of RHA was investigated to clearly bring out its efficiency at 7 and 90 days. It was seen that the efficiency of RHA varied from about 2.5 to 1.5 for the replacement levels of 5%–50% at 28 days. According to these efficiency factors, several concretes resulted in strengths ranging from 70 to 100 MPa at various replacement levels ranging from 5% to 20%. Most of them conformed to the requirements of self-compaction as defined by the European Guidelines for Self-Compacting Concrete (EFNARC, 2005).

From the earlier discussions, it is obvious that the data available is too small, and also the investigators did not look at the efficiency of these pozzolanic materials. Naturally, the concretes could have resulted in strengths that were not as expected. As in the case of the previous materials, the w/p ratio to strength relations of SCCs containing these powders were examined for the limited data available, and how the constituents of these materials are aligned to the different parameters that were investigated earlier was also studied. As already stated, though these materials are expected to be reasonably pozzolanic, they did not exhibit any significant contribution to the strength characteristics of the SCCs. At this stage, the w/c ratio to strength relationships were replaced with the w/p ratio to strength relationships used earlier (Figure 7.11). It can be clearly seen that these SCCs containing RHA and zeolite have shown strengths that are less than w/c ratio relationships that did not even consider the presence of these admixtures, which could because of the dilution effects due to the additional powders. However, they are essentially just below these w/c ratio to strength relationships and still follow the trend that was prescribed for normal concretes. The cementitious material contents to strength relationships plotted next clearly show that the cementitious materials content is far lower than that expected even with the 160 kg/m³ water content that is required. This aspect was also discussed in Chapter 6 both in the fly ash–based SCCs as well as at the end of the discussions on simple powder extenders. Naturally, as the wetting water requirement is still influenced by other materials that are in the system and the volume occupied by these powders has to be accounted for, the presentation in Figure 7.11 is similar to that earlier considering the powder content in the system. Notwithstanding the facts discussed already, it can be seen that concretes containing metakaolin have shown significantly higher strengths (a strength of almost 100 and 70 MPa at the w/c ratio of 0.35 and 0.50 compared to about the expected 75 and 42 MPa from normal concretes. Obviously, this is only because the metakaolin in the system imparted significant strength to the composites. It can also be said that in the case of RHA, the fineness of the grinding plays an important role in the pozzolanic efficiency of the material, as can be seen from the disastrous effects of raw RHA discussed earlier. The earlier discussions show that if appropriately designed, it may be possible to use these materials with a reasonable pozzolanic efficiency (maybe in the region of 2–3 at 5% and 15%) as powder extenders that would impart an effective

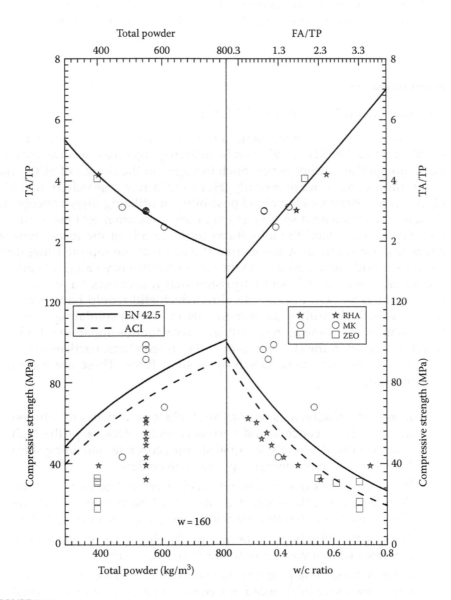

FIGURE 7.11
Correlations in constituents of SCCs with other high-end pozzolans.

cementitious contribution to the concrete matrix. However, Ahmadi is advised that their use be restricted to levels below 15% or 20% to ensure that these very fine powders do not necessitate additional water requirement that will reduce the strength and performance of these SCCs.

7.6 Saturation Concepts and Effects

The investigations on pozzolanic admixtures like fly ash, silica fume, GGBS, metakaolin, RHA, and so on have been going on for several decades now, though the concept is not much in vogue in the design of SCCs that came into existence more recently. However, a few researchers like Liu (2010) were indeed looking at the possibility of utilizing these concepts to design concretes with a specific strength and performance. One need not look far to understand the inhibitions in the minds of the researchers to adopt this approach, as it stems from the fact that the reported literature shows a considerable variation in the strength efficiency values of various pozzolans. Even so, it is still surprising that researchers have not used even a watered-down lower-end value, which still would have been the way to effectively utilize these materials in self-consolidating cementitious composites. Some of the possible reasons for this apparent inhibition could be hidden in the statements of earlier researchers, particularly with reference to the pozzolanic reactivity of silica fume. These are probably the following:

- Increased cohesiveness is the most obvious difference between the silica fume concrete and normal concrete. Because of the high cohesiveness of concrete, at equal slumps, concrete containing silica fume required more energy input for a given flow.
- It was observed that in concrete with a cement content of more than $250 \text{ kg}/\text{m}^3$, the water demand would increase when adding silica fume with no water reducing agents used (Sellevold, 1987).
- The fact that bleeding can be practically eliminated in silica fume concrete makes it vulnerable with regard to cracking.
- In line with general practical experience, it was found that "k" was higher for lower (8%) silica fume content than that for higher (16%) by weight of cement.
- The "k" factors ranged from 2 to 5, increasing for richer mixers and decreasing with higher silica fume contents.
- Silica fume concrete requires protection at early ages to realize its strength potential.

- Cement paste and mortar containing silica fume appear to have a larger shrinkage potential than control.
- Fire exposure tests indicate that silica fume concrete is more vulnerable to spalling than normal concrete.
- The available data indicates that the silica fume in concrete reduces the permeability more than it improves the compressive strength, that is, the efficiency factor is greater with respect to permeability than with respect to compressive strength.

Each one of these statements, however meaningful, were also meant only to be the guiding principles for better construction practices, which have always forced practitioners to be abundantly cautious. It is probably important at this stage to understand the reasons for the varying pozzolanic strength efficiency of these high-end pozzolans, in particular. Naturally having conducted significantly large experimental investigations on the effectiveness and performance of silica fume in concrete composites at the various levels of powder contents, replacements, and water contents, it was felt that there may be an explanation for this variation hidden somewhere on these parameters and their combinations. At the very outset, it was recognized that increasing cement content alone, maybe even at a constant water content of around 200 kg/m³, will certainly not result in concretes of very high strength (say over 80 MPa). Incidentally, this aspect can be explained through the gel/space ratio concept presented by Neville (1995). In a way, the secondary pozzolanic reaction that is supposed to enrich the CSH gel content in the composite is constrained by the pore space that it can occupy in concrete, which keeps reducing as the strength of concrete increases. Just understand this particular dynamics in a system in which this refinement of pore structure has to be accommodated and efficiency values obtained for the different percentage replacements of silica fume at the various total powder contents were first summarized. Figure 7.12 presents such a variation of the pozzolanic efficiency of silica fume at the lower (say around 15%) and the higher (say around 30%) replacement levels with respect to the total powder content. The graph clearly shows that as the percentage replacement level increases, the contribution of silica fume to the pozzolanic strength efficiency appears to be significantly decreasing. In fact, one can even feel that only a part of the silica fume in the system contributed to the efficiency, and probably the higher level in the remaining part of the silica fume was actually a dummy powder acting as a mere filler in the system along the aggregates. This aspect of the efficiency of a pozzolanic admixture in cementitious composites is termed as the "saturation concept" at present. This helps in understanding the need for an effective utilization of such high-end pozzolans in cementitious composites. Thus, it is felt that for generating concrete composites of strength levels up to maybe 120 MPa, it is not necessary to go for replacement levels beyond 15% or 20%. Probably, if one looks at a lower-end pozzolan-like fly ash or GGBS,

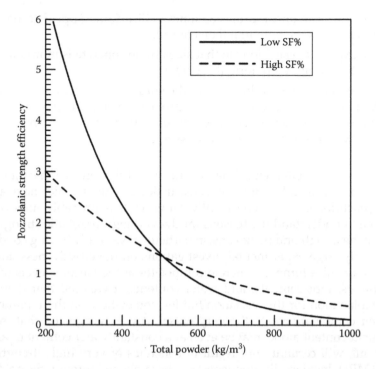

FIGURE 7.12
Effects of saturation on pozzolanic efficiency.

these limits could be different and could be around 40%–50% for the concrete strengths in the range of about 60 MPa. It is obvious that the limits themselves are infinitely related to the strength that is expected of the composites, which will be seen through the explanation on the maximum strength possible at a given percentage replacement presented of fly ash in Figure 5.7.

7.7 SCCs Incorporating Fibrous Constituents

The incorporation of fibrous materials into self-consolidating concretes is actually quite a simple process if only setting some of these precautions were undertaken during the design. The fact emanates from two specific aspects. First, SCCs by their very nature have a substantial quantity of both the finer fractions of aggregate (below 2 mm) and the fines themselves, and second, the use of superplasticizers is obvious. During the initial formulations of fibrous cementitious composites, it was advocated that the sand content be increased by about 10% and also ensured that it has a fluid consistency that will enable the fiber to be distributed in the system, which is indeed what a SCC matrix always provides. The next aspect that might have to be understood is that the length of the fibers and their interaction have always been a problem that needs to be addressed particularly in the case of stiff and inflexible steel fibers, which are sometimes also made with hooked ends or crimped. The stiffness in the fiber, its texture, and the ends specifically make the fibers entangle with each other promoting the formation of fiber balls, which is the sign of the matrix not being able to accommodate the amount of fibers in the system. This problem is relatively or practically nonexistent in inflexible polymeric and carbon fiber inclusions.

To have an understanding of the attempts in making SCCs with fibers is probably in order at this stage. Mazaheripour (2011) studied the effect of polypropylene fibers on the fresh and hardened state characteristics with a lightweight aggregate. The mix contained about 5% silica fume and 15%–20% limestone powder. The variation of slump versus slump flow and the effect of fiber content on slump were discussed. The effect of fibers on other U-box and V-funnel tests in the green state were also studied apart from the strength and modulus on the concretes. El-Dieb (2009) studied concrete mixes containing fibrillated polypropylene and steel fiber at different aspect ratios and diameters. The concretes contained 350, 400, and 500 kg/m^3 of cement and silica fume content of about 40 kg/m^3. The filling characteristics in terms of the slump flow, V-funnel flow, and Kajima box of the concretes containing fibers (3 control mixes, 15 mixes with polypropylene fibers, and 36 mixes containing different steel fibers, a total of 54 mixes) were studied in detail, and the relationships between the various flow characteristics were presented. Studies on the concretes used for the cable-stayed bridge across Yangtze River with the main span of 926 m utilizing a hybrid girder was reported by Li (2011). The concretes contained about 20% fly ash and 10% silica fume. The effects of polypropylene fiber and still fiber were also studied broadly. The effect of 1% polypropylene and polyvinyl alcohol microfibers in self-compacting microconcrete composites containing limestone powder and fly ash was reported by Felekoglu (2009). The flow characteristics as well as the fracture characteristics of 40×40×160 flexure specimens were studied. Caverzan (2011) studied the comparison of the static and dynamic characteristics of steel fiber–reinforced concrete of 70 MPa compressive strength. The concretes contained hooked steel fibers of 35 mm length at a fiber content of 50 kg/m^3. The results are expected to be a better representation of the dynamic characteristics without having to extrapolate from the static characteristics of steel fiber reinforced concrete (SFRC). Akcay (2012) reported the behavior of a hybrid steel fiber–reinforced SCC mix containing 700 kg/m^3 of cement and an additional 15% silica fume at a water–cement ratio of 0.22 resulting in a strength of about 150 MPa for the reference concrete. The concrete was modified with both plain steel fibers of 6 mm length and 0.15 mm diameter attempting a ultra-high performance concrete (UHPC) and also hooked fibers of 30 mm in length and 0.55 mm diameter both normal and high-strength steel. Apart from studying the fiber dispersion, the results of slump flow and J-ring at five different fiber volumes were reported. As can be expected, an increase in the fiber content increases the flow time, though there is no significant increase between the two volume percentages 0.75 and 1.5. The reported strength increase was also not significant, reaching about 124 MPa, but the fracture energy, flexural strength, and ductility ratio improved significantly through the addition of fibers.

As was already discussed, for ensuring a nonsegregating mix to control the viscosity of the paste phase, the method proposed by Saak (2001) and extended by Bui (2002) was used by Ferrara et al. (2007) to accommodate fibers. This concept recognizes the importance of limiting excess paste layer on the aggregate to about twice the thickness, and ensuring only the minimum cementitious paste required to coat the aggregate was used. There are additional fines not only in terms of fine aggregate but also in terms of the cementitious paste. Hence, it is only appropriate to think that the incorporation of fibers into SCCs should not be a serious problem, of course, with the fact that their total quantity is limited to about 2% for the appropriate length of the fibers. Ferrara et al. (2007) presented an extension to the rheology of the paste method based on the investigations on fiber-reinforced concretes containing 50 kg/m³ of Dramix 65/35 type fibers. In this method, the fibers have been treated as equivalent aggregate particles, defined through the surface area equivalence and the density ratio. It is to be noted that the method does not recognize the effects of the length of the fiber in relation to the max size and aggregate, which could be a good parameter that may be of significance in the design of fiber-reinforced concretes. It can also be observed that while the specific surface methodology will indeed cater to the needs of estimating the amount of paste required for coating them, it may not reflect adequately the interactions between the fibers (particularly in the case of longer fibers), and its effect on the workability and the passing ability as desired in the case of SCCs. Maybe by considering the fiber interactions of a triangular mesh of some dimension related to the length of the fiber and the size of the aggregate that could pass through, a more appropriate equivalent aggregate size could be defined for the fiber inclusions in concrete. A similar effort was earlier reported by Grünewald (2009) wherein the fiber length and aggregate diameter were connected to the risk of blocking in fibrous SCCs. For ease of working, it is best to limit the length of unyielding type fibers (mostly steel fibers) to about the maximum size of the aggregate used and the total percentage to about 2% maximum so that there is no excessive loss of self-compactability in the mixture because of the addition of fibers in the system.

Appendix SF-7

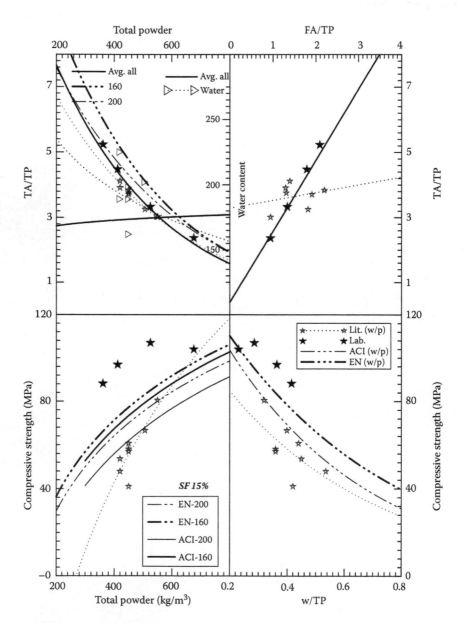

FIGURE SF7.1
Correlations between the constituents in SCCs with 5% silica fume.

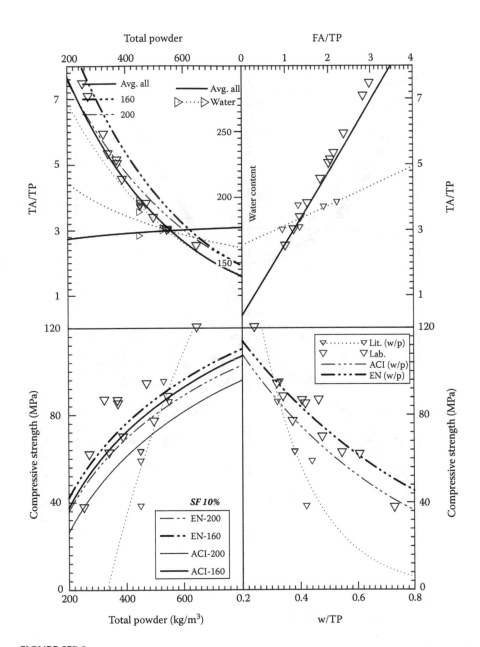

FIGURE SF7.2
Correlations between the constituents in SCCs with 10% silica fume.

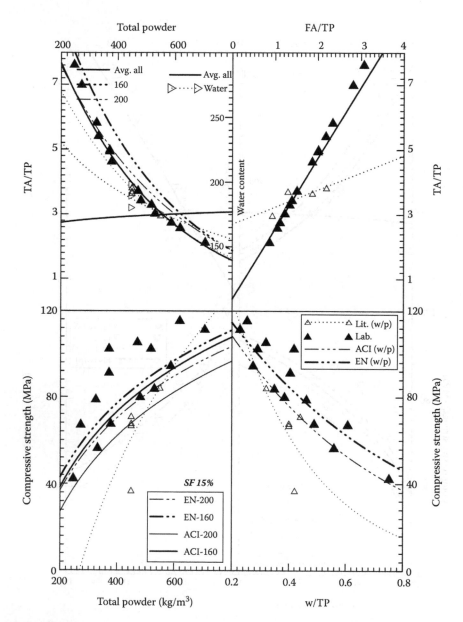

FIGURE SF7.3
Correlations between the constituents in SCCs with 15% silica fume.

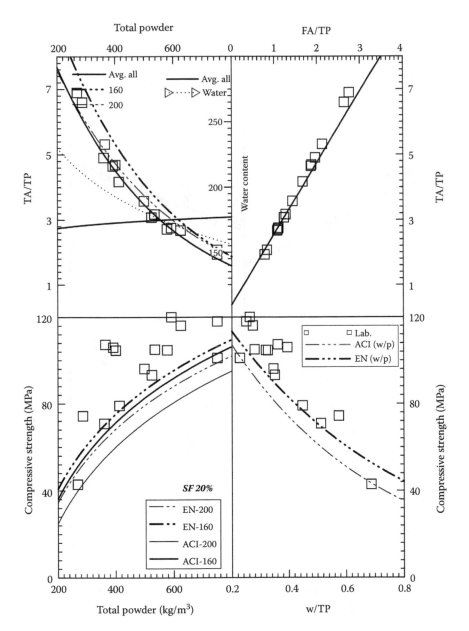

FIGURE SF7.4
Correlations between the constituents in SCCs with 20% silica fume.

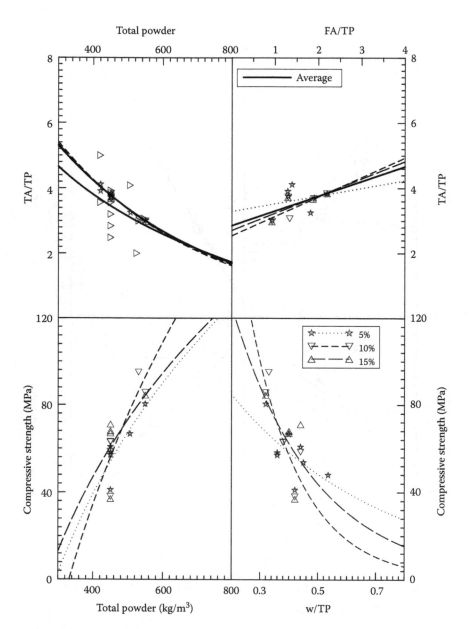

FIGURE SF7.5
Distribution of constituents in SCCs containing silica fume.

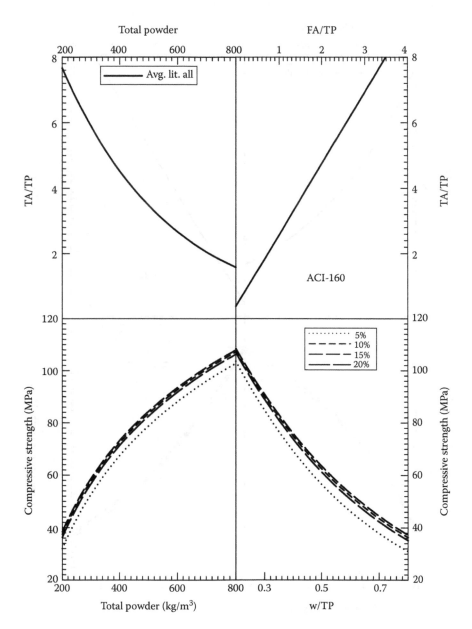

FIGURE SF7.6
Nomogram for design SCCs with ACI at 160 kg/m³ water (cyl.).

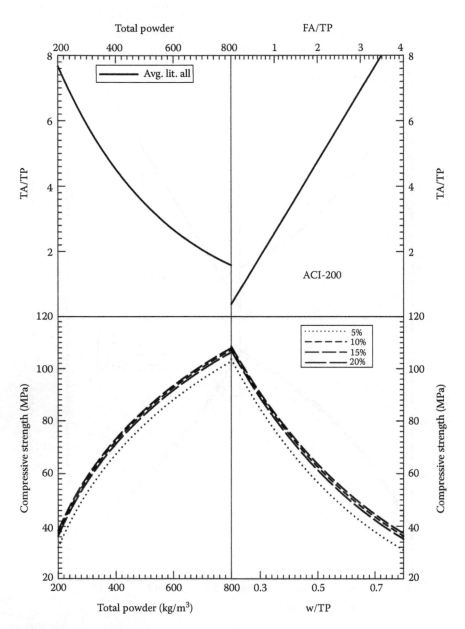

FIGURE SF7.7
Nomogram for design SCCs with ACI at 200 kg/m³ water (cyl.).

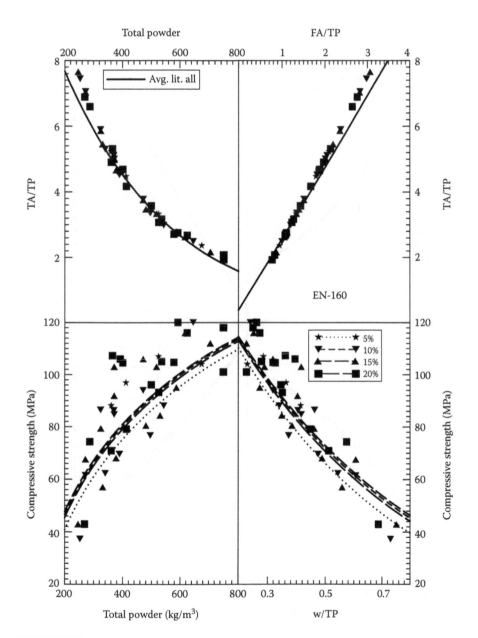

FIGURE SF7.8
Nomogram for design SCCs with EN 42.5 at 160 kg/m³ water (cube).

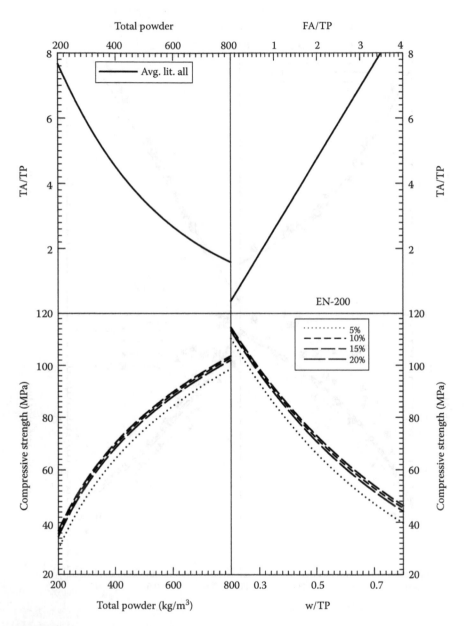

FIGURE SF7.9
Nomogram for design SCCs with EN 42.5 at 200 kg/m^3 water (cube).

References

Ahmadi, M.A., Alidoust, O., Sadrinejad, I., and Nayeri, M., Development of mechanical properties of self compacting concrete contain rice husk ash, *World Academy of Science, Engineering and Technology*, 34, 2007, 168–171.

Akcay, B. and Tasdemir, M.A., Mechanical behaviour and fiber dispersion of hybrid steel fiber reinforced self-compacting concrete, *Construction and Building Materials*, 28, 2012, 287–293.

Appa Rao, C.V., Behaviour of concretes with metakaoline, MS thesis, submitted to Indian Institute of Technology Madras, Chennai, India, 2001.

Atan, M.N. and Awang, H., The mechanical properties of self-compacting concrete incorporating raw rice husk ash, *European Journal of Scientific Research*, 60(1), 2011, 166–176.

Biricik, H. and Sarier, N., Comparative study of the characteristics *of nano silica-, silica* fume- and fly ash-incorporated cement mortars, *Materials Research*, 17, 2014, 570–582.

Bui, V.K., Akkaya, J., and Shah, S.P., Rheological model for self-consolidating concrete, *ACI Materials Journal*, 99(6), 2002, 549–559.

Caverzan, A., Cadoni, E., and Di Prisco, M., Dynamic tensile behaviour of self compacting steel fiber reinforced concrete, *Applied Mechanics and Materials*, 82, 2011, 220–225.

Cohen, M.D., A look at silica fume and its actions in Portland cements concrete, *The Indian Concrete Journal*, September 1990, 64, 429–438.

Collepardi, M., Collepardi, S., Skarp, U., and Troli, R., Optimization of silica fume, fly ash and amorphous nano silica in superplasticized high-performance concretes, in *Proceeding of Eighth CANMET/ACI International Conference on Fly Ash, Silica Fume, Slag and Natural Pozzolans in Concrete, SP-221*, Las Vegas, NV, 2004, pp. 495–506.

Cassagnabère, F., Mouret, M., Escadeillas, G., Broilliard, P., and Bertrand, A., Metakaolin, a solution for the precast industry to limit the clinker content in concrete: Mechanical aspects, *Construction and Building Materials*, 24(7), 2010, 1109–1118.

EFNARC, *The European Guidelines for Self-Compacting Concrete*, EFNARC, Farnham, Surrey, U.K., 2005, p. 68.

El-Dieb, A.S., Mechanical, durability and microstructural characteristics of ultrahigh-strength self-compacting concrete incorporating steel fibers, *Materials and Design*, 30, 2009, 4286–4292.

Felekoglu, B., Tosun, K., and Baradan, B., Effects of fiber type and matrix structure on the mechanical performance of self-compacting micro-concrete composites, *Cement and Concrete Research*, 39, 2009, 1023–1032.

Ferrara, L., Park, Y., and Shah, S.P., A method for mix-design of fiber-reinforced self-compacting concrete, *Cement and Concrete Research*, 37(6), 2007, 957–971.

Ganesh Babu, K. and Appa Rao, C.V., Strength behaviour of concretes containing metakaolin, Paper presented at the *Role of Cement Science in Sustainable Development—Proceedings of the International Symposium Dedicated to Professor Fred Glasser*, University of Aberdeen, Scotland, 2003a, pp. 231–239.

Ganesh Babu, K. and Appa Rao, C.V., Strength characteristics of metakaolin concretes, in R.K. Dhir (Ed.), *International Conference Proceedings*, Dundee, U.K., 2003b.

Ganesh Babu, K. and Narasimhulu, K., Strength efficiency of natural zeolites in concrete composites, Keynote paper in *Proceedings of the International Conference on Advances in Concrete and Construction, ICACC*, Hydrebad, India, February 2008, pp. 789–798.

Ganesh Babu, K. and Prakash, P.V.S., Effective utilisation of lower grade silica fume in the production of high strengtth concretes, in *Fourth NCB International Seminar on Cement and Building Materials*, New Delhi, India, December 1994, Vol. 4, pp. XI, 26–32.

Ganesh Babu, K. and Prakash, P.V.S., Efficiency of silica fume in concrete, *Cement and Concrete Research*, 25(6), 1995, 1273–1283.

Ganesh Babu, K. and Prakash, P.V.S., Sorptivity characteristics of silica fume concretes, in *Proceedings of International Seminar on "Civil Engineering Practices in Twenty First Century"*, Roorkee, India, February 1996, Vol. II, pp. 788–797.

Ganesh Babu, K. and Prakash, P.V.S., Deterioration of silica fume concretes due to sulphate attack, in *International Symposium on Innovative World of Concrete*, Calcutta, India, November 1998, Vol. 2, pp. 6.61–6.69.

Ganesh Babu, K., Rao, G.S.N., and Prakash, P.V.S., Efficiency of pozzolans in cement composites, in R.K. Dhir and R. Jones (Eds.), *Concrete 2000*, Vol. 1, E&FN Spon, Dundee, U.K., 1993, pp. 497–509.

Grünewald, S. and Walraven, J.C., Transporting fibers as reinforcement in self-compacting concrete, *HERON*, 54(2/3), 2009, 101–125.

Guneyisi, E., Gesoglu, M., and Ozbay, E., Strength and drying shrinkage properties of self-compacting concretes incorporating multi-system blended mineral admixtures, *Construction and Building Materials*, 24, 2010a, 1878–1887.

Ioani, A., Domsa, J., Mircea, C., and Szilagyi, H., Durability requirements in self-compacting concrete mix design, *Concrete Repair, Rehabilitation and Retrofitting II—Proceedings of the Second International Conference on Concrete Repair, Rehabilitation and Retrofitting*, Cape Town, South Africa, 2008, pp. 159–165.

Jalal, M., Mansouri, E., Sharifipour, M., and Pouladkhan, A.R., Mechanical, rheological, durability and microstructural properties of high performance self-compacting concrete containing SiO_2 micro and nanoparticles, *Materials and Design*, 34, 2012, 389–400.

Li, B., Guan, A., and Zhou, M., Preparation and performances of self-compacting concrete used in the joint section between steel and concrete box girders of Edong Yangtze River Highway Bridge, *Advanced Materials Research*, 168–170, 2011, 334–340.

Liu, M., Self-compacting concrete with different levels of pulverized fuel ash, *Construction and Building Materials*, 24, 2010, 1245–1252.

Maghsoudi, A.A., Soheil, M.J., and Darbhenz, A., Effect of the nano particles in the new generation of concretes, SCC, *International Journal of Nanoscience and Nanotechnology*, 6(3), September 2010, 137–143.

Mazaheripour, H., Ghanbarpour, S., Mirmoradi, S.H., and Hosseinpour, I., The effect of polypropylene fibers on the properties of fresh and hardened lightweight self-compacting concrete, *Construction and Building Materials*, 25, 2011, 351–358.

Mehta, P.K., Rice husk ash—A unique supplementary cementitious material, in V.M. Malhotra (Ed.), *Proceedings of the International Symbosium on Advances in Concrete Technology*, CANMET, Concord, NH, 1994, pp. 419–443.

Montgomery, J., Abu-Lebdeh, T.M., Hamoush, S.A., and Picornell, M., Effect of nano silica on the compressive strength of harden cement paste at different stages of hydration, *American Journal of Engineering and Applied Sciences*, 9(1), 2016, 166–177.

Narasimhulu, K., Natural and calcined zeolites in concrete, PhD thesis, submitted to Indian Institute of Technology Madras, Chennai, India, 2007.

Neville, A.M., *Properties of Concrete*, 4th ed., Longman, London, U.K., 1995.

Prakash, P.V.S., Development and behavioural characteristics of silica fume concretes for aggressive environment, PhD thesis, submitted to Indian Institute of Technology Madras, Chennai, India, 1996.

Quercia, G., Spiesz, P., Hüsken, G., and Brouwers, H.J.H., SCC modification by use of amorphous nano-silica, *Cement and Concrete Composites*, 45, 2014, 69–81.

Rahmani, Y., Sohrabi, M.R., and Askari, A., Mechanical properties of rubberized self-compacting concrete containing silica fume, *Advanced Materials Research*, 262–263, 2011a, 441–445.

Saak, A.W., Jennings, H.M., and Shah, S.P., New methodology for designing self-compacting concrete, *ACI Materials Journal*, 98(6), 2001, 429–439.

Sellevold, E.J. and Nilsen, T., Condensed silica fume in concrete: A world review, in V.M. Malhotra (Ed.), *Supplementary Cementing Materials for Concrete*, CANMET, Ottawa, ON, Canada, ACI-SP-86, 1987, pp. 167–246.

Su, N., Hsu, K.-C., and Chai, H.-W., A simple mix design method for self-compacting concrete, *Cement and Concrete Research*, 31, 2001, 1799–1807.

Sua-Iam, G. and Makul, N., The use of residual rice husk ash from thermal power plant as cement replacement material in producing self-compacting concrete, *Advanced Materials Research*, 415–417, 2012, 1490–1495.

Surekha, S., Performance of rice husk ash concretes, MS thesis, submitted to Indian Institute of Technology Madras, Chennai, India, 2005.

Turk, K. and Karatas, M., Abrasion resistance and mechanical properties of self-compacting concrete with different dosages of fly ash/silica fume, *Indian Journal of Engineering and Materials Sciences*, 18, 2011a, 49–60.

Turk, K., Turgut, P., Karatas, M., and Benli, A., Mechanical properties of self-compacting concrete with silica fume/fly ash, in *Ninth International Congress on Advances in Civil Engineering*, Trabzon, Turkey, September 27–30, 2010a, pp. 1–7.

Türkmen, I., Oz, A., and Aydin, A.C., Characteristics of workability, strength, and ultrasonic pulse velocity of SCC containing zeolite and slag, *Scientific Research and Essays*, 5(15), 2010, 2055–2064.

Ulucan, Z.C., Turk, K., and Karatas, M., Effect of mineral admixtures on the correlation between ultrasonic velocity and compressive strength for self-compacting concrete, *Russian Journal of Nondestructive Testing*, 44(5), 2008a, 367–374.

Vejmelkova, E., Keppert, M., Grzeszczyk, S., Skalinski, B., and Cerny, R., Properties of self-compacting concrete mixtures containing metakaolin and blast furnace slag, *Construction and Building Materials*, 25, 2011, 1325–1331.

Zia, P., Leming, M.L., and Ahmed, S.H., High performance concretes: A state of the Art report, SHRP/FR 91-103, North Carolia State University, Raleigh, NC, 1991.

References for Evaluations on Silica Fume

Guneyisi, E., Gesoglu, M., and Ozbay, E., Strength and drying shrinkage properties of self-compacting concretes incorporating multi-system blended mineral admixtures, *Construction and Building Materials*, 24, 2010b, 1878–1887.

Ioani, A., Domsa, J., Mircea, C., and Szilagyi, H., Durability requirements in self-compacting concrete mix design, *Concrete Repair, Rehabilitation and Retrofitting II—Proceedings of the Second International Conference on Concrete Repair, Rehabilitation and Retrofitting* II, Cape Town, South Africa, 2008, pp. 159–165.

Jalal, M., Mansouri, E., Sharifipour, M., and Pouladkhan, A.R., Mechanical, rheological, durability and microstructural properties of high performance self-compacting concrete containing SiO_2 micro and nanoparticles, *Materials and Design*, 34, 2012, 389–400.

Rahmani, Y., Sohrabi, M.R., and Askari, A., Mechanical properties of rubberized self compacting concrete containing silica fume, *Advanced Materials Research*, 262–263, 2011b, 441–445.

Sebaibi, N., Benzerzour, M., Sebaibi, Y., and Abriak, N., Composition of self-compacting concrete (SCC) using the compressible packing model, the Chinese method and the European standard, *Construction Building Materials*, 43, 2013, 382–388.

Turk, K. and Karatas, M., Abrasion resistance and mechanical properties of self-compacting concrete with different dosages of fly ash/silica fume, *Indian Journal of Engineering and Materials Sciences*, 18, 2011b, 49–60.

Turk, K., Turgut, P., Karatas, M., and Benli, A., Mechanical properties of self-compacting concrete with silica fume/fly ash, in *Ninth International Congress on Advances in Civil Engineering*, Trabzon, Turkey, September 27–30, 2010b, pp. 1–7.

Ulucan, Z.C., Turk, K., and Karatas, M., Effect of mineral admixtures on the correlation between ultrasonic velocity and compressive strength for self-compacting concrete, *Russian Journal of Nondestructive Testing*, 44(5), 2008b, 367–374.

8

Fresh Concrete Characteristics of SCCs

8.1 Introduction

The concept that compactability through its own self-weight is the fundamental attribute of a self-compacting concrete (SCC) illustrates the importance of the characteristics of fresh concrete and its popularity in the industry. It is to be recognized that while being self-compacting, the concrete should have a uniform consistency to ensure the high performance that is so often attributed to it and expected from it. In view of this, an understanding of the characteristics and the processes of how they are interrelated and the methods by which these can be modulated to suit the needs of a particular application is not only important but is also the very soul of the whole process. The relationship between consistency and compaction, flow and passing ability, and thixotropy and segregation resistance, which are all essential in ensuring that the SCC fills the void and encapsulates the structural reinforcement appropriately to ensure a perfect bond, is understandably a very complex process to be defined by any one single test, let alone a simple one. Naturally, several attempts by various researchers have resulted in a large number of formulations that probably attempted to define and indicate some of the characteristics that are essential for being an SCC.

The second aspect that needs to be kept in mind while examining the consistency and compactability of an SCC is that there are three different classes or ranks that are already accepted by various national bodies. As in the case of normally vibrated concretes, even in the present case it is not appropriate to prescribe a specific group of tests to adequately portray the characteristics of all these different classes. This introduces a further complexity into the needs of the assessment and the relationships between the tests to ensure a better understanding of the SCC on hand.

Apart from an overall perspective of having to define the characteristics of concrete that is probably appropriate for any application, a highly flexible concrete with such low viscosity naturally poses the problem of segregation and bleeding due to the two-phase behavior. Research efforts that looked for synergy of the two phases, mortar and coarse aggregate, have indeed tried to graduate from cement paste to mortar and later to concrete, obviously generating only information regarding the materials on hand and cannot be really holistic. However, these studies will still be useful in finalizing the mixture proportions of a fairly consistent group of materials in an ready-mixed concrete (RMC) plant where they have ensured that the initial work that is required to arrive at an SCC of an appropriate strength and class has already been done. While this is one extreme, the other end obviously is to look at the more scientific and theoretical aspects of the set-tling velocities and viscosity models to ensure that such an approach can result in appropriate solutions through the necessary approximations and assumptions. The fact that in both cases the largest size of aggregate that needs to be addressed is far too dense and is far too large may be far too uneven to arrive at a simple model through experimental investigations based on scientific knowledge. This indeed is one of the reasons that con-crete technology in particular and material science in general have always had to be appropriately modulated and understood to ensure that the theory matches practice in the field. It is also necessary to recognize that there are several singularities in a theoretical approach that are always not understood, because these could be associated with physical characteris-tics of the constituents or they are probably related to the more complex and dynamic chemical characteristics that are to be understood through the changes in the intermediate compounds during the reaction kinetics. It is probably this fact that drives research into the intricacies of concrete or to be more general the entire range of cementitious composites includ-ing polymer and fibrous compositions. Probably this fact will be better understood while discussing the inclusion of fibers into these metrics, spe-cifically steel fibers that are not only thin and long but also have a density several magnitudes higher than the environment of ingredients in which they are situated. A brief overview of some of these concepts, and their interactions in relation to the green state characteristics of a range of SCCs, that are of interest today is attempted below.

8.2 Fundamentals of Consistency and Compaction

Concrete in itself is a bundle of dichotomies, the first and the foremost being the need for additional water to have a high workability while the lower water–cement ratio required for higher strength dictates the significant low-ering of this particular water content. The second aspect that is not recog-nized so directly is the need for a higher cement (or cementitious materials) content ensuring a paste dominant matrix that can be compacted into a void-free structure more easily, while at the same time such a material will have higher drying and thermal shrinkages, which will lower the strength of the cementitious composites significantly. Such dichotomies are indeed common in most engineering practices where the idea is to arrive at a solution with minimum cost and in many scenarios also minimum time and effort. The alleviation of the need for appropriate compaction energy requiring both manpower and finances and the reduction in noise pollution levels during vibratory compaction are also a part of the same story that we have been discussing.

There is indeed a continuous conflict between the need for lower water content to ensure a concrete of the highest strength with minimum cement content and thus cost, and the requirement for a higher amount of water to ensure an appropriate compaction and thus avoid honeycombs or significant air entrapment with minimum energy inputs. The quantity of entrapped air also depends on the other parameters like the size and shape of the member, reinforcement congestion, and ultimately the method of competition. The effect of increasing water content on the consistency of concrete and its effect on concrete quality are well explained by ACI 309 (1984). Figure 8.1 is indeed a broader perception of the same describing

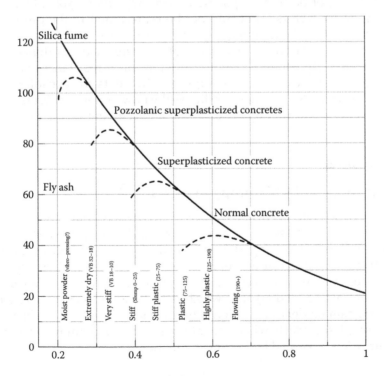

FIGURE 8.1
Effect of competition and material composition needs of concretes in various strength and consistency regimes. (After ACI Committee 309, Recommended practice for consolidation of concrete, ACI 309-72, ACI Manual of Concrete Practice, Part 2, 1984.)

the various consistency characteristics and also showing that they cannot be defined by a single test, in this case either slump or Vebe (VB), appropriately. The reduction in strength of concrete at very low water–cement ratios is in fact well known and the need for superplasticizers becomes obvious for concretes at that stage. The fact that superplasticizers play a major role to ensure concretes of lower water–cement ratio with the same consistency that is possible with the high water content is also represented in this figure. Figure 8.1 also presents, even if indirectly, the effect of the modulation of the internal pore structure through the addition of supplementary cementitious materials, an essential part of the production of very high-strength and even high-performance concretes (which can also be attempted even at lower strengths). The figure clearly explains the effect of water–cement ratio (considering that the water content is essentially constant for a comprehensively designed concrete mixture with a sufficient amount of cementitious materials content) on the consistency on one hand and also the approximate limits of strength that one can comfortably reach with the addition of either superplasticizers or mineral admixtures (the low-end type represented by fly ash and the high-end type represented by silica fume) or both in conjunction with each other. Probably, the only other important factors that could not be represented in the present context are the effects of curing regimes and temperature.

The fundamental flow characteristics of concrete are dependent on properties like viscosity, cohesion, and the inherent shear resistance of concrete. Presenting an overview of the energy requirements for consolidation of concrete, Olsen (1987) defined stability as the flow of fresh concrete without applied forces, measured by the bleeding and segregation characteristics.

8.3 Rheology and Thixotropy of SCCs

Rheology of a material deals with its deformation characteristics under an applied shear stress. In a Newtonian fluid, an applied shear stress produces a shear strain proportional to its magnitude, and the coefficient of proportionality is its shear modulus. However, there are no ideal fluids like this in nature and in most cases there is a certain minimum shear stress (as in the case of starting friction) after which a relation that is characterized by a linear Bingham model or a relation exhibiting shear thinning or shear thickening viscoplastic effects will come into play. The Bingham model depicted by Figure 8.2 is governed by the linear equation:

$$\tau = \tau_0 + \mu\gamma$$

The two other possible non-Newtonian rheological types of shear thickening and shear thinning models are also presented in Figure 8.2. It may not be out

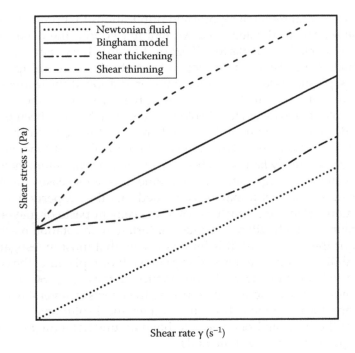

FIGURE 8.2
Bingham and other non-Newtonian models of rheology.

of place to mention that it is indeed possible to visualize some such behavior even in the case of cement and concrete composites if one looks at the various phenomena that occur during the mixing and hydration periods of fresh concrete. One possibility is that when the mixing is relatively slow while the cementitious composites involved are hydrating and forming the calcium silicate and alumni hydrate gels rather fast as in the case of super fine cements, maybe even aided by an internal heat of hydration, the characteristics of the mass slowly tend to thicken naturally and the rheology changes to the situation similar to shear thickening behavior even for cementitious composites. Conversely, if in case the cementitious composites are mixed vigorously through high shear mixing processes particularly in the presence of superplasticizing admixtures, the water quantity locked in the cement agglomerates is released faster due to the action of both the plasticizers and high shear mixing, where a situation similar to shear thinning behavior can also be observed in cement and concrete composites. In fact, some of these processes like reducing the mixing rate throughout the transit period and increasing the mixing rate once the site is reached are in fact very effectively used by the ready mix concrete industry. While the rheological changes in the concrete mass during transport and just before placement are not recorded in detail, it appears that these facts are obviously recognized and appropriately utilized to ensure the best benefit in the case of ready mix operations.

There are several research reports and publications that present an insight into these aspects of rheology and we will not go into the details of these presently (Tatersall, 1991; Bartos, 1992; de Larrard, 1998). Several different test methods have also been developed to study the relationships between yield stress, plastic viscosity, and shear rate for an applied shear stress, mostly through rotational rheometers. The description of some of these methods and their potential to assess the rheology of concrete has also been presented earlier and is also not going to be given here (Ferraris, 2001; Koehler, 2003). However, while it is essential to understand the rheology of concrete the fact that these are seriously hampered because of a matrix containing mostly particulate matter, be it aggregates or even cement, is not so easily amenable to such assessment. In fact it may be construed that the property that is measured is more often the characteristics of the cement paste or maybe at best that the mortar which will not seriously interfere with rotation mechanisms adopted for the assessment. It is for this reason that most investigations on the workability of concretes have always looked for replicating the construction scenario to be more nearer to the practice. In the case of SCCs, researchers have indeed relied more on quantifying the effects in several ways rather than looking for more theoretical and mathematical solutions, even though these were the basis and one needs to have an understanding of the same while adopting the other test methods.

8.4 Critical Evaluation and Comparison of the Test Methods

The one obvious fact that needs to be recognized and ascertained in evaluating the green state characteristics of SCCs is that the different classes or ranks accepted by the various national bodies are never considered a specific parameter in the discussions. This in a way is true even for normal concretes and only the more discerning reader or researcher is keen to look at this before making any assessments. The second factor that makes the consistency evaluations complicated is the enormous variety of pozzolanic admixtures and/or mineral powders that were used by individual researchers and the fact that their physical characteristics could have a significant bearing on the self-compacting characteristics of the system. This while being recognized by some is just relegated to information in passing. Pozzolanic admixtures like fly ash and silica fume coming out of pyroprocessing techniques of their parent materials tend to be spherical and could aid the movement of the aggregates through the ball bearing effects. However, silica fume with its extreme fineness may impart a higher thixotropy to the mix. In contrast, most of the materials like ground granulated blast furnace slag (GGBS), calcined clays, and even rice husk ash have an angular grain structure that helps the sliding or the movement of aggregates through the coating thickness of the paste generated. These two effects are different and come into play in different ways at the different proportioning levels of the cement admixture combinations and can only be discussed for a specific group of mixtures if at all. Apart from this, the type of superplasticizers used, be it the melamine/ naphthalene formaldehyde sulfonates or the more recent polycarboxylate-based materials, could significantly influence the agglomerate characteristics of the cementitious materials, which can also be reflected in the green state characteristics. Needless to mention that the aggregates, aggregate grading, fine aggregate content, powder, and the paste content will all have a significant role. With so many variations, materials, and characteristics, it would not be appropriate to compare the different research efforts unless there are sizeable commonalities that would allow such an effort. Notwithstanding this, a few general comments could always be made. In the present context, it is prudent to select a few specific examples of the research efforts and compare the results to present a broad outline of the trends on these parameters. These discussions are true for the mechanical as well as durability characteristics that form the topics for discussion in the next couple of chapters. In view of this, many of the parameters discussed under various subjects may not be totally broad-based and universal but yet can result in a general understanding of the topics under discussion.

After having seen the enormity of this diversity in materials combinations on test methods just to name a few, one wonders how to get a picture of the more relevant aspects to ensure not only an understanding but also the making of SCCs of high performance. It is also known that as already discussed most investigators looked at a specific dimension of the problem through highly focused and extremely limited investigations, as is necessary. However, correlating several of these different research efforts and making specific recommendations becomes a task that is sometimes not so effective in being inclusive. This should not be construed as a limitation to bring down the importance of any of the investigations, however small they are, as it is only then and only with this diversity that we will be able to make some broader suggestions. Without going further into the depths of the various parameters that influence the characteristics, it may be prudent to look at the broad spectrum of the mosaic to understand the linkages between these materials, combinations, and their effects on the various green state characteristics of SCCs. The attempt is to present a brief outline of the various factors involved in the research efforts and examine any specific aspect that is critical to the understanding of the parameters involved. The next couple of chapters also adopt a very similar methodology with such an outline of some of the investigations into the different aspects based on different materials and methods first presented, and later the various aspects are discussed broadly.

An appropriate point to start the discussion on the characteristics of fresh concrete and the efforts that contributed to them is to look for a reasonably large study that could present an insight into the parameters of interest without considering the effects of materials and methods. After a brief review of the proportioning methods available, Kheder (2010) presented a method for the proportioning of SCC by modifying ACI 211.1 regulations for water requirements of concretes with 75–100 mm slump for different aggregate sizes of 10–20 mm. The fact that the method was attempting to be in line with the already existing concrete mix design procedure of a well-known national body, universally accepted to be one of the simplest methods based on experience and expertise on concretes of that category (normally vibrated concretes with a reasonable slump), should be appreciated and recognized. Incidentally, the fundamental methodology adopted in the formulation of the various discussions presented here is exactly the same philosophy with which instead of looking for the arbitrary mathematical or empirical solutions to achieve the SCCs it is felt that they should be treated more as an extension beyond the collapse slump concretes that are already detailed in the various national recommendations. The necessary modifications required to accommodate the newer materials at their various levels should also be based on such well-known principles of concrete mix design. Incidentally, Kheder also presented the green state characteristics of 30 mixes, namely the slump flow and L-box characteristics along with flow times, for concretes designed according to this method. He also adopted limestone powder as the powder extender in all these SCCs. These were studied comprehensively to arrive at broad guidelines for the interrelations between the various parameters of slump flow and L-box. He also arrived at the strength to water–cement ratio relationship of SCCs based on the results of these mixes and a few others that are available in the literature.

Kovler (2011) presented a very broad overview of the various aspects relating to the behavior of fresh and hardened state concretes in general along with an outline on SCCs, which could be used as a guide to relate SCCs to normally vibrated concretes. SCCs containing higher volumes of fly ash (40%, 50%, and 60% by cement replacement) with water–cementitious materials ratios of 0.35, 0.40, and 0.45 were studied by Bouzoubaa (2001). Even though he recognized the pozzolanic activity index of the fly ash used as part of the material characteristics in his investigations, the effect of such pozzolanic activity was never utilized in the calculations relating to the strength of the concretes. He studied the slump, slump flow, flow time, bleeding, and segregation of these concretes along with the autogenous temperature rise, all of which were in line with the fly ash contents in these mixes. One of the important observations is that the initial and final setting times do get substantially affected due to the increased fly ash content, and at the 60% replacement fly ash concrete (with 0.45 water–cement ratio) the setting times could be as high as 7 and 10 hours, compared to the 4–6 hours exhibited by normal concrete at the 0.5 water–cement ratio. He also observed that the drying shrinkage of the SCCs studied was not very different from that of normal concrete. Saak (2001) suggested a segregation controlled design method by using the falling ball analogy for obtaining segregation resistance. His theory assumes that for the regular aggregate particle size distribution and volume fraction, the rheology and density of the cement paste matrix dictate the fluidity and segregation resistance of concrete. Su (2001) proposed a method to produce an SCC based on the packing factors of both fine and coarse aggregates in terms of the volume ratio of fine aggregate to total aggregate. The method tries to fill the voids in a loosely packed aggregate matrix with the paste of binders that was said to lead to the use of a smaller amount of binder. According to Su et al., the fine aggregate to total aggregate ratios were taken to be ranging from 50% to 57% while the packing factors could vary from 1.12 to 1.18. The cement content was calculated as a function of strength and the water–cement ratio was proposed to be ascertained from ACI or other methods in previous studies. The remaining volume, made up of fly ash (70%) and GGBS (30%), is calculated from the absolute volume method and their water requirements are calculated based on empirical relations to ensure good fluidity for the mixture of cementitious materials. The actual strengths obtained in his investigations were found to be lower than expected.

The studies on plastic settlement of SCC using a laser setup that was earlier developed by Khayat (1997) were reported by Sonebi (2002). Sonebi (2004) also studied 21 concrete mixes having 0.38–0.72 water–powder ratios with cement contents ranging from 183 to 317 kg/m^3 and fly ash contents from 59 to 261 kg/m^3. The green state characteristics, Orimit time, V-funnel, L-box, and J-ring were studied and the results were used in the prediction of a model generated through a factorial design. He also studied concretes through a settlement segregation column apparatus to understand their behavior. In fact the study clearly indicates that with the range of fly ash replacement combinations it is possible to achieve concretes over a wide range, from about 20 to 75 MPa particularly at 90 days. Also, the lower strengths reported at 7 and 28 days obviously indicate that fly ash significantly affects the strength gain rate as already pointed out. The above two aspects of the slower strength gain rate yet contrasting with the possibility of achieving reasonably high-strength concretes at a slightly later age will be more appropriately addressed and discussed in later sections of this chapter.

In another later study, Sonebi (2009a,b) presented the results of a group of concretes containing fly ash at approximately 30%–50% of cementitious content with water–powder ratios of around 0.4–0.65. He uses the various parameters measured as a part of this investigation (including the green state characteristics) for predictions through a genetic algorithm showing a good correlation. The fact that the algorithm trained to the data that was obtained obviously will result in such a good correlation probably proves the effectiveness of the algorithm but not so much its effectiveness in predicting the correlations relating to the behavior of SCCs. Even so the fact that certain correlations could exist should not be lost to the reader. One very broad observation could be that any of the relationships attempting to correlate two parameters should obviously be based on known scientific and engineering facts and also the accepted trends in concrete technology if they were to be proposed for utilization for such concretes in practice. One can also see from the results that the 28-day and 90-day concrete strength variations apparently indicating the effect of the efficiency of fly ash at the different percentages used though such assessments were not attempted in this study.

The effect of the packing density characteristics of aggregates in SCCs was studied by Brouwers (2005). They compared the grading relationships of the aggregates they used as well as others through the Anderson and Anderson model, which considers both the maximum and minimum aggregate sizes in defining the grading. The studies show that the volume of water to volume of powder ratio is directly related to the relative slump and also the fact that these relationships were all different for cement, limestone powder, and fly ash with a decreasing slope. This indicates probably that the Bingham values of the yield stress and plastic viscosity (which probably are an indirect representation of these figures) are decreasing continuously from cement to limestone powder and further to fly ash. In a way this could be anticipated because pure cement compositions will suffer from the early hydration effects, while the other two will have a slower setting and thickening rate. But these are characteristics that depend on the various percentages of the materials in these compositions and need to be correlated accordingly.

Felekoglu (2007) suggested that the adjustment of water–cement ratio and superplasticizer dosage are the two main requirements for an appropriate proportioning of SCC mixtures. He suggested that for avoiding blocking and segregation in the SCC mixtures the water–cement ratio should range between 0.84 and 1.07 by volume. This is similar to that suggested by Okamura (1995) earlier. As already known from a logical point of view, the Bingham properties of SCC can be defined by yield stress and plastic viscosity. It is also recognized by him that slump flow is a poor indicator of the yield stress and is not enough to characterize the fresh concrete behavior of an SCC. He suggested the following relation between the V-funnel times (V_f) to the "T_{50}" values through a small experimental program of only about five mixes.

$$V_f = 2.83 \left[\left(T_{50} \right)^{2.05} \right]$$

He also tried to relate the water–powder ratio by volume with V-funnel time indicating a nonlinear relationship. However, such a relationship is probably not universally applicable with the fact that the maximum aggregate size, shape, and grading distributions and the corresponding loosening effects may vary significantly. Even so one can always look for such correspondence in a tightly controlled RMC scenario to help in modulating the concretes produced.

Another large enough investigation that probably has information on the green state characteristics through slump flow, V-funnel, and L-box results apart from setting times is from the group Guneyisi (2008, 2010) and Gesoglu (2007, 2009). Between these four studies, they reported a total of 65 concrete mixtures at the two water–binder ratios, 0.32 and 0.44, with the first 43 concretes at a total cementitious materials content of 550 and the rest at 450 kg/m^3. The mixes contain both binary and ternary compositions containing fly ash and GGBS with replacement ratios at 20%, 40%, and 60% and silica fume and metakaolin replacements at 5%, 10%, and 15% by weight of the total binder in various combinations. He also adopted the most commonly used EN 42.5 Euro grade cement having 326 m^2/kg Blaine in all these investigations while the FA, GGBS, MK, and SF had a fineness of 287, 418, 12,000, and 21,080 m^2/kg, respectively. Though the fineness of fly ash appears to be a bit lower, these concrete investigations are perhaps the largest database that one could get from any single group of investigators on all these materials. The studies included the green state characteristics apart from the compressive strength and shrinkage characteristics of all these concretes. The conclusions include that the negative effects of fly ash (lower strength gain rates) could be corrected through the use of ternary and quaternary mineral admixtures combinations. The addition of mineral admixtures reduced the shrinkage characteristics of the concretes with silica fume showing the maximum effect, which in any case should be expected with the better pore filling effects of such fine materials apart from their higher pozzolanic efficiency obtaining a highly dense structure and lowering the shrinkage.

The water absorption characteristics of SCCs containing 30% fly ash and 10% silica fume cured for periods of 3, 7, 14, and 28 days under different curing conditions (immersed in water, sealed, and air cured) were reported by Turk (2007). The studies include the compressive and tensile strengths, and the ultrasonic pulse velocities apart from sorptivity coefficients at 28 days. He observed that the sorptivity showed a good correlation with compressive strength regardless of the curing regimes. In a later study, Turk (2010) reported studies on SCCs containing fly ash contents of 25%–40% and silica fume contents of 5%–20%, which are probably the best ranges for their replacement, given the fact that the total cementitious material content was limited to 500 and 450 kg/m^3, respectively. His effort shows an appropriate utilization of the admixtures to arrive at reasonably high-strength concretes of around 70 MPa. He also reported the relationship between compressive strength and modulus of elasticity apart from others. Li (2011) studied concretes containing fly ash along with the effect of superplasticizers and viscosity modifying agents (VMA). He also presented a broad outline of the economics of the system in terms of the electrical energy required for vibratory compaction, indicating benefits in terms of social and environmental parameters even in low-strength concretes.

Georgiadis (2010) attempted to arrive at the relationships between slump flow and V-funnel as well as L-box, specifically looking at the water content variation effects on these consistency parameters. He also attempted to measure the pressure on the formwork experimentally by using a laboratory test with four series of sensors at different levels on a 25×25×100 cm column. He agrees with the fact that concretes during rest periods build up an internal structure, maybe due to the effects of hydration, and the pressure on the formwork decreases as time progresses, which was also observed by Billberg (2001). It was also stated that the classical formwork used in France was able to withstand a formwork pressure of 8–10 t/m², which allows the continuous casting up to a height of 4 m (Cussigh, 2007). As is obvious, the pressure buildup on formwork due to SCC is obviously related to the rate of casting. It is also observed that there have not been any measurements of pressure on SCCs containing a VMA, which anyway was not considered as qualified for vertical casting.

One of the major problems of SCCs in particular and also concretes in general is the possible layering effects arising out of even marginal time lag in the pour sequence or concrete flow distribution in the construction activity of particularly larger depth members. Roussel (2008) critically discusses these distinct casting layer effects of thixotropic materials in terms of their Bingham parameters. An experimental investigation to study that particular effect through shear fracture specimens was conducted. The procedure is to follow a method of casting where the shear fracture surface is cast at different times so that the failure characteristics can be related to the delay in casting these layers. The method of testing appears to be essentially looking into the bond characteristics of the interlayer interface, rather than those due to the distinct layer effect. The distinct layer effect occurring in the field generates minute quantities of bleed water that accumulates even on well-formulated SCCs showing a distinctly identifiable layer in a cut section. If there is a larger accumulation of bleed water, it could be potentially disastrous particularly in huge foundations supporting vibrating machines, wherein the foundation was designed for the mass of the integrated machine and foundation together. But the foundation behaving as two or three separate masses due to layering could lead to not only vibration problems but failure ultimately.

The effects of lightweight aggregates on the absorption characteristics of concrete containing both expanded perlite aggregate (EPA at 5%, 10%, and 15%) and natural aggregates were studied by Turkmen (2007). He also studied the effect of curing conditions in terms of the curing time on these lightweight aggregate concretes. But the addition of silica fume at about 10% necessarily affects the curing times entirely differently, which is very difficult to predict. Mazaheripour (2011) studied the effect of polypropylene fibers on the fresh and hardened state characteristics of self-consolidating concretes containing LECA, a lightweight aggregate. The mixes also contained about 5% silica fume and 15%–20% limestone powder and were all made at a constant water content of 160 kg/m³. The variation of slump versus slump flow and the effect of fiber content on slump were discussed. The effect of fibers on other U-box and V-funnel tests in the green state was also reported along with the strength and modulus on the concretes. The effect of recycled powders was studied by Quan (2011). Recycled powders appear to show a lower elastic modulus and higher drying shrinkage than slag and limestone granules used by him. It is not clear if this is due to any surface pozzolanic effects of these slag and limestone granules. Also the granulated slag improved the superplasticizing effect of the recycled concretes. He also presented the relationships between tensile strength and modulus to the compressive strengths of the concretes studied.

Melo (2010) studied concrete containing metakaolin of varying fineness at 5% and 35% at the different paste volumes ranging from 0.40 to 0.55. The study clearly indicates an obvious increase in water content due to the higher fineness from the metakaolin. The study also indicates that it is possible to establish what he calls a saturation point of the admixture through the mini-slump flow and Marsh cone flow times. However, the use of high water–powder ratios of 0.5 and 0.7 in these mixtures appears to have substantially lowered the strengths of concretes to values of only around 40 MPa even for the finest of metakaolin used while for the coarser variety the strength was only in the region of 20 MPa. It is for this reason that most of the discussions in this particular compilation always emphasize two main factors. The first one is that the comparative normal concrete should not have a strength that is less than 10% of that predicted by the Euro norms for C42.5 grade cement and that the reduction or the increase in strength if any due to the replacement of a pozzolanic admixture should always be related to its strength efficiency. The strength variation between 5% and 35% replacements is also not comparable as they have water–cement ratios of 0.5 and 0.7, respectively, probably because the finer metakaolin at its higher percentage requires a larger quantity of water as well as superplasticizer. In fact there is no sanctity in trying to use such high quantities of water content (as high as 300–350 kg/m^3), which naturally results in very-low-strength concretes that cannot justify the use of the highly expensive and high-efficiency mineral admixtures along with the necessarily higher superplasticizer contents. A further aspect that is also critical is that at 5% such fine materials will simply act as fillers (particularly given that the water–cement ratios are very high) and cannot contribute any measurable pozzolanic effects, while at a high percentage of 35% the pozzolanic strength efficiency of the metakaolin is probably lower than the cement itself, resulting in very-low-strength concretes. Extensive studies on concrete mixes containing fibrillated polypropylene and steel fiber at different aspect ratios and diameters were reported by El-Dieb (2009). The concretes contained 350, 400, and 500 kg/m^3 of cement and a silica fume content of about 40 kg/m^3. The filling characteristics in terms of the slump flow, V-funnel flow, and Kajima box of the concretes containing fibers (3 control mixes, 15 mixes with polypropylene fibers, and 36 mixes containing different steel fibers, a total of 54 mixes) were studied in detail, and the relationships between the various flow characteristics were presented. While the effort involved in such a large experimental investigation of a broad spectrum of fibrous contents is certainly to be appreciated, the green state characteristics of fiber reinforced concretes differ significantly with each quantity as well as aspect ratios of the fibers and to correlate their effects is very difficult to say the least. However, the results probably could be most useful in trying to arrive at the best combination for a particular scenario.

After having looked at some of these research efforts that could give a broad picture of the various combinations in producing SCCs, it is easy to understand the popularity of SCCs. This can also be visualized by looking at the amount of information that is being generated and through the number of national and international conferences on this topic. Some simple facts are that the high powder content, maybe in the range of about 500 kg/m³, is essential for ensuring the required self-compactability. In this regard, cementitious materials and fillers below 125 μm size are generally considered as powders (Skarendahl, 2000). The above review clearly presents the possibility of using several pozzolanic admixtures as reported by a large number of researchers. These were seen to be admirably suitable powder extenders for achieving SCCs. Several efforts were also made to appropriately quantify the various characteristics like filling, passing, and segregation resistance through different test methods designed for this purpose. Modifications in terms of different types of fiber additions to suit specific needs have also been attempted to address improvement required in specific characteristics. In trying to understand the green state characteristics, studies were undertaken not only on flowability aspects but also on the consistency of the material through its depth and length of flow, the bond characteristics with reinforcement, the shrinkage, and temperature effects in general. In most cases, for appropriately designed SCCs, these appear to be in line with the characteristics exhibited by the normally vibrated concretes that are used presently.

A broad understanding of the three different primary characteristics, filling ability, passing ability, and segregation resistance, is available from the earlier discussions. The filling ability is the ability of concrete to flow and compact under its own weight filling the formwork completely, essentially required for thin members and congested reinforcement locations. The passing ability is the ability to flow through confined spaces and narrow openings between reinforcing bars and completely encapsulate the reinforcement. The segregation resistance obviously assumes the highest significance by the fact that the concrete should exhibit a uniform composition throughout the length and the depth of the member during and after the flow. Factors like flow rate generally depicting the viscosity may also have to be approximately ascertained for certain applications.

One observation that can be made at the very start of these discussions regarding the critical evaluation of the test methods is that for a well-trained eye with a reasonable experience on concrete consistency evaluations, the simplest and the most utilized test, slump flow, using the regular slump cone can indeed define almost all these characteristics. To be more specific, one can see the flow time (T_{50}), flow pattern, actual distribution of the various coarse aggregate sizes in the flow spread, possibility of having larger size aggregates remaining in the center (even without the obstruction from a J-ring), thin band of mortar at the periphery of the flow spread, and possibility of also having another small band of bleed water around it, which can positively reveal several of the aspects that we are discussing and put a fix on the characteristics of the SCC.

The second test that attracted attention in the laboratory is the flow characteristics of the SCC from the V-funnel. To describe this observation, one has to closely study not just the flow time but the continuity of the material composite that is coming out of the tapered end, whether it is continuous or discontinuous and if discontinuous how long is the discontinuity. A simple extension to this as already proposed by a few research workers is to have a specific understanding of the flow diameter when the fall from the open end is kept above the base plate by about 100 mm. The observations discussed earlier for slump flow are also quite relevant or probably more relevant in this particular case because of the drop involved in the test, which effectively ensures a definite effect due to the dynamic forces resulting in a better appraisal of the segregation also. There are also a few other ways already reported in the literature in which the V-funnel material is allowed to flow through a channel like in the L-box test or Kajima test to study the horizontal flow rates and passing ability appropriately.

Discussing the guidelines for SCC in line with DIN 1045 (1988), Reinhardt (2001) suggested the introduction of a blocking ring (called the J-ring) consisting of bars intended to hinder the flow of the concrete placed around the slump cone in the slump flow test. The blocking ring is like a circular comb of 300 mm diameter and has round smooth metal bars of 18 mm diameter that are rigidly connected to it. One important aspect that has not yet been fully appreciated in the reported studies on SCCs is that the number of bars suggested by him vary according to the maximum size of the aggregate, namely, 22, 16, and 10 numbers for the aggregate sizes of 8, 16, and 32 mm placed uniformly around the circumference of the ring. The blocking ring obstructs concrete flow and would also result in a height differential between the inside and outside, indicating the passing ability and maybe even the possibility of segregation resistance. The time to flow to a diameter of 500 mm (T_{50}) can also be measured and interpreted similar to the results without the J-ring. It is also suggested that after the withdrawal of the slump cone the SCC must have a slump flow of ≥ 700 mm without the blocking ring and ≥ 650 mm with the blocking ring. Another requirement is that the difference between the slump flow without and with the blocking ring should be no more than 50 mm. A visual check must be made to ensure that the concrete within the blocking ring does not exhibit any increased accumulation of coarse aggregate. He also suggests that an appropriate amount of the coarser fractions in the aggregate should escape through the J-ring and should be distributed uniformly around and that there should be no separation of paste or water at the age of the concrete flow to indicate possible segregation or bleeding. The guidelines also suggest that this property should be verified by the slump flow test with the blocking ring and, in fact, must be verified as the performance test before transfer of the concrete, at least at the first transfer. The T_{50} time should also be checked during the performance testing and at least at the first transfer, which could also serve as a reference value for the consistency of the concrete composition. The guideline contains an escape clause for a deviation from the consistency requirement mentioned if it is shown in an extended performance test that the elements can be concreted without faults. Researchers have also advocated the use of the slump cone inverted, but then it was suggested that the slump cone should not be raised beyond 100 mm, to ensure that the fall is not from a significant height to affect the results, which is actually a difficulty. The methodology of the concept of fixing the cone at a specific height as proposed in the German test discussed earlier (Figure 3.2) is an answer to this particular situation.

In sharp contrast to this, the ASTM recommendations suggest 16 numbers of 16 mm diameter bars in the J-ring for a maximum aggregate size of 25 mm, while the present EN 12350 (2010) retains the earlier recommendation, 16 numbers of 18 mm diameter bars. A critical evaluation of the German recommendations of Reinhardt shows that the maximum size of the aggregates to the clear spacing between the bars works out to be varying between 2.3, 2.5, and 3.1 for the 8, 16, and 32 mm bars suggested. This ratio works out to be 2.2 in the case of ASTM recommendations. It is also known that the spacing of bars in structural members is also governed by the maximum size of the aggregate for this same reason of passing ability. To conclude, the J-ring should have a diameter of 300 mm and height of 120 mm with vertical bars of different diameters and spaced at different intervals in accordance with normal reinforcement considerations, 2.5–3 times the maximum size of the aggregate. In a recent investigation in the laboratory, 14 mm diameter bars of 20 numbers were used for the maximum aggregate size of 12 mm with the spacing between the bars being around 2.75 times the maximum aggregate size. These factors significantly influence the passing ability, the flow spread, and flow time, and comparisons without taking them into consideration may not be totally relevant. It is probably appropriate at this stage that for most concretes containing aggregates up to a maximum size of 20 mm the 16 numbers of the 16 mm diameter ring suggested by ASTM may be more or less appropriate. One further factor that may also be of relevance is that some researchers suggest the use of deformed reinforcing bars for the J-ring, and in such cases one should account for the nominal diameter (approximately a couple of millimeters above) as the obstructing diameter. It may also be noted that for the high-strength high-performance characteristics to be achieved the maximum size of coarse aggregates could be limited to 16 mm as in the case of the DIN specifications.

It is only appropriate at this stage to point out that as already stated no single test is complete in itself to fully portray the different characteristics of an SCC. It will be interesting to see if it is possible to integrate a few of the test methods already reported for establishing the different characteristics of SCCs to arrive at something reasonable. There have already been attempts in this particular direction like the Orimet flow time test proposed by Bartos (1998) and later investigated by Sonebi (2002) with a J-ring to name one that is an integration of the reversed slump cone or the V-funnel discharging into the J-ring. The second one is a simple modification of the slump cone as an extension into something like a Marsh cone for mortars that was proposed by DAfStb (2012), discussed earlier to measure the flow diameter as well as the time to flow through an orifice. Keeping in mind that the flow diameters that have been discussed so far are based on the volume of concrete involved in a slump cone, with or without the J-ring around it to obstruct the flow, Figure 8.3 presents a reversed slump cone with a 10 mm extended tube at the bottom placed exactly 100 mm above the bottom plate in the middle of the J-ring by a specially designed support system. The bottom tube contains a 10 mm reinforcing rod obstruction placed perpendicular to each other at two levels as shown. The opening mechanism for the concrete to flow is situated exactly in between the tube and the cone so that the volume of concrete involved is exactly the same as that in the slump cone study. The apparatus can measure the slump flow with J-ring, the level differential inside and outside the J-ring, and the time for emptying in an obstructed flow path through the tube so that many of the parameters of interest could all be assessed in one go. The two obstructing 10 mm reinforcing rods at the different levels can be replaced with two rods at each level placed perpendicular, for smaller aggregate sizes like 12 mm used in SCCs of higher strengths.

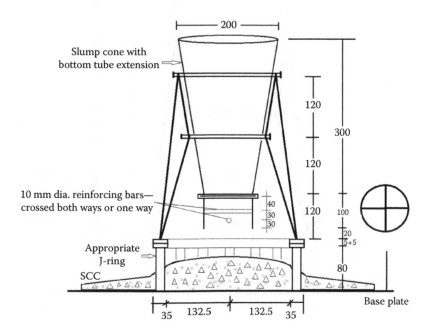

FIGURE 8.3
Proposed inverted slump cone apparatus for comprehensively evaluating SCCs.

Further details about the various assessments that could be done are not discussed as they appear to be absolutely obvious and also because as yet an effective calibration of the apparatus is not available. In sharp contrast to this apparatus, a slightly modified system is presented in Figure 8.4 in which the slump cone is extended at the bottom to reduce the mouth opening to 80 mm. A minor change is that the opening trap is shifted to the bottom of the tube to be able to fill the apparatus without being constrained by the requirement to keep the volume of the concrete constant at that of a slump cone. Here also the reinforcement could be appropriately utilized as discussed earlier. Naturally, the results of this inverted cone apparatus cannot be directly correlated with that of the slump flow or T_{50} but with the constriction due to the smaller opening, the reinforcement obstructions and a slightly larger volume of concrete (which can even be increased further by extending the cone at the top by say another 25 or even 50 mm) will give a much larger time for emptying so that the measurements can be more accurate and the spread and segregation aspects will also be more prominent. In fact, the increase in each of the measurements possible will probably make the recognition of the various

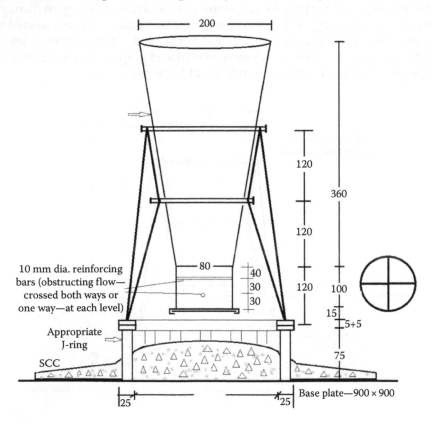

FIGURE 8.4
Proposed modified inverted cone apparatus for comprehensively evaluating SCCs.

facets of the behavior of the SCCs being studied more accurate and recognizable. As was the case earlier, even with this apparatus, further details about the various assessments cannot be presented as an actual calibration of the apparatus has not been made.

In an earlier part, the guidelines on the consistency requirements for the different member types as perceived for the requirements in precast industry were adequately discussed. There have been several scenarios that were comprehensively looked at to prescribe an appropriate rank or class for each of the flow characteristics—slump flow, U-box, T_{50}, L-box, V-funnel, and J-ring (Table 3.6). Maybe at this stage it is only appropriate to recapitulate the broad outline of the test methods and the parameters that they denote be compiled comprehensively. Table 8.1 presents a comprehensive view of the various test methods for establishing the different characteristics along with the appropriate values to qualify in the SCC category of concretes.

The above review and discussions on the test methods, their outcome, and the appropriate range of values to be qualified under an SCC will go a long way in understanding both the results available in the literature and also in putting a fix on the concretes that are utilized in the construction scenario. However, apart from looking at the various provisions to ensure self-compactability of the concrete composites, there is a need to understand broadly the applicability and interrelations between the various tests that are used to assess the level of self-compactability of these composites. It is imperative to understand that the parameters assessed being different, with the tests looking for filling ability, passing ability, and segregation resistance of the composites, a direct comparison of the results is probably not possible. However, a few broad relationships could be arrived at, if and only if the designed composites are appropriately modulated to ensure that the maximum size, proportioning, grading, and distribution of aggregates do not cause serious or perceivable difficulties to the self-compacting ability of the composite and that it also contains sufficient powder content to ensure reasonable control of flow, bleeding, and segregation of the composite.

In this context, the report published by the Nordic Innovation Centre on test methods for SCCs based on the characteristics of the concretes produced by a few companies in the region is an important contribution (Pade, 2005). The report also contains the Nordtest NT BUILD Proposal for Quality control of fresh SCC—workability, air content, density, and casting of test specimens as an attachment. The report tries to find the broad correlations between slump flows and T_{50} with and without J-ring, the J-ring blocking factor to segregation index relation along with an indication of the entrained air. Tables 8.2 through 8.5 summarize a few of these factors as perceived broadly while presenting some of the questions proposed in the report. The report incidentally presents the direct relationships of the various parameters wherever possible, and in places where the effects are highly varying it has shown the trends of the same graphically. The tables offer a look through the maze of data and graphs and help correlate the possible relations between the different parameters with a

TABLE 8.1

Test Methods and Acceptance Criteria for Self-Consolidating Concretes

Parameter	Test Method	Acceptance Criteria	Properties Evaluated
Flowability/ filling ability	Slump flow	600–800 mm, average flow diameter	Filling ability, flowability, segregation, and bleeding
	T_{50} (Slump flow)	2–7 s, time to flow 500 mm	Filling ability, consistency, cohesiveness
	V-funnel	6–12 s, time for emptying	Filling ability, viscosity, segregation
	Orimet	0–5 s, time for emptying	Filling ability, passing ability
	Kajima box	Visual filling, visual flow	Filling ability, passing ability
Passing ability	U-box	0–30 mm, difference in heights	Passing ability, filling ability, blocking effect
	L-box	0.8–1.0, ratio of heights	Passing ability, flowability, blocking effect
	J-ring	0–10 mm, difference in heights	Passing ability, flowability
	Orimet + J-ring	0–5 s, Time for emptying	Passing ability, flowability
	Kajima box	90%–100%	Visual passing and filling
Segregation potential	Settlement column	>0.95, segregation ratio	Shear yield stress, plastic viscosity
	Sieve stability	5%–15% sample passing through 5 mm sieve	Shear yield stress, plastic viscosity
	Mesh test	% passing	Segregation resistance
	Penetration	<8 mm, penetration depth	Aggregate settlement
	V-funnel at 5 min	+3 s, time for emptying	Segregation resistance
Viscosity	T_{50} (Slump flow)	2–7 s, time to flow 500 mm	Filling ability, consistency, cohesiveness
	T_{20} (L-box)	2–7 s, time to flow 200 mm	Flowability, viscosity
	T_{40} (L-box)	2–7 s, time to flow 400 mm	Flowability, viscosity
	V-funnel	6–12 s, time for emptying	Segregation resistance
	Rheometer	Torque resistance	Viscosity
Bleed test	Bleeding	Bleed water and time	Bleeding/rate
Shrinkage	Bar/ring	Length/stress	Drying
Site acceptance	Flow	Flow resistance	Passing ability through narrow spaces

TABLE 8.2

Parameters Relating to Slump Flow

Parameter[a]	Slump Flow (Normal Cone) (mm)							
	<500	500	550	600	650	700	750	800
Slump flow (inverted cone), mm	—	500	530	570	610	650	—	—
				($y = x$)				
Slump flow (with J-ring), mm		475	525	575	625	675	725	775
		($y = 1.2x - 180$; maximum 50 mm less)						
No. blows (wooden mallet)[a]	25	(15)	10	(7)	5	(2)	0	0

Source: After Pade, C., Test methods for SCC, Nordic Innovation Centre, project number: 02128, December 2005, 56p.

Note: The values in the tables are the ones proposed after looking at the corresponding graphs. However, the equations given are the ones from the report.

[a] Values in brackets are the expected interpolated values.

TABLE 8.3

Parameters Relating to Slump Flow T_{50}

Parameter	Slump Flow T_{50} (Normal Cone) (s)				
	2	3	4	5	6
Slump flow T_{50} (inverted cone) (s)	3	4	5	6	7
			($y = x$)		
Slump flow T_{50} (with J-ring) (s)	3		6		9
			($y = 1.5x$)		

Source: After Pade, C., Test methods for SCC, Nordic Innovation Centre, project number: 02128, December 2005, 56p.

TABLE 8.4

Parameters Relating to Slump Flow With J-Ring

Parameter	Slump Flow (with J-Ring)						
	500	550	600	650	700	750	800
J-ring block (mm)	40		25		15		0
			(Substantial variation)				
Segregation index (%)	0	0	0	10	20	50	90

Source: After Pade, C., Test methods for SCC, Nordic Innovation Centre, project number: 02128, December 2005, 56p.

TABLE 8.5

Parameters Relating to J-Ring Block

Parameter	J-Ring Block (mm)			
	10	20	30	40
Segregation index (%)	50–70	20–30	0–10	0

Source: After Pade, C., Test methods for SCC, Nordic Innovation Centre, project number: 02128, December 2005, 56p.

broad physical understanding of the phenomena involved. It should also be said that while taking a fix on these parameters related to slump flow with or without J-ring and the blocking effect of J-ring, the results that are seen to be essentially outliers of the system (maybe due to the effect of aggregate size, grading, and distribution as well as the paste and powder content deviations involved) are avoided to present a slightly different version of the same.

Table 8.2 presents the slump flow relationships between the normal: the inverted cone in the first place. One important aspect that needs to be kept in mind is that the inverted slump cone technique was first introduced for fibrous concretes with time of flow as the parameter, and with the cone lifted by about 100 mm. Its relevance to SCCs can at best be said to be limited, the effect being limited to the fact that the initial surface from where the concrete has to spread or flow reduces from 200 to 100 mm, so long as it is not left to fall from any height. While it was felt that there is not much difference between the two, the data indicates slightly lower slump flow values, with the reduction increasing with decreasing fluidity. In case a J-ring is introduced into the normal slump flow measurement, the restricted area flow at the ring periphery will reduce the slump flow. Neglecting the outliers, it is seen that the reduction in the slump flow of diameter is about 25 mm for most SCCs. It is also a fact that the viscous nature of SCCs can entrain a larger quantity of air particularly at higher concentrations of superplasticizers and VMAs. Nevertheless, a simple tapping by a wooden mallet on the side of the container could release a part of this air in SCC, and the number of blows for this, as a measure of this additional air which could escape, is indicated by no further subsidence. This is also presented in the table, which can be seen to be decreasing and becoming nil as the fluidity increases from 500 to 750 mm slump flow. Table 8.3 indicates that though the report suggested that there is no major effect due to an inverted slump cone test, a close look indicates that the T_{50} values could increase by about a second or two. However, the T_{50} values will increase by 1.5 times when the J-ring is introduced, and these effects are due to the reasons that were discussed earlier in Table 8.2. Table 8.4 presents the possible blocking heights in the J-ring at the different slump flow levels with the J-ring in place, which, as can be expected, was decreasing with increasing fluidity. However, it is important to note that the blocking effect is highly influenced by the maximum size, shape, amount, grading, and distribution of coarse aggregates. The substantial scatter in the test results reported is naturally because of this. For the same reasons, the segregation index was also showing a very large scatter. It is important to note that as was reported the greatest challenge in SCC production is to avoid segregation and there is no reasonably acceptable method for assessing the tendency to segregate. The sieve test presenting the amount of concrete passing a 5 mm sieve is probably not ideal. In this project, the segregation index was evaluated based on the difference in blocking step between successive J-ring tests on SCC at the top and bottom of a bucket that has been resting for 2 minutes, unlike that in the earlier table (Table 8.5). The segregation indicator is the relative difference in blocking step between the two J-ring measurements, which appears to be more reasonable. These results also show a significant scatter for reasons already explained earlier as can be understood. Table 8.4 presents the same segregation index in relation to the J-ring blocking parameter that was discussed earlier. The two tables indicate that segregation is critical for concretes of high fluidity as can be expected.

In a very similar effort, Hwang (2006) studied SCC mixtures with air entrainment for the interrelations between the various consistency parameters like slump flow, V-funnel, L-box, and filling capacity apparatus. The mixes contained 475 kg/m³ of cementitious materials with 25% Class F fly ash and 5% silica fume and the two water–cement ratios of 0.35 and 0.42. Also, the maximum aggregate size was limited to about 10 mm with the high-range water-reducing admixture (HRWRA) and air-entraining admixture (AEA) content adjusted to an initial slump flow of 660. The material was indeed found to be highly suitable for repair application. They also presented typical relations between the various consistency measurements and based on these proposed combined test methods and recommended workability values to effectively address SCCs for repair applications.

The consistency characteristics that are required for different applications are really a complex mix of the filling, passing, and segregation resistance as was already recognized by the EFNARC. The preferred choice of combinations for some construction requirements like ramps, walls and piles, tall and slender structures, and floors and slabs have been presented earlier by Walraven (2003). A general correlation of these with the actual values expected for these classes as defined (EFNARC, 2005) has already been presented in Figure 3.10. It may not be out of place to show actual relationships as they are seen in this interpretation to have a clear idea of the broader correlation that is possible to a discerning reader. It should be noted that these are not based on direct experimental results but do have a strong background of the committee that has gone through several iterations before suggesting these limits. The present contribution is an attempt only to the extent that the intermediate values were filled in to arrive at a certain set of possible correlating relationships as presented in Figure 8.5. A close look at Figure 8.5 clearly indicates that the estimated or proposed intermediate values for the different characteristics like

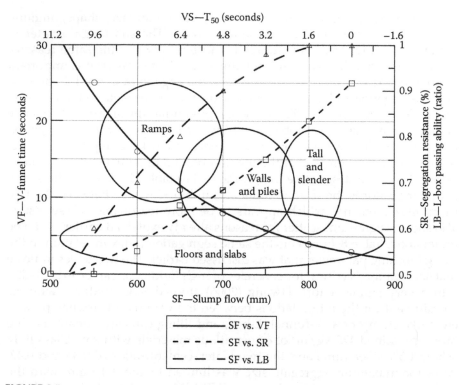

FIGURE 8.5
Correlation between consistency characteristics of SCCs. (After Walraven, J., Structural applications of self-compacting concrete, *Third RILEM and International Symposium on Self-Compacting Concrete*, Reykjavik, Iceland, 2003, pp. 15–22.)

slump flow, T_{50}, V-funnel time, segregation resistance, and L-box ratios have certainly resulted in acceptably continuous variations on these parameters to justify the intermediate variations in the parameters that are under discussion.

As was already discussed, Kheder (2010) reported the fresh state characteristics of a total of 30 concrete mixes containing limestone powder at various replacement levels. The study presents the results of slump flow and L-box characteristics systematically in terms of slump flow, T_{50} time in slump flow, L-box ratio, and also the T_{20}, T_{40} times of the L-box apart from other parameters like strength that will be addressed separately at a later stage. An attempt to examine the relationships between the various green state characteristics as reported by him resulted in Figure 8.6. It is obvious from Figure 8.6 that the slump flow is related to the T_{50} time in slump flow through a linear relationship while it is related to the L-box height ratios in a marginally nonlinear way. The L-box height ratio in itself is related almost linearly to the flow time T_{20} while the flow time T_{40} appears to be more scattered and nonlinear. This particular difference between the flow times T_{20} and T_{40} in the case of the L-box can be easily understood by looking at

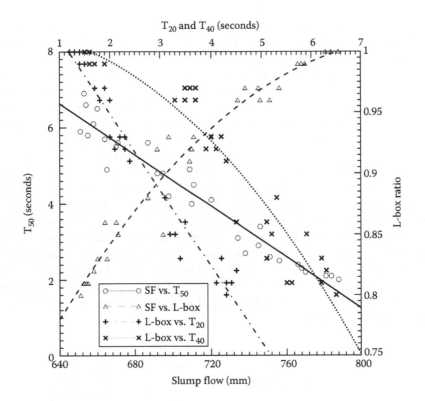

FIGURE 8.6
Relationships between slump flow and L-box characteristics of SCC.

the consistency levels, which indicate clearly that for L-box height ratio up to 0.97 the T_{40} time is also linear and is almost parallel to the T_{20} relation (a difference of approximately 1.8 seconds) but for high-fluidity concretes the difference between them keeps reducing, ultimately becoming negligible. This means that the difference between T_{20} and T_{40} is about 2 seconds for the operative L-box ratio and the corresponding slump flow ranges of 0.75–0.97 and 640–750 mm.

A fact that should be remembered about the previous discussions and the relationships that were presented is that they are primarily the results of investigations relating to SCCs utilizing limestone powder as the powder extender. As was already observed, physical characteristics play a vital role in these flow characteristics and the behavior of physically angular materials like limestone powder and GGBS obtained through crushing and powdering in roll presses or ball mills could be significantly different from that of the essentially spherical materials obtained during the pyrocondensation of flue gases. In view of this, it is felt that the accuracy of these relationships should be verified with data from other investigations, particularly with other materials if possible.

In line with this objective of establishing how far these relationships are valid for other materials used in SCCs or for other studies in different environments, the results of the green state characteristics from the studies by the group Guneyisi (2008, 2010) and Gesoglu (2007, 2009) discussed earlier were superimposed on these relationships individually. It is known that the studies contain a total of 65 concrete mixtures at the two water–binder ratios, 0.32 and 0.44, with the total cementitious materials contents of 550 and the rest at 450 kg/m³. The mixes used both binary and ternary mixes of fly ash and GGBS with replacement ratios at 20%, 40%, and 60% along with silica fume and metakaolin at 5%, 10%, and 15% by weight of the total binder in various combinations. Figure 8.7 presents a picture of these slump flow and T_{50} results superimposed on the earlier linear relationship of the SCCs made with limestone powder additions. The relationship between the two parameters was also presented in the figure. It is clear from the figure that the earlier relationship is essentially an upper bound solution and most of the finer pozzolanic admixtures like silica fume and metakaolin or even fly ash show lower T_{50} values at any specific slump flow. To be more broad-based, an approximate band with the upper and lower bound possibilities was also presented in Figure 8.6.

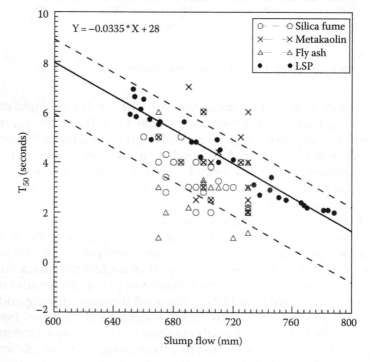

FIGURE 8.7
Variation of T_{50} with slump flow.

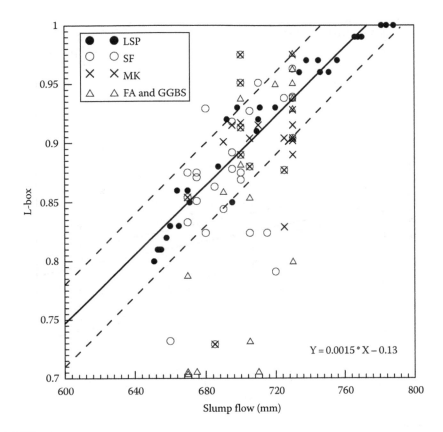

FIGURE 8.8
Variation of L-box ratio with slump flow.

Similar to this the results of the L-box ratio variation with slump flow are also presented in Figure 8.8. To simplify the earlier nonlinear relationship assumed was replaced by a linear relationship along with the approximate upper and lower bound limits as was done earlier. The superimposed results of Guneyisi (2008, 2010) and Gesoglu (2007, 2009) appear to be almost in line with the predictions of the limestone powder SCCs of Kheder (2010). The actual relationship as predicted by the figure is also given in Figure 8.8 for easier access.

As could be the case with most experimental investigations by different researchers, the parameters chosen for the assessment of the self-compacting capabilities of concretes could be different as in the present case. The earlier two figures could be generated with the slump flow, T_{50}, and L-box height results observed for the concretes investigated by both the programs. However, Kheder's study did not contain the V-funnel results for these concretes and thus cannot be compared with the present investigation. Even so, the possible V-funnel variation with slump flow

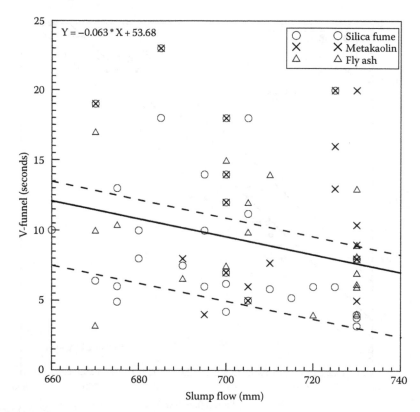

FIGURE 8.9
Variation of V-funnel time with slump flow.

along with the broader limits is presented in Figure 8.9 from the results of Guneyisi (2008, 2010) and Gesoglu (2007, 2009). Though the experimental values show a considerable scatter, it can be seen that the band proposed is kept essentially at the lower bound region so that the results of the more highly thixotropic concretes are avoided. For at least having an approximate estimate, the corresponding relation is also included in the figure.

Similar to the variation of V-funnel time with slump flow presented above, the variation of the V-funnel time with both T_{50} and L-box heights (Figures 8.10 and 8.11) also shows a large scatter, but to have a general assessment of the trend, the lower bound band was chosen for the relationships that could be relevant in these cases also. It can be seen that the bands in all these three figures encompass almost about 80% of the results, suggesting that a reasonable level of confidence can be placed on these relationships. It is possible that with the availability of more comprehensive investigations on different materials some of these relationships could be refined to give better and more reliable predictions.

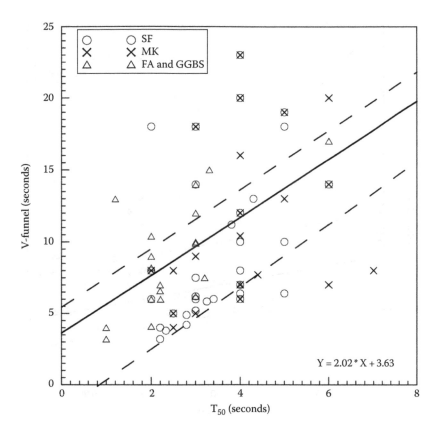

FIGURE 8.10
Variation of V-funnel time with T_{50}.

8.5 Effects of Quality and Quantity of Cementitious Materials

The quantity of cementitious material needed to ensure the required paste for coating the aggregates adequately and for filling the void spaces in the aggregate skeleton is an important parameter. This should be marginally higher than the minimum required ensuring adequate flexibility of the matrix yet should not be so high that it will lead to substantial amount of shrinkage and maybe even segregation and bleeding. The proposed range for this as given by EFNARC (2005) is 380–600 kg/m³ for low-end pozzolanic admixtures or simple powder extenders particularly for achieving concretes of low- and medium-strength grades. For enabling high- and very-high-strength concretes also the same range could be acceptable if one uses high-end pozzolanic admixtures such as silica fume. At this stage, one can still observe that while investigating concretes containing silica fume

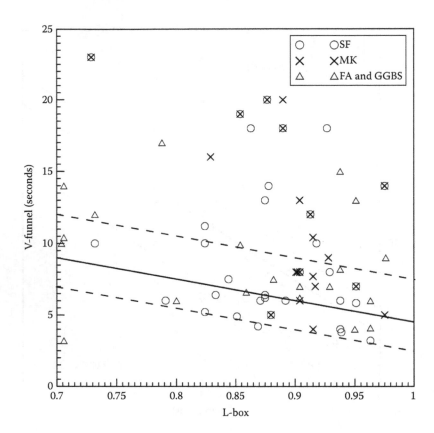

FIGURE 8.11
Variation of V-funnel time with L-box.

Sonebi (2009a,b) proposed cement contents as low as almost 200 kg/m³. The ACI recommendations also suggest cement contents of the same range. To be more appropriate, it is felt that the range of cement contents should also be looked at in conjunction with the other parameters such as water requirement and strength characteristics of the concrete to be produced. Some of these aspects have already been discussed at length in one of the previous chapters while trying to look for the appropriate water content that is required for ensuring the desired strength characteristics. Needless to say that some of the research efforts in trying to be more assured of the self-compactability have used much higher cementitious material contents, which may be a concern in terms of the creep, shrinkage, and temperature effects of such concretes. Additional aspects regarding this are discussed in the following sections as it is convenient to explain the same in relation to the materials, compositions, and methodologies utilized in their experimental investigations.

8.6 Wetting Water Requirements of Powder Materials

The need for additional water for wetting the increased fines in the case of SCCs was always understood though it was never essentially quantified, until probably an explicit method for this particular aspect was proposed by Marquardt (2002). In fact some of the methods suggest that an appropriate water content as suggested by ACI 211.1 (Su, 2001) be adopted in design. It is also a fact that with the available high-efficiency polycarboxylate-based superplasticizers the water content can be modulated to suit the requirements of strength and the cementitious materials content within reasonable limits. However, the experience is that very low water contents require an extremely large amount of superplasticizers (over 2%–3%) making the concrete highly economical. It is also not clear if it is a superplasticizer or the water in the superplasticizer that makes this difference toward achieving the required flow as superplasticizers contain only 35%–37% solids and the rest is water. For this reason, it is suggested that the quantity of the superplasticizer as a liquid can simply be added to the water content for determining the water–cement ratios required for strength calculations.

Through a study on SCCs containing pulverized fuel ash or fly ash at levels up to 80%, Liu (2010) observed that while achieving the same filling ability the superplasticizer dosage reduced considerably and this was attributed to the lubrication effect of fly ash. Also the superplasticizer's repulsive force effects may not be acting on fly ash. While this is partly true, at least to the extent that the spherical particles of fly ash invariably promote the ball bearing effect promoting lubrication and that there may not be a major influence on the deflocculation of the cement agglomerates through ionic repulsions and steric hindrance effects of the more modern polycarboxylic ester based superplasticizers, the fact that the cementitious material contents reduced from 539 to 439 kg/m^3 and the cement content reduced from 539 to 115 kg/m^3 at 80% replacement is certainly noteworthy though there is a small reduction in the water content from about 184 to 170 kg/m^3. In view of this, a direct comparison as is being attempted is probably not justifiable. It is enough to say that the cementitious material contents of 450 to 650 kg/m^3 and water content in the range of 180 kg/m^3 with about 1% superplasticizer will indeed be adequate in ensuring SCCs with a slump flow of over 700 mm. He also compared the superplasticizer dosage variation with the fly ash replacement ratios of Sukumar et al. (2008), showing that they also follow a very similar trend.

Apart from this, several attempts have been made to understand the superplasticizer compatibility and requirements through studies on cement pastes and mortars by a few researchers. The development of high-strength high-performance concrete needs a specific understanding of the behavior of the primary cementitious composites consisting of cement and pozzolanic admixtures at different percentages of replacements. Specific studies on cement pastes and mortars consisting of cementitious compositions are criti-cally analyzed and their effects are discussed. In fact, these are a part of a much larger investigation into the behavior of such cementitious composites to have a comprehensive idea about the behavior of fly ash and silica fume as cementitious materials in concrete.

The additional powder requirement of SCCs being obvious, the additional wetting water requirements of the increased fines content are always a mat-ter of concern. It is also well understood that these increased fines or for that matter a fairly large proportion of the fines could be made up of differ-ent pozzolanic admixtures of varying physical and chemical characteristics like particle size distributions, silica alumina and calcium oxides, and amor-phous glass content, to name a few. While the physical characteristics directly influence the wetting water requirements, the chemical characteristics play a large role in defining the superplasticizer compatibility and requirement, flow retention, setting, and finally the strength and strength gain characteris-tics of the cementitious mass.

Studies on pastes and mortars are probably the simplest to be done to under-stand some of the basic properties and effects of cementitious compositions con-taining pozzolanic admixtures. These are very effective in assessing the physical modifications like the effect of water content on consistency and the chemical effects like setting times as well as the effects on strength. The tests basically need the smallest amount of materials and simple gadgets like a mortar flow table and mini-slump cone or even a flat glass plate to study the flow characteristics. The mini-slump cone test, as the name indicates, is a smaller-scale version of the Abrams truncated cone and has a top diameter of 19 mm, a base diameter of 38 mm, and a height of 57 mm needing only a sample of 40 mL per test; this was proposed earlier by Kantro (1980) for studying the influence of water-reducing admixtures on properties of cement pastes. The average spread diameter of the paste measured in two perpendicular directions is taken as a measure of the flow behavior of the paste. These types of tests are obviously very simple but require good titration skills with a burette for accurate and reliable results.

The need for understanding the effects of different pozzolanic admixtures at different proportions in the cementitious matrix particularly with the enormous variety of these constituents is always a fascinating subject and one that has eluded a simple or even a very complex solution in more ways than one. Also the enormous amount of research literature that is available on concrete composites presently has not really clarified the significantly dif-ferent behavior of the low-end pozzolanic admixtures like fly ash and GGBS and the higher-end counterparts like silica fume and metakaolin, particu-larly their applicability and effectiveness in ensuring high-performance con-crete composites of various grades. This is not to say that either the research is lacking or that there is no understanding of the behavior of these cementi-tious materials in concrete in any way, but the need and the economy of their use in different situations still require some clarity. It is absolutely obvious that lower-end pozzolans are a great help in enhancing the powder content of normal-strength concretes that are required for general-purpose applications reaching up to the levels of marginal strengths without much difficulty with the aid of superplasticizers. In contrast, higher-end pozzolans with their better cementitious efficiency could significantly reduce the cement and water contents required for a particular strength, making them most appropriate for the production of high- to very-high-strength concretes. While these are facts that are already known, their application in enabling SCCs with the entire range of possibilities is not seen in the research efforts of today. The second factor that is of interest here is that there are specific limitations associated with the use of pozzolanic admixtures in cementitious composites, which were also not appropriately recognized and addressed in ensuring SCCs of either high strength or higher performance. By using about 30% fly ash and a higher amount of powder content, the resulting compact mass was assumed to be of high performance automatically. In view of this, a few fundamental and simple studies that project both the limitations and advantages with pozzolanic admixtures, particularly with reference to fly ash and silica fume (to represent both extremes of efficiency), are examined here.

8.6.1 Superplasticizer Requirements of Pozzolanic Cementitious Mixtures

In the first place, it is understood that the behavior of neither cement mixtures nor mortars containing them is in any way a perfect representation of the behavior of concrete. The varying mass distribution of these in actual concrete with the corresponding shrinkage and temperature effects apart from the effects of the interfacial transition zone behavior and even the water lenses that could be associated with the larger coarse aggregates in the system are obviously significant differences. Even so there is certainly a broad understanding that can be had, which cannot be and may not be a direct representation or correspondence to the field concrete. In view of this, the paste and mortar studies are designed in such a way that they represent a correspondence within themselves and can be compared appropriately without much difficulty, which probably will be obvious if one looks through the entire experimental procedure adopted for such an evaluation. The factors that are studied in principle are the effect of the pozzolanic admixture content in the cementitious material in terms of the plasticizer requirement and water demand, flow characteristics, initial and final setting times, and finally strengths of both the cement mixtures and mortar combinations. It is obvious that such a well-directed, simple, and comprehensive investigation could lead to a significant understanding of the problems associated with pozzolanic admixtures, though not in the best quantitative terms as would be needed in the final concrete mixtures. However, such a broad understanding will go a long way in ensuring an appropriate utilization of these materials in structural concrete and in modulating the same for applications of both cast in situ and precast applications proposed.

In the first place the varying fineness of these pozzolanic admixtures, with the fly ash having a fineness just a little above that of cement (with an average particle size of about 10 µm resulting in the fineness of around 4000–4500 cm^2/g compared to that of around 3000 cm^2/g of cement) and the silica fume or microsilica exhibiting a fineness of about 20,000–30,000 cm^2/g with an average particle size below 1.0 µm, will indeed need significantly different wetting water requirements to start with. In one such effort, paste studies were conducted on fly ash at different percentages in a C53 grade cement available in India, which in the present case was comparable to C42.5 of the Euro grade (Rao, 1996). To ensure an appropriate and reasonable comparison the water to total cementitious materials ratio was kept not only a constant but also at a significantly low value of 0.25, and the fly ash (or silica fume later) was replaced on an equal weight basis. To maintain the workability of the paste approximately constant, a naphthalene formaldehyde sulfonate–based superplasticizer of 50% concentration (the near highest concentration that was possible at the laboratory temperature) was used in all the studies. The low water–cementitious materials ratio and the highest concentration of the superplasticizer ensure that the parameters that are compared are indeed being studied at almost a constant water to the total cementitious materials ratio of 0.25.

The workability measured in terms of flow on a mortar flow table (ASTM C 33-84) without using any drops of the table was kept generally in the range of 150–200 mm by trial and error. It may be noted that the cementitious mass that was initially a sticky matrix up to a certain point changes suddenly to a flowing mass by just the addition of a couple of drops of the superplasticizer from the burette or pipette as the case may be. At this stage to ensure a good and almost equal flow with all mixtures, a couple of additional drops of the superplasticizer were added and the corresponding flow was measured. The fly ash replacements varied from 0% to 50% and efforts to have higher replacements resulted in difficulties with consistency and bleeding. The parameters studied were flow table spread, superplasticizer required for flow, initial and final setting times, density, and compressive strength at different ages on the 50 mm cube specimens. The flow table spread was generally in the range of 160–190 mm and the superplasticizer required at the w/c ratio of 0.25 for obtaining flow ranged from 0.3% to 0.5% of the cementitious materials (Figure 8.12). The pastes containing fly ash required only marginally higher dosages of the superplasticizer to obtain the flow regimes expected. It can also be seen from this figure that the variation of the flow spread at the different replacement levels of fly ash was fairly uniform over the entire range. These observations could be easily explained with the fact that the fly ash utilized was only marginally finer than the cement. This spherical nature of fly ash could have also helped in ensuring that there are no major surprises in terms of the superplasticizer requirements on the flow characteristics.

A very similar investigation was also undertaken at the appropriate levels of replacement of silica fume ranging from 0% to 25% in increments of 5% by

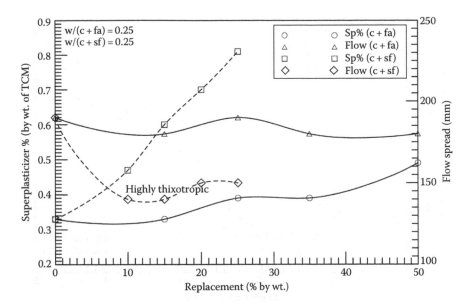

FIGURE 8.12
Effect of fly ash and silica fume on the plasticizer requirement.

Surya Prakash (1996) in line with the presently known practice of concrete technology. The methodologies adopted for the testing of both the cement and mortar mixtures were essentially the same as those reported earlier, thus enabling a broader understanding though not exactly at the same replacement levels for obvious reasons. The complete set of parameters as were investigated in the case of fly ash was also studied with these silica fume replaced cementitious compositions. Two specific factors stand out in the behavior of cementitious materials with silica fume compared to that associated with fly ash. First, even though the silica fume replacements were significantly smaller than those of fly ash, there is a very steep increase in the superplasticizer requirements as the percentage of silica fume replacements increased, with the value reaching almost about 0.8% of the cementitious materials at 25% as can be seen. The other significant observation that can be made is the obvious fact that the material becomes highly thixotropic or viscous reducing from an initial flow of about 200 mm to about 140 mm.

8.6.2 Effect of Pozzolanic Admixtures on Setting Times

It is also obvious that the higher dilution effects of larger replacement levels of fly ash resulted in a substantial increase in the initial and final setting times of the cementitious materials containing fly ash. It was observed that both the initial and final setting times increased continuously from about 245 and 350 minutes for the C53 grade cement to about 505 to 580 minutes,

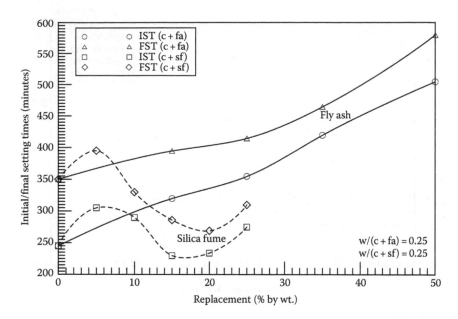

FIGURE 8.13
Effect of fly ash and silica fume on initial and final setting times.

with the fly ash replacements increasing from 0% to 50% (Figure 8.13). The cementitious materials containing silica fume presented a different picture, maybe due to the reason that the replacement percentages were significantly lower (only 0%–20%), thus not having any substantial dilution effects and also due to the fact that the higher pozzolanic reactivity of the significantly smaller particles with amorphous silica sites could indeed reduce the setting times at the optimum percentage levels of replacement showing the highest reactivity. The initial as well as final setting times increased marginally for the lower replacement levels of about 5% and 10% but dropped well below at percentages around 15%–20% where the reactivity is apparently significant. At the higher percentage of about 25% studied, the setting times seem to get back to those for the normal cement alone (Figure 8.13).

8.6.3 Effect of Pozzolanic Admixtures on Strength Characteristics

At the very outset it should be stated that this parameter is more relevant to the discussions in Chapter 9 on strength characteristics. However for the sake of continuity and comparison in discussions, these results are presented directly here. The compressive strengths of the 50 mm cement paste cubes made at different fly ash replacements show a continuous increase in strength with age, with the higher replacements showing a lowering of strength (Figure 8.14). It is also clear that the highest strength was obtained at 15% replacement level and

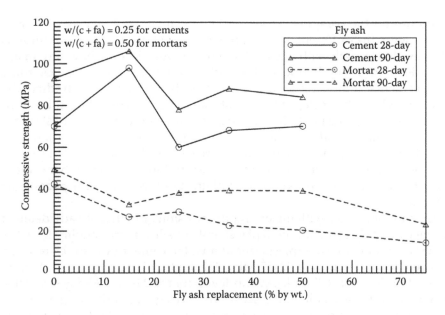

FIGURE 8.14
Effect of fly ash on compressive strengths.

at all percentages beyond 25% the strengths were essentially lower than those of the plane cement cubes, indicating the validity of efficiency factors as was described in the previous chapters. The mortar studies adopted the same C53 grade cement as was done earlier in the cement paste studies. The grading of sand proposed earlier by Ganesh Babu (Prakash, 1996; Rao, 1996) was adopted because it was resulting in higher strengths and densities than the other standard gradings in practice. The mix proportions used were 1:3 [(cement + fly ash) : sand] at a water to cementitious materials ratio of 0.50 as recommended earlier to have a specific understanding of the strength of cement in relation to the Euro standards. Cube specimens of 50 mm size compacted as per ASTM standards were adopted in the investigation. As in the case of cement pastes, mortars were also studied at the various replacements of fly ash on the basis of equal mass but over a wider range of replacements (0%–75%). All the constituents of these mortars were mixed in a mortar mixer. First the cement, sand, and fly ash were dry mixed thoroughly for about 90 seconds and then the water was added. No superplasticizer was used in any of these mixes. It was obvious from these investigations that in the case of mortars the simple replacement may not give an equivalent 28-day strength as that of normal mortars without fly ash. Even though fly ash mortars show lower early strengths, there is strength gain after 28 days. These studies on cements and mortars essentially indicate that an assessment of the efficiencies at the different percentages through them is indeed not the way forward, and studies on actual well-designed concretes are essential to get a complete, clear, and comprehensive picture of the behavior.

Studies on the compressive strengths of cement paste cubes of 50 mm made with the different silica fume replacements show a continuous increase in strength with age as it should be but did not show any major variations in strength due to the different percentages of silica fume replacements, indicating an efficiency of only 1.0, which contradicts the known higher efficiencies of these materials (Figure 8.15). Only at the minimum 5% SF replacement, there appears to be a slight lowering of strength indicating that the silica fume was used up in pore filling and there was very little effect of the later pozzolanic reactions. A more detailed discussion on the lower efficiencies due to the saturation effects at higher cement contents was presented earlier to explain these aspects with pozzolanic admixtures.

The mortar studies with mixer proportions similar to that already detailed in the case of fly ash with the replacements of silica fume ranging from 0% to 25% without any superplasticizer at a water–cement ratio of 0.5 resulted in strength values around 42–57 MPa. The highest efficiency was exhibited around the 5% and 10% replacements levels for the 28- and 90-day strengths, respectively. These results and discussions once again clearly bring home the fact that apart from the percentage of replacement, the water to cementitious materials ratio, grading, and compaction will all have a significant influence on the strength and the corresponding cementitious efficiencies of silica fume or for that matter pozzolanic materials in general.

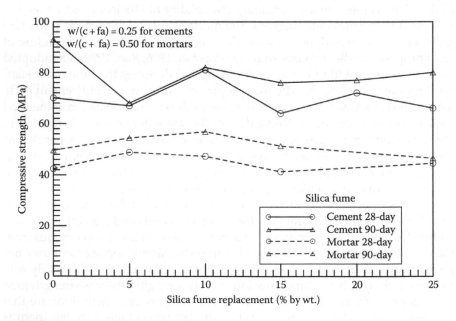

FIGURE 8.15
Effect of silica fume on compressive strengths.

8.6.4 Effect of Superplasticizer on the Water Requirement of Pozzolanic Cementitious Mixtures

The earlier paragraphs looked at some of the influences through a certain perspective. Now a study on the water requirements at different super-plasticizer levels (Ganesh Babu, 2005a) is considered. In brief, the water requirements of a specific quantity of cementitious material containing low-end mineral admixtures (both fly ash and GGBS at 0%, 10%, 30%, 50%, 70%, 85% by weight) and a high-end mineral admixture (silica fume at 0%, 5%, 10%, 15%, 20%) were investigated at the two different superplasticizer dosages of 1% and 2% by weight of cementitious materials. The studies were done with the help of a mini-slump cone described earlier. Initially, about half the quantity of water that is required for the standard consistency of cement along with the chosen amount of the superplasticizer is mixed in a small porcelain or glass bowl, and the additional water required for flow is determined generally by continuing to add water from a burette, which can be controlled dropwise particularly toward the end. As often repeated, a couple of drops actually render the sticky mass of cementitious material to its flowing consistency at the end and it is necessary to control only the total time for testing to ensure that water is not lost through drying during the test. The mini-slump cone, a smaller-scale version of the Abrams cone, consisting of a mold in the shape of a truncated cone with a top diameter of 19 mm, a base diameter of 38 mm, and height of 57 mm is filled with a sample of just 40 mL. The average diameter measured across two perpendicular directions on a smooth wet glass plate is a measure of the flow of the mixture. While it is very difficult to control the flow to be exactly of a particular value, it is generally ensured that in all the tests the mixture is free flowing and results in a spread of about 150–180 mm. It is possible to arrive at an accurate measure of the water required for a specific flow by repeating the tests a few times and plotting the results of the flow versus water content. In most cases, just trying to arrive at the point where the cementitious mass turns from a reasonably stiff mix to totally flowing consistency at the addition of a couple of drops is enough to define the water content required for a flow.

The water requirements at the various percentages of pozzolan in cementitious materials (fly ash and GGBS at 0%–85%; silica fume at 0%–20%) at both 1% and 2% dosages of the superplasticizer are presented in Figure 8.16. The figure shows that as the pozzolan content increases the corresponding water requirement increases only marginally for both fly ash and GGBS. However, the increase is very steep in the case of silica fume even at the much lower percentages, easily attributable to the very high specific surface as seen earlier. Even Figure 8.17, which shows the actual amount of water required for flow, presents a very similar trend for the different pozzolans discussed. One important observation is that even a twofold increase in superplasticizer content has not made any significant

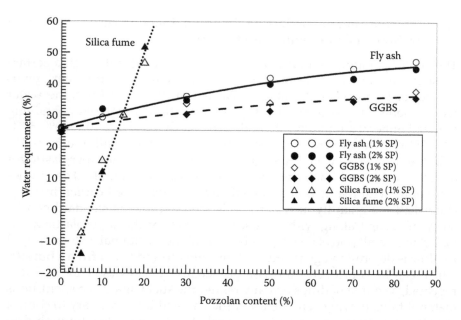

FIGURE 8.16
Water requirements at the various percentages of pozzolan.

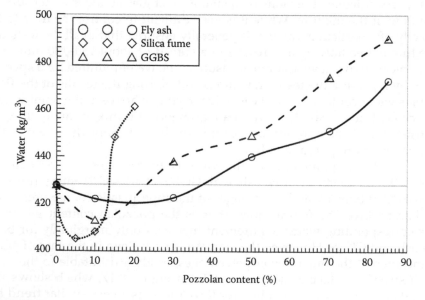

FIGURE 8.17
Amount of water required for flow.

difference to the water required to achieve the flowing consistency. This obviously is the proof for the often stated observation in several places that there is a minimum water content required for achieving flowing consistency given the constituents. Efforts by researchers to modulate this with larger quantities of the superplasticizer are not effective. In fact as already stated, it is the water in the superplasticizer that has been doing the needful and not the additional percentage of superplasticizer in any way. The final result of such efforts in using higher dosages of superplasticizers will only make the concrete more expensive. It is also important to understand that contribution to the water requirement of concrete from all the other constituents is at best marginal. The compressive strength variation with the water–binder ratio for the three different cementitious systems (at 2% superplasticizer dosage) can be seen in Figure 8.18. The perception of a few that such a study on cementitious components or on mortars can be used to depict to a scale the strength to water–cement ratio relations of concrete or for the evaluations of the efficiency is totally farfetched. This is clearly seen from the strength distributions of the three different cementitious matrices presented, clearly indicating that they are specific to the matrix on hand. However, the compressive strength variation with the percentage replacement (Figure 8.19) appears to present a reasonable picture of the percentages at which the equivalent and maximum strengths for the corresponding cementitious mixtures occur.

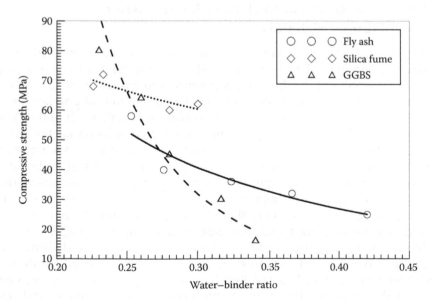

FIGURE 8.18
Compressive strength variation with water–binder ratio.

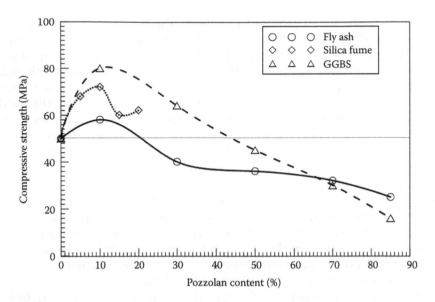

FIGURE 8.19
Compressive strength variation with the percentage replacement.

8.7 Effects of Granular Skeleton Characteristics and Fibrous Materials

The methodologies that were discussed earlier contain an explicit or implicit requirement that the granular skeleton modulated to have a suitable losing effect will be the most appropriate to ensure an SCC with the highest aggregate content necessitating the least amount of powder material and thus the least water content. This obviously will result in the most economical and also the most performance-effective self-compacting composite in its grade. While addressing packing density and packing characteristics as the way forward in most of these discussions till today, it is felt that the most effective, simple, and readily implementable alternative would be to ensure that the aggregate grading follows a well-defined combined grading curve. Even here there have been several alternatives suggested in terms of using a specific exponent to define percentages in terms of the maximum aggregate size (D/d ratio) or through the definition of both the maximum and minimum aggregate sizes being used (Anderson and Anderson model). It is felt that for most purposes with cement contents in the range of 500 kg/m³ a combined grading of coarse and fine aggregates conforming to the average of DIN-A and DIN-B (DIN AB) as suggested earlier (Marquardt, 2002) will ensure a satisfactory performance for all practical purposes. At this stage, it is absolutely important to point out that the use of 20 and 12 mm combinations for coarse aggregates as the normal practice is certainly not going to lead to an effective solution as it would require the aggregates between 6 and 2 mm sizes to fill in the space between the coarse aggregate and sand (almost invariably below 2 mm and in most cases around only 1 mm). The experience in the laboratory clearly shows that it is this fraction and this fraction alone that makes the difference between a nonsegregating SCC and that which could readily segregate in a mortar-rich concrete environment.

Apart from this, considering fibrous additions as a part of the coarse aggregate may be effective in addressing the volume constituents of the concrete mixture but cannot adequately address their interaction and potential blocking, balling, and even water requirements beyond a certain limit. In general, well-designed SCC appears to have the capacity to absorb about 1% fiber content without serious difficulty and could also be pushed to accommodate percentages up to 2% by marginal adjustments in fines and water content. One specific exception could always be steel fibers with hooked ends or heavily crimped fibers, which are very effective but should be limited to percentages below 1% unless explicitly designed for. One more important factor is that longer fibers beyond the maximum size of the aggregate will always pose specific problems and the way to address such situations where longer fibers are incorporated with a specific purpose is to ensure that their proportion is only a small fraction of the total fiber content while using the smaller fiber to supplement the additional needs in the matrix design. In the case of extremely thin fibers like carbon filaments, basalt fibers, or even polymer fibers of polypropylene, the addition by weight factions will mean an enormous surface area as the fibers are very thin and long, which will require additional wetting water as in the case of fine materials though they are highly flexible and may not interfere in the mixing processes. Even so at marginally higher percentages for their applications, they tend to inhibit flowability (slump flow) as was observed in a few experimental investigations reported earlier.

8.8 Segregation and Bleeding

Various aspects of segregation and bleeding problems have been discussed at different stages. In fact, these two aspects are essential prerequisites for the formulation, making, and use of self-consolidating concretes. It is enough to say that the additional powder content invariably ensures a certain viscous thickening of the paste phase that will help in supporting the aggregates in suspension. However, in case the powder quantity is not adequate particularly at a content below 400 kg/m³, the need for additional assistance through VMAs may be necessary. Also, for SCCs with slump flow beyond 700 mm, even concretes containing 500 kg/m³ powder content may still exhibit segregation and bleeding without the assistance of VMAs. Another factor that changes this dynamics is the type of powder extender itself. Materials like silica fume and some calcined clays, which are several times finer than cement and other powder extenders like fly ash, will impart a significant thixotropy to the system even at their commonly used level of only about 10%–15%. It is obvious that in such cases the need for a VMA may almost be nonexistent or even if required for very high slump flow regimes it will be very small. Many of these aspects can be judged or perceived only if the powder constituents and the required slump flow are available to the designer.

8.9 Shrinkage and Heat of Hydration

The need for a specific understanding of the shrinkage characteristics of SCCs emanates from the fact that the additional paste content along with its additional water requirements could certainly lead to higher drying shrinkage characteristics. However, considering the fact that in most cases the additional fines required are supplemented mostly through pozzolanic admixtures, which in any case will use the additional water in the system for producing secondary cementitious compounds, the effect of shrinkage at later ages after the permanent set may not be significant. Notwithstanding this one is advised to be vigilant about the possible settlement effects during the flexible matrix state. A further word of caution is that it may be required to ensure appropriate curing to replenish the loss of water due to the drying period to support the formation of the secondary cementitious compounds unhindered.

It is also to be recognized that SCCs are considered to be inherently of a higher performance not only due to the effectiveness of the well-compacted mass but, as indicated by several others, also due to the use of pozzolanic admixtures. The greatest advantage of having these materials in the cementitious system is that they reduce the initial peak of the heat of hydration due to the lower cement content possible and also help in spreading the hydration reaction for a much longer period, which automatically means that the thermal effects are also spread out helping lower thermal differentials that are actually sensitive for the performance of concrete. In a nutshell, most investigators working in the region of 400–600 kg/m³ cementitious materials in concrete have found that there is no appreciable effect in SCCs due to the problems associated with shrinkage.

8.10 Transport, Placement, and Finishing

The effectiveness of any material in a structural application finally revolves around its final characteristics in place. Obviously, the operations before transferring even a well-designed SCC, namely, the transport, placement, and finishing, significantly influence the performance of the material and also the structure. With the several lessons learnt, the procedures for transporting and placing the highly flexible normally vibrated concretes required to be pumped to significant heights are well known and are also understood. Even these normally vibrated concretes require a slump in excess of 130 mm to be effectively pumped at greater heights. There have been several methodologies suggested to ensure that the aggregates are uniformly suspended in the matrix such as slightly reducing the total water requirements for self-compatibility (including the superplasticizers required) through the transport process (with the transit mixer rotating at a lower speed) and ensuring the remaining water is mixed through vigorous mixing at site to ensure SCCs. This is a method that is often adopted. One serious drawback of the system is that the excellent control possible at the RMC plant is compromised, sometimes quite seriously, for ensuring operational convenience. If the system is to be approved, there should be a complete record and excellent quality control of the process and the various quantities of water and admixture added at various places to ensure that the resulting concrete is of the quality and strength that were proposed. It has also been known in admixture circles that for long haul and city operations, where the time consumed in transport is large, superplasticizers in the form of granulated powders (Melgran, a melamine formaldehyde sulfonate normally known as Melment, by SKW, Trostberg, Germany) were also suggested and used effectively. However, the need for additional quality control measures in all these methods has always been difficult to implement even in significantly large constructions. One word of caution in any case is that quality control and quality assurance are an absolute must for effective implementation of SCCs in large construction scenarios.

The filling and passing ability being the essential characteristics of an SCC, the ability to flow for the concrete is ensured. Reports suggest that the concrete can flow up to about 5 m without any difficulty for most well-designed concretes. But it also depends on the depth of the member, the obstructions or tortuosity in the flow path particularly in thin members, and the density of reinforcement. It is important to remember that time lapse between successive flows, if ever they were to be there, has to be appropriately addressed to ensure that there is no layering effect in the final product. There is also a need to understand that being a highly fluid mix, even minor openings in the formwork can lead to the loss of the paste matrix creating a honeycombed location. The problem of entrapped air or sometimes water pockets if there is excessive bleeding could also cause defects in the structure, which is very difficult to repair later. Pumping of SCC in most cases is not a serious problem. However, mixes with very high fines content and with a high differential in aggregate gradations could still cause problems of segregation and bleeding due to the pumping pressure. The joint bends in the pipeline system should be adequately cared for to ensure a smooth and uninterrupted flow of concrete during the pumping process so that the earlier mentioned problems of segregation and bleeding do not take place. It is of course possible to use a skip and crane for intermittent construction pockets effectively to ensure economy.

8.11 Formwork and Pressure on Formwork

The ability for self-compaction and free flow of SCCs has given rise to the feeling that with increasing depth of the member the possibility of exerting the full hydrostatic pressure is obviously a fact. However, concrete composites with the distributed particulate material suspended inside a highly thixotropic mass do not have the capacity to generate such pressures on the formwork. There have been several investigations to look at the parameters that influence the pressures on formwork, apart from the effects on other characteristics like the bond with reinforcement and the aggregate grading distributions across the depth. The pressure on the formwork was measured experimentally by using a laboratory test with a series of sensors at different levels on a 25×25×100 cm column by Georgiadis (2010). They also attempted to look at the water content variation effects on these consistency parameters of filling and passing ability. It was observed that SCCs build up an internal structure during the rest periods, maybe due to the effects of hydration, which helps decreasing the pressure on the formwork as the time progresses. Similar observations were made by Billberg (2001) earlier. In general, the pressure buildup on formwork due to SCC is obviously related to the rate of casting height and the rest periods during the traversing of the concrete pour along the structural member. However, for slender column members, this could still be a problem to be aware of and guarded against. Another factor that has not received so much attention is the effect of the reinforcement grill and the spacing of reinforcement, as this could generate a small amount of confinement effect, which will be significantly higher with time. It is also difficult to assess the effects in all their entirety as the material combinations, effects of superplasticizers, and effectiveness of VMAs over time are different with different cements and other components in the concrete mix. There is a need for establishing these characteristics at least briefly before a large construction activity is undertaken. The time gap for mixing to pouring at site, with the intermediate operations of transport and pumping, and so on, will significantly influence the initial and final setting times of cementitious composites, and these will also have a bearing on the effects of internal stiffening in the concrete mass. As can be seen, these transient phenomena need to be addressed individually for each particular project, which obviously means that quality control authority is required at site. In fact, such an independent body that looks after the various activities not encumbered with the daily routine of the construction process is advocated for all major constructions.

8.12 Setting Times and Removal of Forms

The use of pozzolanic mineral admixtures at their different percentages in the varying amounts of cementitious components in SCCs naturally makes a clear and specific discussion on this particular topic difficult. With the possibility of adding higher percentages of low-end pozzolans (more than 50% of fly ash) with a limited cement content, the reaction kinetics of the system will slow down considerably compared to that in which these pozzolans are at a much lower percentage even at the same cement content (about 20%–30% fly ash). The effect of the higher contents of these low-end pozzolans was seen earlier reflecting in the higher setting times and lower strength gain rates (Figure 8.13). While the earlier information was generated on cement pastes and mortars, Figure 8.20 shows the same in a different format and that too on the actual concretes. The figure presents observations on concretes containing 0%–85% fly ash in terms of the time required before attempting to demold the

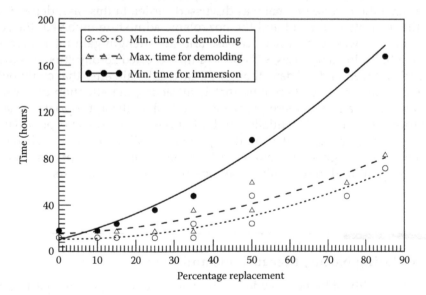

FIGURE 8.20
Effect of pozzolan content on hardening and curing needs.

specimens. Even at 50% replacement, the minimum normal demolding time of about 12 hours jumps to over 32–48 hours. The minimum time required for demolding is to see that the material is still soft in the mold, which would be observed if by pressing a thumb an impression can be made. Demolding at this point of time is certainly not possible. The difficulty of demolding could be reflected in the fact that the specimens spall, the sharp edges break off, and many a time the mortar sticks to the sides and bottom of the mold. The maximum for demolding is when there is no such difficulty as observed earlier and the specimens are all firm and exhibit sharp edges. A further observation at this point is that after demolding the specimens are either stored under wet gunny bags or placed in the curing tank for testing at the appropriate age. At higher percentages of replacement particularly beyond 35%, specimens literally crumble and appear to dissolve in water if placed in the curing tank immediately or we lose the edges under the wet gunny bags without further moist air curing for a further length of time as the case may be. This in the present case is termed as the minimum time for immersion in the curing tank. Figure 8.20 clearly shows that with about 75% fly ash content this time for immersion could be as high as 6–7 days. If these factors are not understood appropriately, the use of high-volume fly ash even in the prefabrication industry is going to be extremely difficult. The hydration dynamics and strength gain rate can be appropriately modulated through the use of higher temperature or even steam curing in the prefabrication industry.

In the case of high-end pozzolans like silica fume that start to react much faster due to their large surface area and distribution, and also due to the large quantity of amorphous silica that is present, the observed effects would be exactly the reverse of what was discussed earlier. In this case, depending on the quantity of cement and the pozzolanic admixture involved, the setting times as was seen earlier decrease and the strength gain rates increase substantially. In many cases, 90%–95% of the expected 28-day strengths may be realized well under 7 days though the strength gain after that could only be marginal. The one alternative that is attracting the attention of several researchers to alleviate these problems is to look at the appropriate ternary mixtures of cements. Essentially, the limitations of the low-end pozzolanic admixtures are compensated by using a very small quantity of their more expensive and higher-end counterparts. Some of these concepts have been effectively utilized in some high-rise constructions in recent years.

8.13 Curing Needs, Precautions, and Best Practices

The availability of higher powder content and the corresponding paste content requires that for addressing the possibility of drying and shrinkage problems an appropriate curing is used. With such powder contents and in cases where the SCCs contain higher quantities of additional powders or even pozzolans, it is essential not to cure through ponding techniques for reasons already discussed. Apart from this to avoid shrinkage and crazing cracks, early moist curing (even misting) or ensuring that the surface is covered is mostly essential, particularly in the case of large exposed surfaces like slabs. Typically all highly flowing concretes that have a larger water content will experience considerably large shrinkage stresses while drying that needs to be appropriately addressed through specific curing methodologies.

8.14 Effect of Accelerated Curing and Maturity Concepts

The traditional 28-day strength of the concrete is still the yardstick to define the quality of the concrete and many a time also its performance. However, it is necessary to evaluate the strength at the earliest possible point in time (maybe even before casting) for an appropriate corrective action, if required, as remedial measures to redress any decrease are expensive, time consuming, and often not totally perfect. Such an evaluation will not only provide guidance for the construction process but will also aid in predicting the long-term behavior. There have been several attempts to assess the 28-day compressive strength of concrete through accelerated curing methods for a long time. The fundamental principle in most of these attempts is to look at either a direct comparison of the 28-day strength with a specified accelerated curing procedure or to look for a more general methodology like the maturity concept. The product of time and temperature is termed as the maturity of concrete, and it is expected that concretes of the same maturity will have similar strengths. Exhaustive literature survey shows that there have been several researchers who worked on these aspects and some of them led to the various existing national provisions. Without going into the various research studies and their findings, limitations, and applicability, it is sufficient to say that the investigations were done to assess the combined effect of time and temperature on the hardening process and the strength development in concrete. Temperatures ranging from 35°C to 105°C have been investigated to have a broader understanding based on which a few national recommendations have been made. Table 8.6 presents a comprehensive overview of the different provisions in the American [5], British [6], and Indian [7] standards. These concrete standards suggest methodologies that can be used to provide an indication of the 28-day strength of concrete through its own heat of hydration or by thermal acceleration.

While information regarding the effects of accelerated curing on SCCs is scanty, the recommendations of the various codes could be used as guidelines. A word of caution is that the different mineral admixtures at their various percentages are not helpful in establishing definitive relationships.

TABLE 8.6

Specifications for the Accelerated Curing of Concrete

National Standard	Test Method	Curing Medium	Curing Start (After Casting)	Curing Duration/Temperature	Age at Test
ASTM C 684-99 (2003) (20°C–30°C)	Warm water	Hot water	Immediate	23.5 h ± 0.5 h (35°C ± 3°CC)	24 h ± 0.25 h
	Boiling water	Boiling water	23 h ± 0.5 h	3.5 h ± 0.08 h (100°C)	28.5 h ± 0.25 h
	Autogenous	Heat of hydration	Immediate	48 h ± 0.25 h (45°C ± 4°C)	49 h ± 0.25 h
	Autoclaving	Steam and pressure	Immediate	5 h ± 0.08 h (149°C)	5.25 h ± 0.08 h
BS 1881: Part 112: 1983 (20°C ± 2°C)	35°C method	Insulating water	Immediate	24 h ± 0.25 h (35°C ± 2°C)	25 h ± 0.25 h
	55°C method	Heating by water	1.5 h	19 h 50 m ± 0.33 h (55°C ± 2°C)	22 h 30 m ± 0.25 h
	82°C method	Heating by water	1 h	14 h ± 0.25 h (82°C ± 2°C)	16 h ± 0.25 h
IS: 9013-1978 1998 (27°C ± 2°C)	Warm water	Heating by water	1.5 h to 3.5 h	19 h 50 m (55°C ± 2°C)	22 h 50 m
	Boiling water	Heating by water	23 h ± 0.25 h	3.5 h ± 0.25 h (100°C ± 3°C)	29.5 h ± 0.25 h

However, the methodologies enumerated can be stated to be totally appropriate for use in establishing the necessary relationships for SCCs also. Such assessments are site and materials specific and have a high value in large construction sites or for the prefabrication industry. However, looking at the simplicity of maintaining the bath temperature at 100°C the boiling water method or steam curing method is more suitable for field applications.

8.15 Quality Assurance and Control

It is obvious that quality assurance and quality control are considered an essential prerequisite for all major constructions in recent years, particularly given the inadequate performance of several constructed facilities in recent years. SCCs in particular stand at the crossroads between high-fluidity, collapse slump flowing concretes and those that are now termed as self-compacting with a thin knife edge as the difference between the two if at all. Apart from the flowability, which is common to almost both these concretes, the passing ability and segregation resistance need to be more critically evaluated in the case of SCCs. Thus, most recommendations, of national or individual researchers, mention the need for a specific understanding and implementation of the quality assurance criteria to avoid performance failures in SCC. The fact that at the very start of this chapter on green state characteristics a broad outline of "Test methods and acceptance criteria for self-consolidating concretes" (Table 8.1) is introduced shows clearly the importance associated with quality assurance and control. It is obvious that the acceptance criteria proposed have to be implemented appropriately by a third party that has no association with the site activities in particular and is directly and independently responsible to the local governmental administrative authority through the owner of the constructed facility. Without going through this huge discussion on what constitutes this quality assurance and control protocol for SCCs, the entire sequence of what could be the various parameters of its operation is presented in Table 8.7. It is to be specifically noted that this table is only an indicator of the various tasks and the procedures involved in a broad sense, and for each constructed facility depending on the need, importance, and appropriateness a suitable formulation of the same has to be done. Even so, it is important to reiterate that part of the implementation protocol in terms of inspection and acceptance by authorities shall be appropriately ensured. These two components finally decide the resulting product and should have the capacity and competence to be entrusted with that particular responsibility, after looking at the specific competence for the role that they are to play.

TABLE 8.7

Quality Assurance and Control Protocol for SCCs

Task	Operations
Material selection	Cementitious components
	Cement—type, grade, compatibility
	Pozzolanic admixture—fly ash, silica fume, LSP
	Aggregates—size, gradation
	Superplasticizer—compatibility
	Viscosity modifier—if required
Mixture design	Proportions
	Cementitious—mineral admixtures percentage
	Self-compactability class
	Mixing sequence
	Transport and placement
Acceptance criteria	Green state
	Class—flowability, passing, and segregation tests
	Limits of variability for acceptance
	Hardened state
	Strength—standard, accelerated
	Performance
	Durability—absorption, permeability
	Corrosion—resistivity, potentials, RCPT
Implementation protocol	Sequence
	Inspection—plant, site
	Approval—plant, site

References

ACI Committee 309, *Recommended Practice for Consolidation of Concrete*, ACI 309-72, ACI Manual of Concrete Practice, Part 2, American Concrete Institute, Detroit, MI, 1984.

ASTM C 684-99, *Standard Method of Making, Accelerated Curing, and Testing for Concrete Compression Test Specimens*, American Society for Testing and Materials, West Conshohocken, PA, 2003.

Bartos, P., *Fresh Concrete: Properties and Tests*, Elsevier, 1992.

Bartos, P.J.M., An appraisal of the Orimet Test as a method for on-site assessment of fresh SCC concrete, *Proceedings of International Workshop on Self-Compacting Concrete*, Kochi University of Technology, Japan, 1998, pp. 121–135.

Billberg, P., Influence of filler characteristics on SCC rheology and early hydration, in *Proceedings of Second International Symposium on Self-Compacting Concrete*, Ozawa, K. and Ouchi, M. (Eds.), COMS Engineering Corporation, Tokyo, Japan, 2001, pp. 285–294.

Bouzoubaa, N. and Lachemi, M., Self-compacting concrete incorporating high volumes of class F fly ash: Preliminary results, *Cement and Concrete Research*, 31, 2001, 413–420.

Brouwers, H.J.H. and Radix, H.J., Self-compacting concrete: Theoretical and experimental study, *Cement and Concrete Research*, 35, 2005, 2116–2136.

BS 1881: Part 112, *Methods of Accelerated Curing of Test Cubes*, British Standards Institution, London, U.K., 1983.

Cussigh, F., SCC in practice: Opportunities and bottlenecks. In *Proceedings of the Fifth International RILEM Symposium on SCC*, De Schutter, G. and Boel, V. (Eds.), Ghent, Belgium, 2007, pp. 21–28.

DAfStb, DAfStb-Richtlinie Selbstverdichtender Beton (SVB-Richtline)–Teile 1, 2 und 3, Entwurf, DAfStb, Berlin, September 2012, pp. 1–18.

de Larrard, F., Ferraris, C.F., and Sedran, T., Fresh concrete: A Herschel-Bulkley material, *Materials and Structures*, 31, 1998, 494–498.

DIN 1045, *Beton und Stahlbeton*, Beton Verlag GMBH, Koln, Germany, 1988.

EFNARC, The European guidelines for self-compacting concrete, EFNARC, Farnham, Surrey, U.K., 2005, p. 68.

El-Dieb, A.S., Mechanical, durability and microstructural characteristics of ultra-high-strength self-compacting concrete incorporating steel fibers, *Materials and Design*, 30, 2009, 4286–4292.

EN 12350-1, Testing fresh concrete, Part 1. Sampling fresh concrete, European committee for standardization, Brussels, Belgium, 2000, pp. 1–6.

Felekoglu, B., Turkel, S., and Baradan, B., Effect of water/cement ratio on the fresh and hardened properties of self-compacting concrete, *Building and Environment*, 42, 2007, 1795–1802.

Ferraris, C. and Brower, L., Comparison of concrete rheometers: International tests at LCPC (Nantes, France) in October 2000, National Institute of Standards and Technology (NISTIR) 6819, Gaithersburg, MD, September 2001.

Ganesh Babu, K. et al., Design of self-compacting concretes with fly ash, in *Second North American Conference on the Design and Use of Self-Consolidating Concrete*, Center for Advanced Cement-Based Materials, Chicago, IL, 2005a.

Ganesh Babu, K. et al., Fly ash based high performance cementitious composites, material science of concrete, in *Indo-US Workshop on High-Performance Cement-Based Concrete Composites*, American Ceramic Society, Chennai, India, 2005b, pp. 171–182.

Ganesh Babu, K. et al., Self-compacting concrete with fly ash, in *ACI Sponsored Second International Symposium on Concrete Technology for Sustainable Development, with Emphasis on Infrastructure*, Hyderabad, India, 2005c, pp. 605–614.

Georgiadis, A.S., Sideris, K.K., and Anagnostopoulos, N.S., Properties of SCC produced with limestone filler or viscosity modifying admixture, *Journal of Materials in Civil Engineering*, 22(4), 2009, 352–360.

Gesoglu, M., Guneyisi, E., and Ozbay, E., Properties of self-compacting concretes made with binary, ternary, and quaternary cementitious blends of fly ash, blast furnace slag, and silica fume, *Construction and Building Materials*, 23, 2009, 1847–1854.

Gesoglu, M. and Ozbay, E., Effects of mineral admixtures on fresh and hardened properties of self-compacting concretes: Binary, ternary and quaternary systems, *Materials and Structures/Materiaux et Constructions*, 40, 2007, 923–937.

Guneyisi, E. and Gesoglu, M., Properties of self-compacting mortars with binary and ternary cementitious blends of fly ash and metakaolin, *Materials and Structures*, 41, 2008, 1519–1531.

Guneyisi, E., Gesoglu, M., and Ozbay, E., Strength and drying shrinkage properties of self-compacting concretes incorporating multi-system blended mineral admixtures, *Construction and Building Materials*, 24, 2010, 1878–1887.

Hwang, S., Khayat, K.H., and Bonneau, O., Performance-based specifications of self-consolidating concrete used in structural applications, *ACI Materials Journal*, 103(2), 2006, 121–129.

IS: 9013-1978, *Method of Making, Curing and Determining Compressive Strength of Accelerated Cured Concrete Test Specimens*, Bureau of Indian Standards, 1998.

Kantro, D.L., Influence of water reducing admixtures on properties of cement pastes—A miniature slump test, *Cement Concrete Aggregates*, 2, 1980, 95–102.

Khayat, K.H., Manai, K., and Trudel, A., In situ mechanical properties of wall elements cast using self-consolidating concrete, *ACI Materials Journal*, 94, 1997, 491–500.

Kheder, G.F. and Al Jadiri, R.A., New method for proportioning self-consolidating concrete based on compressive strength requirements, *ACI Materials Journal*, 107(5), 2010, 490–497.

Koehler, E.P. and Fowler, D.W., Summary of concrete workability test methods, ICAR Report 105.1, International Center for Aggregates Research, Austin, TX, August 2003, 76pp.

Kovler, K. and Roussel, N., Properties of fresh and hardened concrete, *Cement and Concrete Research*, 41, 2011, 775–792.

Li, J., Configuration of low-strength self-compacting concrete adapted to local raw materials, *Advanced Materials Research*, 335–336, 2011, 1159–1162.

Liu, M., Self-compacting concrete with different levels of pulverized fuel ash, *Construction and Building Materials*, 24, 2010, 1245–1252.

Marquardt, I., Diederichs, U., and Vala, J., Determination of the composition of self compacting concretes on the basis of the water requirements of the constituents materials, *Betonwerk + Fertigteil-Technik*, 68(11), 2002, 22–29 (in German and English).

Mazaheripour, H., Ghanbarpour, S., Mirmoradi, S.H., and Hosseinpour, I., The effect of polypropylene fibers on the properties of fresh and hardened lightweight self-compacting concrete, *Construction and Building Materials*, 25, 2011, 351–358.

Melo, K.A. and Carneiro, A.M.P., Effect of metakaolin's finesses and content in self-consolidating concrete, *Construction and Building Materials*, 24, 2010, 1529–1535.

Okamura, H. and Ozawa, K., Mix-design for self-compacting concrete, *Concrete Library, JSCE*, 25, 1995, 107–120.

Olsen, M.P.J., Energy requirements for consolidation of concrete during internal vibration, Consolidation of concrete, ACI SP-86, American Concrete Institute, Detroit, 1987, pp. 179–196.

Pade, C., Test methods for SCC, Nordic Innovation Centre, project number: 02128, December 2005, 56p.

Prakash, P.V.S., Development and behavioural characteristics of silica fume concretes for aggressive environment, PhD thesis, submitted to IIT Madras, Chennai, India, May 1996.

Quan, H., Study on properties of self-compacting concrete with recycled powder, *Advanced Materials Research*, 250–253, 2011, 866–869.

Rao, G.S.N., Effective utilization of fly ash in concretes for aggressive environment, PhD thesis, submitted to IIT Madras, Chennai, India, May 1996.

Reinhardt, H.W., DAfStb guideline on self compacting concrete, *Betonwerk und Fertigteil-Technik/Concrete Precasting Plant and Technology*, 67(12), 2001, 54–58 + 60–62.

Roussel, N. and Cussigh, F., Distinct-layer casting of SCC: The mechanical consequences of thixotropy, *Cement and Concrete Research*, 38, 2008, 624–632.

Saak, A.W., Jennings, H.M., and Shah, S.P., New methodology for designing self-compacting concrete, *ACI Materials Journal*, 98(6), 2001, 429–439.

Skarendahl, A. and Petersson, O., Self compacting concrete, state of the art report of RILEM TC 174-SCC, Report 23, RILEM Publications, Cachan, France, 2000.

Sonebi, M., Medium strength self-compacting concrete containing fly ash: Modelling using factorial experimental plans, *Cement and Concrete Research*, 34, 2004, 1199–1208.

Sonebi, M. and Bartos, P.J.M., Filling ability and plastic settlement of self-compacting concrete, *Materials and Structures*, 35, 2002, 462–469.

Sonebi, M. and Cevik, A., Prediction of fresh and hardened properties of self-consolidating concrete using neurofuzzy approach, *Journal of Materials in Civil Engineering*, 21(11), 2009a, 672–679.

Sonebi, M. and Cevik, A., Genetic programming based formulation for fresh and hardened properties of self-compacting concrete containing pulverised fuel ash, *Construction and Building Materials*, 23, 2009b, 2614–2622.

Su, N., Hsu, K.-C., and Chai, H.-W., A simple mix design method for self-compacting concrete, *Cement and Concrete Research*, 31, 2001, 1799–1807.

Sukumar, B., Nagamani, K., and Srinivasa Raghavan, R., Evaluation of strength at the early ages of self-compacting concrete with high volume fly ash, *Construction and Building Materials*, 22(7), 2008, 1394–1401.

Tatersall, G.H., *Workability and Quality-Control of Concrete*, E & FN SPON, London, U.K., 1991.

Turk, K., Caliskan, S., and Yazicioglu, S., Capillary water absorption of self-compacting concrete under different curing conditions, *Indian Journal of Engineering and Materials Sciences*, 14, 2007, 365–372.

Turk, K., Turgut, P., Karatas, M., and Benli, A., Mechanical properties of self-compacting concrete with silica fume/fly ash, in *Ninth International Congress on Advances in Civil Engineering*, September 27–30, 2010, pp. 1–7.

Turkmen, I. and Kantarci, A., Effects of expanded perlite aggregate and different curing conditions on the physical and mechanical properties of self-compacting concrete, *Building and Environment*, 42, 2007, 2378–2383.

Walraven, J., Structural applications of self-compacting concrete, in *Third RILEM and International symposium on self-compacting concrete*, Reykjavik, Iceland, 2003, pp. 15–22.

9

Mechanical Characteristics of SCCs

9.1 Introduction

The fact that self-compacting concretes (SCCs) can be produced with reasonably different consistencies with different flow characteristics is already accepted through various national recommendations, and the fact that they can be produced in three different ways by using fines, viscosity modifying agents (VMAs), or both appropriately makes it very difficult for a conclusive prediction of their mechanical characteristics. Another important consideration that needs careful attention is that the additional fines that are required for free flow of the SCCs are generally augmented by the use of simple powder extenders at different percentages through various pozzolanic materials like fly ash, ground granulated blast furnace slag (GGBS), silica fume, and others. It is well known that each of these pozzolanic admixtures have their own reaction kinetics in the presence of cement, which can also be affected by the temperatures particularly in the body of the larger members. Some chemical admixtures can have issues of compatibility both with the cement and the pozzolan. Even the use of VMAs and chemical admixtures can interfere with the strength gain rate or strength of concrete particularly at early ages.

9.2 Physical Properties and Microstructural Effects

The microstructural changes that occur inside concrete have a special bearing on the strength and more importantly durability characteristics of concretes in general. SCCs are essentially produced by different approaches, namely high fines approach, VMA approach, and combination approach. Increasing the viscosity by additional fines, particularly pozzolanic admixtures, understandably improves not only the green state characteristics but can also impart strength as well as durability and performance to concrete. However, the pore filling and pore refinement effects are defined by the fineness as well as pozzolanic characteristics of the admixtures utilized. The microstructural changes that can be modulated are basically the parameters that allow the making of such a wide range of structural cementitious composites, ranging from the normal low-strength concrete (20–30 MPa) with an additional powder material to facilitate an SCC with high-strength, very-high-strength and ultra-high-performance concretes (UHPCs) (up to and beyond 200 MPa), or even nano cementitious composites depending on the type of pozzolanic and chemical admixtures along with fibrous inclusions. The use of superplasticizers in such a case is essential to ensure that the water content required for the mix is reduced, correspondingly reducing the cement content. Such increased strength that was not envisaged through the mix design has already been recognized as reported in the discussions earlier, particularly when silica fume, a high-efficiency pozzolan, was used in the mix. The VMAs, essentially consisting of welan gum–based polysaccharides, and even sometimes the ones that were used earlier like methyl cellulose–based materials often exhibit a specific retardation during the early ages. Some of these characteristics have to be appropriately addressed to arrive at and to take advantage of the variations that are possible in the strengths of the SCCs to the benefit of the construction sequence and structural performance.

In the first place, it is best to look at the effects of physical properties of the constituent materials to have an understanding of their influence on the resulting concrete. The strength and performance of concrete are decided primarily by the hydration characteristics of the cementitious materials, which will ultimately decide the porosity and pore structure. Also, an understanding of cement hydration along with its effects on secondary cementitious materials is of importance. Among the physical characteristics of cement that affect the performance of the concrete, strength, fineness, and setting time are parameters that need attention. Presently, there are three different grades of Ordinary Portland Cement (OPC) based on strength (C32.5, C42.5, and C52.5) available as proposed in the European norms. The American standards in fact do not talk about any such specific strength of cement in particular. In this regard, it can be seen that the water–cement ratios adopted for the evaluation of cement strength by the American and German standards are 0.485 and 0.5, respectively. Furthermore, the British standards consider cement strengths in an indirect way, through the compressive strength of concrete at a water–cement ratio of 0.5, for concrete mix design. Cement strengths are overestimated in Indian standards, as they still use the old consistency approach for fixing the water–cement ratio to be used for the assessment of cement strength, which allows a much lower water–cement ratio of about 0.35–0.45. Considering all these aspects, it is probably best to establish cement strength at a water–cement ratio of 0.5. The strength of a representative concrete mix at a water–cement ratio of 0.5, similar to the British standards, with an appropriately graded aggregate and a reasonable amount of cement content (may be about 400 kg/m^3, giving a water content of 200 kg/m^3) for non-superplasticized concretes is considered appropriate. If superplasticizers and other chemical admixtures are involved, the water content can be appropriately reduced to about 160, however, keeping the water–cement ratio constant at 0.5, thus allowing for a reduction of cement to 320 kg/m^3. It is also strongly suggested that the minimum and maximum strength limitations for each grade of cement may also be recognized and specified as was adopted in the German or the present Euro standards.

The fineness of cement affects the rate of hydration, workability, bleeding, cracking, and the amount of entrained air. It also influences consistency and thus strength evaluations as already discussed. From this consideration, the earlier recommendation of a unique value (preferably 0.5) for the water–cement ratio for strength evaluation of cement (through concrete rather than mortar) is more appropriate. Also, a finer grinding of cement influences the early strengths but the gain in strength at later ages may only be marginal. It is felt that a proper grinding to obtain an effective grain size distribution in the cement may help in modulating both the early strength gain rate and the later strength developments after 28 days. Thus, the rate of strength gain should also be a characteristic, needing specific recommendations, preferably up to the 90- or 180-day value specifically in the case of pozzolanic cements. Furthermore, finer grinding of cement also influences the early heat of hydration and may influence the characteristics of the concrete produced due to the shrinkage and temperature effects. The water requirement for a specified workability may also increase disproportionately due to this. It may also influence the requirements of a superplasticizer for obtaining higher workabilities.

In fact even adopting the best packing density methodology like Fuller's curve for spheres possible, there is no way to ensure the highest packing of the aggregate mass. Various attempts have been made to arrive at a particular grading relationship for aggregates in general in an attempt to have an appropriate packing desired by the design practice. Further the attempt to combine two or more aggregates is only to ensure that the aggregates proportioned are as desired and some of the difficulties in trying to ensure the required gradation through a single mixture of a specific maximum size are addressed appropriately. The several options and possibilities that could be used to arrive at an optimum packing density with three aggregates were discussed in detail by Roy (1995) and (Scheetz, 1993) earlier. They also suggest a method of even proportioning the cementitious material combination for cement, fly ash, and silica fume in a very similar fashion. It is only appropriate to mention at this stage that apart from the physical combination of these materials to arrive at an optimum packing density, the possibility of achieving the highest strength through an insight into the hydration dynamics of the system was discussed in detail by Scheetz (1993). The effectiveness of fly ash and silica fume (calling them as cementitious waste forms) in addressing the fundamental aspects of the durability of concrete was explained in specific detail. In fact if only one can optimize both the physical and chemical systems together along with the minimum necessary cementitious materials and with an optimal aggregate packing, the resultant concrete will have the highest strength and even performance. This in fact is the aim of this book and in this the need for powder extenders without increasing the cement content is the most important part in arriving at the highest performance in any concrete, particularly the SCCs that require such powder extenders.

In practice, these concepts have led to modifications in cements through two specific routes to arrive at such cements that are modulated for ensuring durability and economy. In the first case, the cements are ground finer to achieve the high early strength due to the faster hydration of the larger surface area available. However, this had only limited success as faster hydration obviously means larger heat of hydration within that period, which cannot be dissipated, leading to thermal cracking. This problem was attempted to be redressed through the adoption of effective packing. The effective packing of the various size grains has been recognized as a means of improving the microstructural integrity of cement, particularly while using blended cements. The use of packing density diagrams for this is discussed here. In an effort to present a method for the control of the intrinsic properties of cementitious materials, researchers suggest two specific aspects: control over the physical behavior of the cementitious matrix and control over the chemical behavior of the hydrating system. Control over the physical properties of the cementitious materials involves a careful control of particle size distributions of all the systems components. Figure 9.1 depicts a typical packing density diagram for a couple of reactive mineral admixtures and cement (Scheetz, 1993).

Control over the chemical aspects of hydrating cementitious materials involves a careful consideration of all components based on bulk chemistry, phase composition, and bulk chemical reactivity. Synergistic effects are common and can often be used to enhance performance. The phase assemblage and porosities of OPC as a function of curing time have been discussed by Scheetz (1993). The amount and size distribution of porosity is linked to the components of the constituent cementitious system, with the water–cement ratio having the most significant influence upon the mature cement product. Therefore, any processing aid or step that can be taken to minimize the amount of water in the mix will result in reduced porosity. Likewise, the use of mineral admixtures in the formulation offers an opportunity to tailor the nature and amount of hydration product formed. Thus, control over the porosity and pore size distribution can be achieved through this understanding of the chemical and mechanical parameters in a cementitious system. Until now the discussions were all directed toward the cement and water phase in the matrix. This is because the intrinsic properties of the matrix are decided only by this phase and the filler will not in any way influence the behavior of the mixture. However, it is obvious that the water requirement for workability is a function of the total wetted surface of all the constituents and as such can have a significant influence on the final water–cement ratio that is possible. Naturally, this aspect will make it essential to look at mixture proportioning in a much broader perspective.

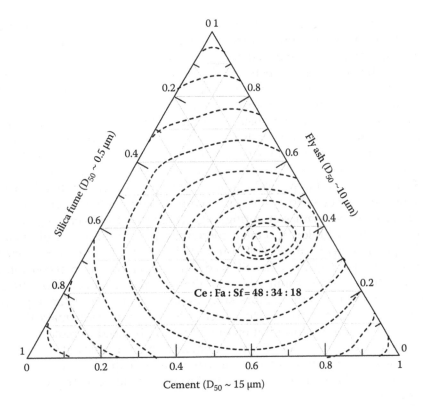

FIGURE 9.1
Concept of packing density in cements. (After Scheetz, B.E. et al., Fundamental aspects of the durability of cementitious waste forms, *Proceedings of the Indo-US Workshop on Durability of Concrete Structures*, Indian concrete Institute, Bangalore, India, 1993.)

It is well recognized that concrete is a porous material, which is an indicator of its strength and also the deciding factor of its vulnerability to the external environment. The interconnected capillary pore structure resulting out of the water trying to escape is primarily responsible for this. The porosity of the cement paste varies from the air bubbles in the range of millimeters to the microporosity that exists between the hydrating layers of cement (Glasser, 1993).

The mechanisms involved in the cementitious participation of silica fume in concrete are generally explained through filler and pozzolanic effects. The effect of SF on the rheology of fresh concrete is generally viewed as a stabilizing effect in the sense that the addition of very fine particles to a concrete mixture tends to reduce segregation and bleeding tendencies. Without the SF the finest particles in concrete are those of Portland cement, which are normally in the 1–80 microns size range. Since the fine and coarse aggregate particles are much bigger than the cement particles, the latter act as stabilizers by reducing the size of channels through which bleed water rises to the surface of concrete. When very fine particles of SF are added to concrete, the size of flow channels is greatly reduced because these particles are able to find their way into the empty spaces between the two cement grains, causing drastic segmentation of the bleed-water flow channels. It is this effect of breaking the continuity of the capillary porosity through pore filling that is responsible for the thixotropy and lower bleeding potential of the concretes with silica fume. This mechanism was earlier explained by Mehta (1986). It is already known that concrete mixtures containing silica fume can be proportioned based on an "efficiency factor – k," which was also defined earlier (Jahren, 1983; Sellevold and Radjy, 1983). Jahren (1983) also concluded that the activity index (efficiency factor) of SF varies considerably with SF dosage, curing conditions (higher with wet curing), age (higher with lower age), cement content (higher with lower cement content), type and dosage of plasticizer, and type of cement. Jahren (1983) reported that it was often necessary to increase slump by 50 mm when SF is added compared to normal concrete to maintain the same workability for both the concretes. From all these discussions it is obvious that the compressive strength of concrete can be improved by incorporating SF into concrete. Jahren (1983) illustrated schematically that there is an optimum concentration of SF that can be used to replace cement to achieve maximum strength improvement. The dynamics of the various effects associated with silica fume in concrete are explained in Figure 9.2, which is a modification of the conceptual representation earlier proposed by Jahren (1983).

Without going much further into the discussions regarding various aspects of the distribution of the void space, in the first place the strength of the cementitious composites, including the ones like SCCs, is primarily related to the water–cementitious materials ratio. The methodology of effectively addressing these factors through the efficiency of pozzolanic materials has already been presented. The earlier paragraphs further reinforce these discussions and help in understanding the methods of achieving high-strength high-performance concretes. The fact that the water–cementitious materials ratio could be kept low using more modern superplasticizers and an effective combination of high-efficiency pozzolans like silica fume or even nano silica provided avenues for producing ultra-high-strength cementitious composites of strength nearing that of steel in at least compression.

Having looked at the microstructural aspects and their effects on the physical properties it may be relevant to just discuss the causes for concern, if any, that need a specific mention in the case of SCC. One of the aspects that immediately comes to mind is whether the available relations between the various mechanical parameters from conventionally vibrated concretes are valid in the case of SCCs also. While agreeing that SCCs are in no way different from the conventional vibrated concretes, the fact that there is a much smaller fraction of coarse aggregate and in some cases there is an excessive thrust toward higher fines than probably justifiable to simply alleviate the

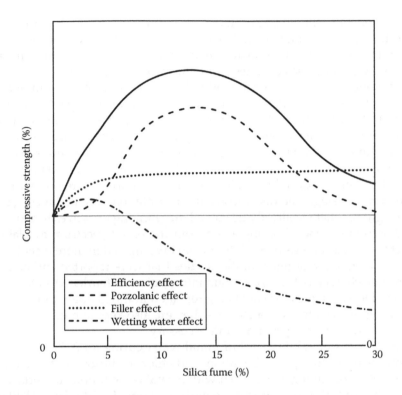

FIGURE 9.2
Schematic on effects of silica fume in concrete. (After Jahren, P., Use of silica fume in concrete, ACI SP-79, Vol. II, 1983, pp. 625–642.)

problems of compactability causes concern. This emanates from the fact that the coarse aggregates in a way participate through sheer transfer forces that are concentrated at specific points like in the case of end zones of prestressed concrete members. Even the fact that the complete development of interfacial share transfer is probably not possible in such mixes is well acknowledged.

After this brief glimpse into the intricacies of the microstructural characteristics of cementitious composites and their possible modifications to pozzolanic materials and superplasticizers, one can understand that in general investigations into the various facets of the behavior of SCCs have been understood in their proper perspective. As in the case of green state characteristics each of these investigations has to be first looked at through the prism of materials, compositions, and interactions both in terms of the chemical and physical perspectives apart from the aggregates involved. In a way such an individual perspective on each of the reported study is essential before summarizing what was understood and what could be the way forward in terms of the performance of SCCs. It is to be reiterated that this is in no way an effort to underrate the importance of the investigations so far and is in fact the platform on which broad consensus is being built to ensure high-performance SCCs. It is also to be pointed out that a large amount of data on the mechanical strength behavior of SCCs is available in the past couple of decades. In fact a brief overview of these concretes and their strength results was already presented while trying to understand individually the effect of different pozzolanic admixtures at different percentages and how they stack up when assessed for their efficiency. Later these results were used to present various nomograms for ensuring a reasonable starting point for the design of concretes of a specific strength. At that stage but for referring to the investigations that were used in those efficiency assessments the actual characteristics of specific concretes that were produced in the investigations reported were not discussed. At the present stage, the discussions are basically on specific investigations in the light of what we already discussed, not limiting ourselves to those that are chosen but in fact choosing a cue from the broad spectrum available to depict the characteristics of SCCs that were investigated in more recent times. It is also be noted that many of the studies that were reported earlier under green state characteristics also contain information on hardened state characteristics. However, these are not being repeated in this chapter, though while discussing the results and conclusions they are also kept in mind and will be referred to wherever appropriate and necessary.

Zhao (2012) studied the effect of initial water curing for 3, 7, and 14 days on compressive strength, carbonation, and chloride diffusion characteristics in concretes containing 20% fly ash with a total cementitious materials content of 460 kg/m^3, resulting in a strength of about 60 MPa. The NT-BUILD 355 test was used to establish the chloride migration coefficient. Accelerated carbonation studies indicated carbonation depths ranging from 6.5 to 3 mm in concretes of strength ranging from 58 to 63 MPa. Concretes with fly ash replacement levels of 20%, 40%, 60%, and 80% at a constant cementitious materials content of 500 kg/m^3 and a constant water content of 180 kg/m^3 resulting in a water–binder ratio of 0.36 were studied by Khatib (2008). It is surprising to see that SCC without fly ash showed strength of 85 MPa at 56 days while the maximum strength in SCCs containing fly ash obtained for 40% replacement was only about 68 MPa. The water absorption increased with increasing the level of fly ash as was also observed in studies by Zhao (2012). The drying shrinkage however was seen to be reducing with increasing fly ash percentage. He also presented the relationships between compressive strength and ultrasonic pulse velocity (UPV) as well as shrinkage to fly ash content in his study, which are useful guidelines. A range of concretes containing fly ash and limestone powder (LSP) with a cement content of around 400 kg/m^3 and at fly ash percentages varying from 20% to 34% were studied by Gheorghe (2011). He also studied the effect of LSP in such concretes at 20% and 30% for a comparison. The important aspect of the study is that he also recognized the efficiency concept for fly ash, though the value suggested appears to be much lower than what was presented by Ganesh Babu (1996) earlier.

Filho (2010) studied three concretes containing about 350 kg/m³ cement and 100 kg/m³ of LSP as the additional filler. His investigations concentrated on the statistical distributions of concrete strength, moduli, and tensile strengths in general from a total of almost 25 specimens in each case. He also presented theoretical equations relating the modulus of elasticity to the compressive and tensile strengths as well as the density along with the statistical distributions. With such a limited investigation on almost similar concretes of very limited strength range (40–50 MPa), the relationships between the different parameters studied are difficult to justify, though it may be relevant for the design of a specific concrete structure in which the strengths could only vary in such a limited range. Bosiljkov (2003) reported concretes containing 15% slag cements supplemented with almost 50% limestone filler at the water–cementitious materials ratios of around 0.22–0.25 and water content of around 160–180 kg/m³, resulting in a strength of around 50–70 MPa. The studies include the tensile strength and UPV for these concretes. He also studied the behavior of a reinforced concrete beam, which clearly established the suitability of the concrete for structural applications. The behavior of what was considered as self-compacting sand concretes (mortars) with a cement content of 350 kg/m³ in combination with 150, 250, and 350 kg/m³ of marble powder at a water–powder ratio of 0.44, resulting in water contents of 220, 264, and 308 kg/m³, was studied by Tayeb (2011). He investigated fresh concrete characteristics using the mini-slump cone (cone used with the mortar flow table), V-funnel, and viscosity tests. The compressive strengths range from 35 to 55 MPa. In a way such mortar mixtures may be quite useful for repair and rehabilitation of spalled cover concrete of corrosion-affected structural members. Tittarelli (2010) studied the strength characteristics of Portland limestone blended cement containing glass reinforced plastic (GRP) powder as the additive at 10% and 20% in 1:3 cements and mortars. Compressive and flexural strength studies were done on 40×40×160 mm specimens apart from capillary absorption and drying shrinkage. The compressive strength reduced from 40 MPa to around 25 MPa due to replacements with GRP powder. The reason for this could be the dilution effect of the cementitious capacity of the binder wherein the glass reinforced plastic powder was only supplementing the powder requirements and cannot be considered as a part of the cementitious system. However, the capillary absorption of water reduced in all the specimens containing GRP powder. This obviously is a reflection of the pore filling effects of such fine powders in a cementitious system though they may not really be contributing to the strength characteristics of the concrete composites investigated. Dehwah (2012) reported a total of four concretes containing quarry dust powder with cementitious materials limited to 400 kg/m³. One of the mixes contained silica fume and another contained fly ash. The resulting concretes appear to have strength in the region of 50–70 MPa. In brief he studied the split and flexure tensile strengths apart from the UPV of these concretes.

The use of metakaolin as a clinker replacement material of 12.5%–25% in cement was suggested by Cassagnabere (2010). His studies in a precast factory suggest that for slip forming applications and self-compacting applications the optimum percentages were 25% and 17.5%, respectively, with the compressive strength and porosity not being affected. This obviously is not really appropriate as it is possible to achieve much higher strengths and performance through appropriate use of metakaolin as a cement replacement material in the production of concrete rather than mixing it at the clinker stage. In any case, the strength characteristics of the cement produced might have altered significantly and these would be responsible for the changes that take place in the green state or the hardened state performance of concrete. It is easier to account for the strength as well as the performance improvements accruing from the replacements effected through the earlier suggested efficiency approach. Sua-Iam and Makul (2012) studied concretes containing rice husk ash (RHA) at 10%, 20%, and 40% of the two water–cement ratios 0.28 and 0.33 with a cementitious materials content of 550 kg/m^3. The studies included slump flow, compressive strength, and UPV, and the concretes show strengths varying from about 35 to 60 MPa. The maximum strength was obtained at the 20% replacement level.

The effects of fly ash and silica fume in SCCs under air and water curing were studied by Mohamed (2011). The fly ash percentages varied from 10% to 50% and silica fume varied from 10% to 20% with a few ternaries included. The optimum percentage of fly ash was 30% and that of silica fume was 15%, and for ternary a 10+10 percentage was found to be the optimum through his studies. The water absorption characteristics of SCCs containing 30% fly ash and 10% silica fume cured for periods of 3, 7, 14, and 28 days under different curing conditions (immersed in water, sealed, and air cured) were reported by Turk (2007). The studies include compressive and tensile strengths and UPV apart from sorptivity coefficients at 28 days. He observed that sorptivity showed a good correlation with compressive strength regardless of the curing regimes. In a later study, Turk (2010) reported studies on SCCs containing fly ash content of 25%–40% and silica fume content of 5%–20%, which are probably the best ranges for the replacement, given the fact that the total cementitious material content was limited to 500 and 450 kg/m^3, respectively. His effort shows an appropriate utilization of the admixtures to arrive at reasonably high-strength concretes of around 70 MPa. He also reported the relationship between compressive strength and modulus of elasticity apart from others.

A significantly large investigation that has information on the various green and hardened state characteristics was reported by the group Güneyisi (2008, 2010) and Gesoğlu (2007, 2009). Between these four studies, they reported a total of 65 concrete mixtures at two water–binder ratios, 0.32 and 0.44, having a total cementitious materials content of 550 and 450 kg/m^3. The concretes contain both binary and ternary mixes containing fly ash, GGBS, silica fume, and metakaolin. Individual publications discussed here have clarity of what each one of them is directed at along with their conclusions. Gesoğlu (2007) reported the sump flow time, L-box height ratio, and V-funnel flow time of 22 mixes. He also studied the initial and final setting times of these compositions. The more important projection of the results is his depiction of the changes in viscosity with time for the different combinations in these 22 mixes. He also reported the UPV and electrical resistivity of these concretes in relation to the compressive strengths. His studies indicate that a reasonably high strength of about 80 MPa can be easily obtained with the use of these pozzolanic materials and their combinations. In the next report that came out of this group, Güneyisi (2008) presented studies on the consistencies of mortars containing supplementary cementitious materials (fly ash up to 60% and metakaolin up to 15% apart from ternary cements containing both) at a constant water–cementitious materials ratio of 0.4. The studies include flow characteristics through a mini-slump cone of 100 and 70 mm diameter with a height of 50 mm, apart from flow through a mini V-funnel and the setting times. He also observed that the use of fly ash significantly prolonged the initial and final setting times while the mortars containing metakaolin only increased the setting times marginally, increasing with the increasing metakaolin percentage. These studies were very similar to those by Siva Nageswara Rao (1996) and Surya Prakash (1995) for fly ash and silica fume, which also presented similar results. He also reported the strength of these binary and ternary mortars. Later Güneyisi (2010) presented his studies on a total of 65 concrete mixtures at the two water–binder ratios, 0.32 and 0.44, with the first 43 concretes at a total cementitious materials content of 550 and the rest at 450 kg/m³. The mixes contain both binary and ternary compositions containing fly ash and GGBS with replacement ratios at 20%, 40%, and 60% and silica fume and metakaolin replacements at 5%, 10%, and 15% by weight of the total binder in various combinations. He also used the most commonly adopted EN 42.5 euro grade cement that has 326 m²/kg Blaine in all these investigations while the FA, GGBS, MK, and SF had finenesses of 287, 418, 12,000, and 21,080 m²/kg, respectively. Though the fineness of fly ash appears to be a bit lower, these concrete investigations are perhaps the largest database one could get from any single group of investigators. The studies included the green state characteristics of slump flow, V-funnel, and L-box results apart from the compressive strength and the complete shrinkage characteristics of all these concretes. The conclusions include that the negative effects of fly ash could be corrected with the use of ternary and quaternary mineral admixture combinations. The addition of mineral admixtures reduced the shrinkage characteristics of the concretes, with silica fume showing the maximum effect, which in any case should be expected with the better pore filling effects of such fine materials apart from its higher pozzolanic efficiency obtaining a highly dense structure and lowering the shrinkage. In the next study, Gesoğlu (2009) discusses 22 concrete mixtures having a binder content of 450 kg/m³ at a water–binder ratio of 0.44, the binder itself containing several combinations of fly ash, GGBS, and silica fume. He studied both the fresh and hardened state characteristics. The fresh state characteristics have already been reported earlier. He used a multiobjective optimization technique to predict the various characteristics studied. Information regarding the UPV and resistivity characteristics apart from sorptivity, chloride permeability, and shrinkage was discussed. The conclusions indicate that ternary mixtures containing slag and silica fume performed best in terms of durability characteristics.

The behavior of normal and SCCs containing 0%, 10%, and 20% RHA at constant water–binder ratios of 0.4 and 0.35, with water contents of 184 and 161 kg/m^3, resulting in the cementitious material content of 460 kg/m^3 was reported by Ahmadi (2007). Ordinary concretes containing RHA resulted in strengths of 60–75 MPa while the SCCs appear to have achieved 80–95 MPa at the end of the 180-day period. The significant change in the compressive strength observed is hard to explain for the SCCs. He also studied the flexural strengths and modulus for these concretes. Studies on the mortar phase of SCCs using the mineral additives fly ash, LSP, and granite filler with two different types of cements of 32.5 and 42.5 grades were attempted by Nepomuceno (2012). He also tried to correlate the relative spread area of a flow cone with the relative flow velocity through a V-funnel for these mortars. The mortar strengths obtained reach almost 65 and 90 MPa at 0.25 water–cement ratio in the curves presented for the two cements. Based on these mortar studies, he tried to extrapolate and arrive at the compressive strength of concretes. Parra (2011) studied both normally vibrated and SCCs with 32.5 and 42.5 cements for their strengths and modulus. The concretes contained LSP with total cementitious materials content of around 500–600 kg/m^3, at the water content of around 180 kg/m^3, resulting in a strength of around 30–60 MPa at 90 days.

The studies by Domone at the UCL are the most extensive, totaling to about 70 efforts toward an understanding of the SCC behavior in their various forms. One of the obvious results of this huge effort is the UCL method for SCCs, which is based on an appreciation of the effects of the different parameters associated with all the concrete studied. The article (Domone, 2007) presents a comprehensive summary of all these investigations on the various characteristics, particularly the water–cement ratio to strength relations and cube to cylinder strength ratios apart from the other relations for modulus and the effect of bond strengths of SCCs at different depths. He also presented typical transmission lengths applicable to the prestressed concrete members in practice. Jalal (2012) reported studies on fresh, mechanical, and durability characteristics of SCCs containing micro- and nanosilica at 10% and 2% individually as well as together. The concretes were made with a cement content of about 400–500 kg/m^3 with these materials along with 177 kg/m^3 of LSP as filler at a water–cementitious materials ratio of 0.38. The compressive strengths of concretes ranged between 50 and 90 MPa. The water absorption and capillary absorption were also measured apart from the chloride ion concentrations at different depths and resistivity to represent the durability characteristics. The characteristics of SCCs and mortars with both polycarboxylate (1.6%) and naphthalene formaldehyde (1.8%) based superplasticizers containing GGBS were investigated by Boukendakdji (2012). The concretes had a total cementitious material content of about 465 kg/m³, with slag content ranging from 0% to 25%, at a water content of 183 kg/m³. He also studied the various green state characteristics like slump flow, T50, J-ring, V-funnel, U-box, and stability through the 5 mm sieve test to characterize the concretes as self-compacting. The compressive strength of concretes varied from about 40 to 70 MPa at the age of 7 to 90 days. The change in spread for the slag SCCs appeared to decrease with slag, with the spread of 15% slag SCC being the highest at all ages.

The effect of different chemical admixtures (polycarboxylates, melamine formaldehyde, aqueous dispersion of microsilica, and high-molecular-weight hydroxylated polymer) in conjunction with mineral admixtures (fly ash, brick powder, LSP, and kaolinite) was reported by Sahmaran (2006). The water–powder ratio selected was 0.40 with a total powder content of 650 kg/m³ for all the tests. The green state characteristics included slump flow and V-funnel apart from setting times. The UPV and strength values were also studied. Akcay (2012) reported the behavior of a hybrid steel fiber reinforced SCC mix containing 700 kg/m³ of cement and an additional 15% silica fume at a water–cement ratio of 0.22 resulting in a strength of about 150 MPa for the reference concrete. The concrete was modified with both plain steel fibers of 6 mm length and 0.15 mm diameter (as in the case of attempting UHPCs) and also hooked fibers of 30 mm length and 0.55 mm diameter of both normal and high-strength steel. Apart from studying fiber dispersion, he reported the results of slump flow and J-ring at different fiber volumes. As can be expected an increase in the fiber content increases the flow time. Though there is no significant strength increase between the two volume percentages of fiber 0.75% and 1.5%, reaching about 124 MPa, the fracture energy, flexural strength, and ductility ratio improved significantly through the addition of fibers.

Vejmelkova (2011) studied two concretes containing 40% metakaolin and a blast furnace slag cement with 56% slag to establish the fresh and hardened state characteristics of the concretes. It is surprising that he feels that the slag concretes shows zero yield stress. The observation that the strength gain rate of metakaolin concretes is far higher than the other is understandable from the fact that the efficiency of metakaolin is high. The characteristics like specific heat capacity and thermal conductivity for these two concretes were also reported. It was also observed that the metakaolin concretes performed better in terms of the freeze–thaw resistance because of the finer pore structure compared to the slag concrete, with a mass loss of 0.396 kg/m² after 56 cycles (considered to be in the good category), while the slack concrete exhibited mass loss in the satisfactory category as prescribed by the Polish standards. The effects of LSPs with varying specific surface areas in SCCs were studied by Esping (2008) through slump flow, autogenous deformation measurement, and ring test for plastic shrinkage. The Bingham characteristics were also measured. The effect of additional water required for the increased surface area of LSP was found to be decreasing the compressive strength as can be expected, which also reflected in the plastic shrinkage characteristics. However, the minute variations that are being discussed may or may not have the specific bearing on characteristics of the volume of concrete that is involved in structural constructions, and this is one reason for the thrust to examine more comprehensive information on the actual concretes. Even so with the complications involved in the void spaces created by the different aggregates and the distributions of mortar and aggregates in the mix, a highly trained eye is required to look at the exact effect of any such variation in fineness. But from experience it may be said that the conclusions that were drawn by the author are certainly in line with the behavior of concretes with such materials. Altin (2008) studied the strength characteristics of four SCC mixes with cement content ranging from 450 to 550 kg/m^3 and water content ranging from 204 to 224 kg/m^3 in a ready mix plant. The 28-day compressive strength of each of the concrete mixes was established through 24 samples, and the data gathered from these results was used in an artificial neural network model. A statistical analysis was further used to study the reliability of the model. The study indicates essentially the usefulness of neural networks and statistical modelling for RMC plants to perfect a specific mix design for a coherent, well-defined, and consistent set of materials in such an industry with consistent materials. Studies on concretes containing binary and ternary blends of fly ash and silica fume were reported by Askari (2011). The concretes had a total binder content of 550 kg/m^3, with the fly ash and silica fume as powders. The studies show that a cementitious mixture containing 30% fly ash along with 5% silica fume will have a compressive strength similar to that of normal concrete at the same cement content. This is probably obvious from the fact that 5% silica fume will essentially not be able to influence cementitious characteristics substantially as it would be absorbed in the pore filling, and whatever strength gain was obtained from it will only help in compensating the minor strength reduction at 30% fly ash.

A study on the mechanical and durability characteristics of high-strength SCCs and their counterparts containing steel fiber (HELIX 5-25, twisted fiber) at the different dosages of 6, 20, and 40 kg/m^3 corresponding to the volume fractions of 0.08%, 0.12%, and 0.52% was reported by Deib (2009). The plain concrete mixers contained 15 and 17.5% silica fume. But with the cementitious materials totaling 775 and 900 kg/m^3, which is almost the extreme end of the cementitious content if ever they can be permitted, the drying shrinkage and heat of hydration characteristics and the curing regimes of such concretes are always a matter of concern. The resulting concretes developed a strength of about 100–130 MPa, while the concretes with fiber additions reached strengths of around 110–150 MPa. A comparison of the static and dynamic characteristics of steel fiber reinforced concrete (SFRC) of 70 MPa compressive strength containing hooked steel fibers, 35 mm long, at a fiber content of 50 kg/m³ was presented by Caverzan (2011). The results were expected to be a better representation of the dynamic characteristics without having to extrapolate from the static characteristics of SFRC as is normally attempted. The workability and mechanical properties of fly ash–based SCCs containing monofilament polypropylene fibers were investigated by Gencel (2011). The specific concretes with cement content of 350 and 450 kg/m³ and 120 kg/m³ of fly ash at a water content of 140 and 180 kg/m³ were investigated at 3, 6, 9, and 12 kg/m³ of fiber. Tests on slump flow, J-ring, V-funnel, and air content were related to the fiber content showing a corresponding linear increase in the thixotropy of the material due to fiber. There is only a marginal increase of about 5–10 MPa in the compressive strengths, while the tensile strengths improved by 2–3 times at the highest percentage of fiber. Felekoglu et al. (2007) studied the effect of the quantity of fines in sand on the strength characteristics along with the water absorption characteristics in concretes containing fly ash at about 30% arriving at strengths ranging from 25 to 45 MPa. In a later study, Felekoglu et al. (2007) also studied the effect of one percent polypropylene and polyvinyl alcohol microfibers in self-compacting micro-concrete composites containing LSP and fly ash. The flow characteristics as well as the fracture characteristics of 40×40×160 flexure specimens were studied. The flexure specimens were tested under three-point loading at a constant displacement rate of 0.02 and 0.2 mm/min for plain and fiber reinforced specimens. The first crack strength and the mid-point displacement at first crack were all compared for the different matrices considered. He also compared the toughness characteristics of the specimens containing the different fiber compositions at a displacement of 1.5 mm typically showing improvements and ductility characteristics of the fibrous composites developed.

Li (2011) reported studies on the concretes used for the cable-stayed bridge across the Yangtze river with a main span of 926 m using a hybrid girder. The concretes contain a binder content of 550 kg/m³, with the water content of about 170 kg/m³, using about 20% fly ash and 10% silica fume. The effects of polypropylene fiber and steel fiber were also studied broadly. As a part of the overall investigations, he studied the chloride ion permeability and frost resistance of these concretes to look at the suitability of the concrete for riverine and waterfront structures. In fact this type of concrete with 30% fly ash and 10% silica fume at a maximum of 550 kg/m³ of the cementitious materials at a slightly lower water content of 160 kg/m³ or even lower with steel fibers at about 1% minimum could result in a concrete that will have the best performance in terms of the environmental residents and also be reasonably high in compressive strength.

The applicability of the current codal provisions for estimating the mechanical properties, tensile strength and modulus, to SCC was studied by Vilanova (2011) based on 627 mixtures from a total of 138 references. A closer look at the variation of modulus with compressive strength indicates that there is a substantial scatter particularly at the higher compressive strengths. Li (2011a) studied a concrete containing recycled coarse aggregate with about 500 kg/m³ of cement and 195 kg/m³ of water. The apparent increase in water content from a general maximum of 180 kg/m³ could be due to the absorption characteristics of the recycled coarse aggregate. However, the strength of about 20 MPa obtained appears to be rather low considering the significant quantity of cement in the mix. The use of 50–150 mm for the aggregate size also appears to be an anomaly even though the test cube size happens to be 450 mm. In a later study, Li (2012) presented the effect of the same concrete in unreinforced concrete beams of size 250 × 500 × 1700 mm with 16 mm diameter bars as reinforcement. The deflection and failure characteristics studied appear to be broadly acceptable.

Pandurangan (2010) studied the bond characteristics of lap splices in tension reinforcement in concrete beams with specifically incorporated crack initiators in the form of a notch at the start of the lap splice. The beam cross-section was 200 × 250 mm with an effective span of 2100 mm containing two numbers of 16 mm bars with a cover of 25 mm from both edges and having an overlapping splice length of 300 mm at the center. The top two hanger bars were made up of two numbers of 10 mm bars and the beams were tested with a four-point bending test with a 700 mm constant moment zone. The stirrups were made up of 8 mm reinforcements kept at three different spacings of 75, 150, and 300 mm resulting in different confinement effects. The concrete consisted of an SCC with cement and fly ash content of 400 and 135 kg/m³ with a water–binder ratio of 0.33 and a water content of 178 kg/m³. Graded aggregates of 20 and 10 mm were used, resulting in a slump flow of about 660 mm and a compressive strength of around 36 MPa. The confinement effect was quite evident from the ultimate load, low deflection, and bond stress to slip relationships studied apart from the crack widths and crack spacing. Finally, beams with a confinement index less than 2.0 exhibited poor ductility and increasing the confinement index to 3.0 resulted in improved ductility.

Najim (2012) studied the mechanical and dynamic properties of self-compacting crumb rubber modified concretes with the crumb rubber in the form of coarse, fine, and combined aggregates replacing the fine and coarse aggregates at 5%, 10%, and 15%. His studies on compressive strength indicate that for the 5%, 10%, and 15% replacement of crumb rubber the compressive strength of the concrete reduced from around 55 MPa to around 45, 35, and 25 MPa, respectively. The flexural and split tensile strengths also reduced correspondingly. He studied the fracture characteristics in terms of I_5, I_{10}, and I_{30} values of the fracture toughness of these materials. More importantly, he looked at the dynamic characteristics in terms of both longitudinal and transverse UPVs resulting in the dynamic modulus values for these materials. He also studied the vibration damping capabilities of these materials, which probably are more relevant to crumb rubber concretes, but the substantial lowering of strengths by the addition of crumb rubber really does not give enough impetus to its use extensively as there are several other avenues for achieving the same result particularly with the addition of different fibers. Kou (2009) reported studies on mortars and concretes containing glass as both fine aggregate and coarse aggregate replacement in SCC as a measure of recycling glass in concrete matrix. The mortar mixes contain 15%, 30%, and 45% recycled glass while the concrete mixes contain 5% and 15% recycled glass as 10 mm aggregates. The concrete compressive strengths of about 75 MPa at a water–cement ratio of 0.37 with the powder content of 500 kg/m³ have not varied very much with the replacements studied. As can be expected with inert crystalline glass materials in the form of aggregate and not powder, the other characteristics like tensile strength modulus and chloride ion penetration have not been affected by these replacements. The drying shrinkage as well as the ASR expansions were also not affected significantly in any way.

The behavior of low-calcium fly ash–based self-compacting geo-polymer concrete at a fly ash content of 400 kg/m³, activated by 57 kg/m³ of 12 M sodium hydroxide and 143 kg/m³ of sodium silicate with 7% superplasticizer, was reported by Ahmed (2011). The additional water required ranging from 40 to 80 kg/m³ has resulted in concretes of strength varying from 22 to 54 MPa depending on the additional water content utilized. The workability characteristics of slump flow (about 700 mm), T_{50}, V-funnel, L-box, and J-ring values were also presented. The curing time varying from 24 to 96 hours and curing temperature varying from 60°C to 90°C did not appreciably change the compressive strengths. The behavior of SCCs containing micro encapsulated phase change materials (PCM) based on paraffin in concrete at 1%, 3%, and 5% by mass or 2.5%–12.4% by volume was reported by Hunger (2009). He also used grated aggregates, with the Anderson and Anderson model, using 16 mm maximum and 0.22 mm minimum sizes. The hydration temperatures of these concretes were also studied through an experimental setup modelled to present a semi-adiabatic environment. The fresh concrete characteristics in terms of V-funnel, slump flow, and J-ring studies were also reported. However, the strength reduction from 74 MPa of the reference mix continuously to 52, 35, and 21 MPa at the 1%, 3%, and 5% PCM content certainly makes the use of PCM highly questionable, though in the literature some of these materials are utilized particularly as water-repellent concretes.

Valcuende (2009) studied eight different concretes to understand the bond strength characteristics of SCCs. The concretes were produced using both EN 32.5 and EN 42.5 grade cements at water–cementitious materials ratios of 0.45 and 0.65 with cement content varying from 275 to 400 kg/m³. Essentially the first part of the investigation consisted of 200 mm cube specimens with 16 mm bars having an anchorage length of 80 mm perpendicular to the direction of casting to study the mixes. To study the effect of the depth of casting (with the possible segregation effects to be included), 1500 mm column specimens were made with bars embedded across at various depths. The concretes investigated have no additional powder content other than the cements and exhibited strengths ranging from about 30 to 70 MPa. The studies indicate that the bond strengths decreased significantly as the height increased and were reducing to almost 40% of that at the bottom in the 1500 mm height of SCC columns studied. The reduction was much higher (about 80%) in the case of normally vibrated concrete. However, with the very low cement content and maybe even the compaction methodologies being difficult for such concretes particularly in the case of normally vibrated concretes, the utility of the results cannot be gauged adequately. However, the methodology adopted for the study and the implementation mechanism in the experimentation are certainly to be appreciated. This clearly shows the need for an understanding of SCC in terms of the granular distribution in the aggregates and the proportioning of SCC. The bond characteristics of 12–40 mm high-strength deformed bars of yield strengths decreasing from 629 to 570 MPa as the bar diameter increases were studied by Desnerck (2010). The concrete itself was an SCC with 600 kg/m^3 of cementitious materials content containing 240 and 300 kg/m^3 of limestone filler with the water content of 165 kg/m^3 resulting in a cylinder strength of around 60 MPa apart from a control concrete with 360 kg/m^3 of cement. The bond strengths are studied through a standard RILEM beam test. The studies indicate that the SCC beams performed similar to those with normal concrete.

Maghsoudi (2011) looked for the development of lightweight SCCs containing light expanded clay aggregates. The two concretes studied contain cement content of 360 in 450 kg/m^3 along with 40 and 50 kg/m^3 of silica fume and 150 and 50 kg/m^3 of LSP at a water content of 256 and 240 kg/m^3. The concretes achieved a strength of about 25–30 MPa. This is obviously significantly low for concretes containing silica fume but can be readily seen to be the result of the very high water content utilized as well as the fact that they used lightweight aggregates. The slump flow, V-funnel, L-box, and J-ring tests show a very good performance in terms of self-compaction. The stress–strain characteristics of concrete in compression show a very similar behavior to normally vibrated concretes. Atan (2011) attempted to look at the behavior of six SCCs primarily incorporating RHA along with other fine materials like LSP, fly ash, and silica fume. The concretes containing only about 475 kg/m^3 of cementitious materials used water content ranging from 185 to 255 kg/m^3 resulting in compressive strengths ranging from 20 to 45 MPa. His studies included other characteristics like flexural strength, UPV, and water absorption. Mohammadhassani (2011) studied the failure behavior and serviceability of six high-strength SCC deep beams of size 200×500×1300 mm reinforced on the face appropriately. The concrete itself was an SCC with a characteristic cube strength of 75 MPa design, resulting in a strength of around 90 MPa in the beams tested. The first crack and ultimate load capacities were reported apart from the failure characteristics of the beams tested. As the brittle failure reduces the strength of the structural elements below the flexural capacity, the performance of the beams was assessed through the energy absorbed, calculated from the area under the load deflection curve.

This review on a broad spectrum of studies shows the variety of powder extenders in particular and the corresponding strengths and mechanical characteristics broadly. While the studies present the individual compressive strengths of the concretes studied, it is obvious in most cases that the initial proportioning was probably done for a corresponding normal concrete and mineral admixtures were either added or replaced. The corresponding compressive strengths of the resulting concretes were investigated for the remaining mechanical and durability characteristics in most cases. It looks obvious thus that while the material combinations in the making of these SCCs were generally not a matter of concern for the mechanical behavior in particular most of the relationships follow essentially the trend of the corresponding normally vibrated concretes of the same strength. Insofar as the mechanical characteristics are concerned, this is mostly acceptable and most research has found that there could be minor variations from the relationships proposed by the various national codes. Even so the various factors that may be a major concern in the design of structures are broadly discussed to ensure that SCCs can be used with confidence. One obvious limitation that could be suggested as an overriding fact in all these discussions on mechanical behavior of SCCs is that the relationships as well as the discussions are probably valid for only concretes that do not contain very large quantities of powder content and are probably produced only to ensure self-compactability and not mechanical performance in one extreme end and on the other side those that do not contain sufficient fines but yet modulated through plasticizer–VMA combinations.

9.3 Compressive Strength and Strength Gain Rate

The compressive strength and strength gain rate are parameters that depend on the water and the cementitious constituents and their reactivity is modulated by time and temperature. It is also known that VMAs, essentially consisting of welan gum or others based on polysaccharides, often exhibit retardation during the early ages. These can be suitably addressed to take advantage of the retardation in the strengths in construction practice. Also, the compressive strength of SCC appears to have not been given the same level of importance as in normally vibrated concretes, basically in an effort to ensure that the mix in the first place is self-consolidating. Some of the mix designs reported earlier do start with the premise of obtaining a particular strength. Su (2001) while proposing a method to design SCCs based on the packing factors suggested that the cement content be calculated as a function of strength. He suggested that the water to cement ratio be ascertained from ACI or other methods in previous studies yet giving an equation for cement content in terms of the compressive strength, which obviously means that the specific water content was presupposed. Sebaibi (2013) proposed a mix design similar to that of Su with the same relation for the cement content, while the water–cement ratio strength relation used was from the Euro norms. These and other such facts were already discussed in the mix design procedures discussed at length previously.

The reported data in the literature indicates that in spite of the significant amounts of filler in SCCs the compressive strengths can be best understood through their porosity as in the case of normal concretes (Bonen and Shah, 2004). The investigations also indicate that similar to the gel-space ratio concept used for conventional vibrated concretes (Neville, 1995), the compressive strength was seen to be represented by a parameter called the binder-space ratio. They observe that the binder–space ratio calculated as the binder volume consisting of gel volume + filler volume + solid volume of the superplasticizers and VMA over the volume of the binder + capillary porosity + air will be similar to that in conventionally vibrated concrete gel-space ratio approach. The data from 10 different series shows that the binder–space ratio shows an adequately comfortable exponential relationship with the compressive strength. This in a way gives the confidence that for all practical purposes SCCs are indeed no different in their behavior from the conventionally vibrated concretes, assuming that they are designed with an understanding that abnormal fines content is avoided and an appropriate distribution of aggregate grading is assured.

There have also been discussions about the compressive strength to water–cement ratio relationships of SCCs that consider the effects of pozzolanic admixtures. Domone (2007) suggests that the effect of the pozzolanic admixture (additives) on strength can be analyzed by determining the cementitious strength (cement) efficiency factor (k) enabling the calculation of an equivalent water/cement ratio. He states that the powders adopted as additives are generally LSP, pulverized fuel ash (PFA) (fly ash), GGBS, and/or condensed silica fume (microsilica) and that the strength tends to be governed as much by the type and proportion of powder addition than by the water/powder ratio. The effect of an addition on strength can be analyzed by determining its cement efficiency factor (k). This is the factor by which the quantity of addition is multiplied to give an equivalent amount of cement, which is then added to the actual amount of cement for calculation of an equivalent water/cement ratio. Estimates of the k-factors could be made by the following:

- Assuming a typical variation of strength with the water/cement ratio for the cement
- Determining the equivalent water/cement ratio for the concrete with the addition
- Obtaining the k-factor by simple calculation

The values of k-factors for the 28-day strengths obtained by him in the analysis of the results from about 30 references for different powder extenders are as follows:

Limestone powder (LSP)	–0.29
Chalk powder	–0.23
Fly ash (PFA)	–0.56
Blast furnace slag (GGBS)	–0.86

It may be significant and highly appropriate at this stage to reiterate that this is the exact philosophy with which the entire assessment of the available data on SCCs was made presently. In fact the k-factors assessed for a reasonably high-slump well-compacted concretes containing aggregates below 25 mm were first examined. With the understanding that the dynamics of the chemical kinetics will not be in any way different between these and the well-formulated SCCs, the data available was assessed and presented at the different percentages of replacement for the various pozzolanic cementitious materials used hitherto in the earlier chapters. This proves that the assessments made and the mix design charts containing the water–cementitious materials ratio to strengthen relationships presented could all be used with confidence.

In an effort to develop a method capable of proportioning SCC mixtures with specified compressive strength, contrary to previous SCC proportioning methods that emphasized the fulfillment of fresh properties requirements more than strength requirements as described by them, Kheder et al. (2010) presented a proportioning method that combines the ACI 211.1 and EFNARC method to design concretes covering a wider range of 15–75 MPa. Their investigations concentrated on SCCs with LSP as a filler. It was observed that these mixtures fail to reach the required design strength level by about 8%, attributed to the lower 32.5 MPa grade Type I cement used. Kheder presented the green state characteristics of 30 mixes with LSP as the powder extender designed according to this method. He arrived at the strength to water–cement ratio relationship of SCCs based on the results of these mixes and a few others that are available in the literature (Druta, 2003; Al Salami, 2008; Rahman, 2008). Additional literature data was used to show that the proposed relation is in good agreement with others results. This strength to water–cement ratio relationship presented was similar to the original Abrams relation and was given as

$$w/c = \left(16/f_c'\right)^{5/6}$$

with the value of f_c' being in MPa. The one important factor specifically reported was that the study utilized an ordinary Portland cement, Type I, complying with ASTM C150 (Grade 32.5 MPa) and that the SCC cylinder specimens of 152 × 304 mm (6 × 12 in.) were cast without vibration. It was also explicitly stated "these mixtures failed to reach the required design strength level by approximately 8%. This may be attributed to the intrinsic properties of the cement used in this work (ordinary Portland cement, Type I, Grade 32.5 MPa)." It was also stated that to adjust this problem data from the literature (Druta, 2003; Al Salami, 2008; Rahman, 2008) was also used to construct a more reliable relationship between concrete compressive strength and w/c. Also, LSP in these investigations was considered just a simple filler. However, it is known from a few earlier reported studies that there is indeed a certain amount of reactivity that is exhibited by LSP, which was not considered by him. The pozzolanic reactivity of limestone in terms of its efficiency has already been discussed and presented earlier. The results of the present investigation were not included in those assessments because the cement used was specifically of a lower grade.

Figure 9.3 shows the date replotted along with the best fit relation and the proposed equation superimposed on it. The figure also contains the

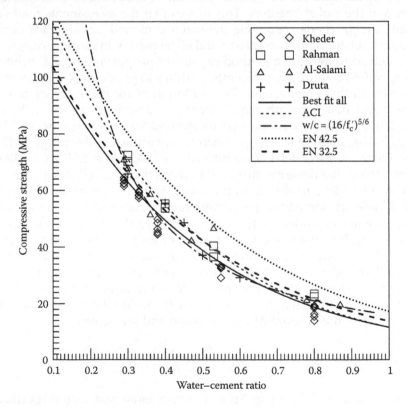

FIGURE 9.3
Water–cement ratio strength relation for SCCs. (After Kheder, G.F. and Al Jadiri, R.A., *ACI Mater. J.*, 490, 2010.)

water–cement ratio to strength relationships of ACI and also the EN 42.5, already discussed in earlier chapters. Apart from these, the EN 32.5 relation was also imposed on the graph to give a complete picture. As can be seen, the EN 32.5 relation is very close to the best fit. However, it is easy to recognize that the ACI code relationship also appears to be perfectly reasonable if only the upper bound values are considered. The power relationship suggested obviously tends to predict very high values for compressive strengths at low water–cement ratios (below 0.3 where there are no experimental results). It also underestimates the strength in the region between 0.3 and 0.8). Without going further it is best to conclude that the ACI relationships that are essentially valid for the cylinder strengths appear to be the most appropriate representation of the compressive strength to water–cement ratio relationship as can be seen even in these comparisons.

It can also be seen from the figure that the proposed power relationship is also acceptable within the range of concrete water–cement ratios available but can lead to much higher strength estimation and water–cement ratio below 0.25. Though the LSP has a very small effective cementitious reactivity, it is likely that the lower-grade cement that was used could not effectively bring out its full potential. This only goes to prove that in the production of SCCs the need for a better cement like the EN 42.5 MPa grade cement is considered a minimum. If one can use the EN 52.5 MPa grade, there could be several benefits like earlier strength gain, effective pozzolanic efficiency utilization, and even possible lower cement requirements for a particular strength grade of concrete.

Another such effort on concretes containing pulverized fuel ash from 20%, 40%, 60%, to 80% by volume (16%, 32%, 52%, 75% by weight) with powder content ranging from 439 to 539 kg/m^3 was studied by Liu (2010). A sand to mortar volume ratio of 45% and a coarse aggregate to concrete volume ratio of 35.5% were adopted in all the mixes. The water content varied between 160 and 180 kg/m^3, with the concretes achieving strengths of 16–70 MPa at 28 days and finally 37–83 MPa at 180 days. It was observed that the efficiency factors (k) evaluated by him for the 7-, 28-, and 90-day strengths of these concretes containing replacements up to 80% were fairly in line with what was already presented by Ganesh Babu (1996) for fly ash earlier. This is the proof that conclusively shows that the estimates of the efficiency of fly ash in concrete based on a large volume of data from several research investigations on normally vibrated concretes used by Ganesh Babu (1996) were indeed applicable for SCCs as well. In fact this should not come as a surprise as the factors that are discussed are the cementitious strength efficiency factors, which depend only on the reaction kinetics of the water–cementitious material system, which, if at all, is only more homogeneous in the case of SCCs because of the fluidity involved. His conclusions that the k-values presented earlier were a little lower than those observed for the SCCs studied by Domone (2007), who only indicates the efficacy of the methodologies that were proposed earlier. The fact also remains that the suggested k-values for fly ash in particular were actually lowered intentionally to ensure that the resulting strength of the concretes is always within reach even with the variations that are possible in the cement and fly ash characteristics from various sources. The equivalent water–cement ratio to strength relationship presented by him is almost in line with the relationship given by the Euro norms for the C 42.5 grade cement, which is the relationship proposed for all the evaluations of the pozzolanic efficiencies of the different materials considered.

As already stated, concrete mixtures containing silica fume can be proportioned based on an "efficiency factor—k" (Jahren, 1983; Sellevold and Radjy, 1983). Jahren also concluded that the activity index (efficiency factor) of SF varies considerably with SF dosage, curing conditions (higher with wet curing), age (higher with lower age), cement content (higher with lower cement content), type and dosage of plasticizer, and type of cement. He also recognized the need for additional water content as it was often necessary to increase slump by 50 mm when SF is added compared to normal concrete to maintain the same workability for both the concretes. From all these discussions, it is obvious that the compressive strength of concrete can be appropriately predicted through the efficiency of the different pozzolans.

At this stage, it is important to realize that with the increased fines and adoption of the cement content provisions and the codal recommendations, even values as low as 200 kg/m³ at about an additional 100% fly ash strengths of the order of 40–50 MPa can be easily achieved by the modulation of water and superplasticizer. In a way, it only proves that the hitherto held belief of cement content of over 300 kg/m³ for even 20 MPa concrete is no more valid. In fact it is easy to arrive at a medium-strength concrete through simple SCC formulations using fly ash or GGBS and even LSP. The possibility of achieving high-strength concretes of almost 80–100 MPa using silica fume should pose no difficulty. The strength gain rate of these concretes with pozzolanic cementitious constituents is very similar to their counterparts in the normally vibrated concrete systems. One of the important observations that can be reported from experience with the compression testing of both normally vibrated and SCCs is that the failure surfaces in the case of SCCs do not show very clean cleavage surfaces even in the case of high-strength concretes. The broken surface of an SCC specimen has a lot of powder associated with the fractured surfaces, which is certainly not the case of normally vibrated concretes or even medium strengths. This clearly shows that higher powder content and lower aggregate volumes have an effect on compression failure of these specimens. However, the split tensile fracture surfaces show a very clean cleavage and splitting without any such powder material observed on the fractured surfaces.

It may be prudent at this stage to recapitulate a few facts regarding the strength estimation of cementitious composites containing pozzolanic admixtures as already seen. In the first place, the idea of a cementitious strength efficiency factor (different from an efficiency factor that can be computed in terms of the various other parameters relating to the physical parameters or durability and corrosion) was originally suggested by Smith (1967). Starting then or maybe even a few years earlier, several attempts have been made to establish the strength efficiency of fly ash in concrete in particular by several researchers, and their recommendations have also been reflected in the specifications and codes. In fact the British code presently accepts a value as high as 0.5 for even replacements up to 50%. The same philosophy has been extended a little to specifically delineate this efficiency factor in two components, as the general efficiency factor dependent on the age and the percentage efficiency factor dependent on the percentage, and it is the same at all ages (Prakash, 1996; Rao, 1996; Sree Rama Kumar, 1999). A similar methodology was also adopted for the assessment of the efficiency of other pozzolanic admixtures later (Sree Rama Kumar, 1999; Appa Rao, 2001; Surekha, 2005; Narasimhulu, 2007). However, in most of these investigations for establishing the efficiency of the different cementitious composites, the basic strength to water–cement ratio relationships were individually established from the data available on the concretes containing no mineral admixtures as a part of their study. This in fact is probably the best way, as cementitious composites containing pozzolanic admixtures will certainly be influenced by the cement in particular and probably even the aggregates and their distribution in general. However, in the first place it is absolutely essential to be clear if the cements and other materials and their combinations produce concretes of acceptable quality as delineated by the well-accepted provisions in the standards available presently. While looking at some of these aspects, two equations representing cube and cylinder strengths proposed in the Euro code (for EN 42.5 grade cement) and the ACI recommendations for high-strength concretes (ACI 211.4, for 20 mm aggregate with high-range water-reducer (HRWR)) were proposed to be the guidelines against which most investigations will be compared. In fact it is probably most appropriate to clearly see how the assessments that were done initially in evaluating the efficiency of fly ash after correcting for the cementitious efficiency compare with the strengths that are expected by these relationships. Figure 9.4 presents such a view on the comparisons.

It is clear from this figure that almost all the concrete strengths are above the expected ACI values and almost 50% of them even surpass the values expected by the Euro norms. This indicates that the proposed basic strength to water–cement ratio relationships are indeed conservative in their expectations and could be reasonably guaranteed with most cementitious compositions containing fly ash even for strengths up to about 80 MPa. Not satisfied with this assessment, it was felt necessary to look at a similar comparison of concretes containing the high-efficiency pozzolanic admixture, silica fume, to have a reasonable confidence in what is being proposed. Figure 9.5 shows a similar comparison and one should remember that these concretes certainly did not contain cementitious materials in such high volumes or pozzolanic admixture replacements at such high levels. Even in the case of these silica fume

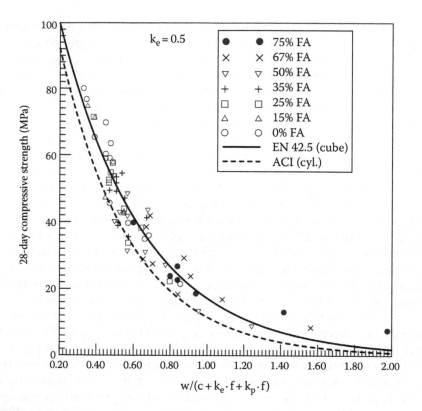

FIGURE 9.4
Comparison of fly ash concrete strengths with Euro and ACI norms.

concretes, the relationships appear to be reasonably adequate for strengths of about 80 MPa or water–cement ratios of about 0.25. In fact it is prudent to use this particular relationship for strengths ranging between 20 and 80 MPa or water to the effective cementitious materials ratios between 0.25 and 0.75 only, which is actually the range of strengths used for most concrete constructions.

9.4 Near-Surface Characteristics

One of the primary observations associated with SCCs containing VMAs in particular is that it is difficult to ensure that the entrapped air is expelled from the system. The type of mixer may also contribute to some of these problems. An obvious reflection of the entrapped air is that a good proportion of it adheres to the side surfaces and may be a cause for concern both aesthetically and also in terms of performance in certain cases. It is also possible that

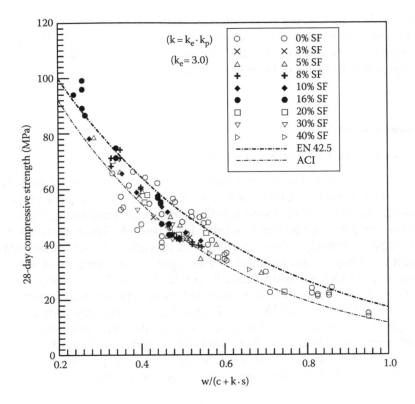

FIGURE 9.5
Comparison of silica fume concrete strengths with Euro and ACI norms.

this phenomenon could be observed with certain types of superplasticizers and also some specific powder extenders used. In a way it is obvious that pozzolanic admixtures like silica fume, which impart a very large thixotropy to the system, can also exhibit such surface air bubbles. Apart from this, internal air entrainment and distribution are also a matter of concern for the performance of the concrete in general. It is obvious that such problems are associated with specific constituents and maybe also due to the type of mold surfaces, and release agents have to be addressed locally by a suitable adjustment of the constituents or by changing the superplasticizer and VMA types as required. Another important factor that is very specific to such powder-rich materials requiring a much larger water content is the drying shrinkage problem, particularly at the surface leading to the hexagonal crazing cracks. The environmental effects due to the rate of drying could seriously accentuate these problems. Last but not the least of the problems is that with large quantities of other materials like fly ash, LSP, GGBS, and so on, which tend to absorb moisture through capillary rise. This can be a serious concern both for moisture migration from ground as well as surface absorption and walls, which will ultimately permeate into the interior making them significantly damp and giving rise to mold growth. Suitable measures both in terms of appropriate curing and probably the use of water-repellent treatments through admixtures are options in such cases.

9.5 Tensile and Shear Strengths

In an effort to study the effect of pulverized fuel ash or fly ash at levels up to 80%, Liu (2010) attempted to analyze the results of a few such investigations reported earlier (Dehn, 2000; Holschemacher, 2002; Leemann, 2005; Sukumar, 2008) along with his own data to understand the behavior of concrete. The evaluations of the data on the relationship between compressive strength and split tensile strength though according to him indicates that the split tensile strengths of SCCs are in the range of CEB-FIP model code 90 predictions for the normally vibrated concretes, clearly shows a distinct variation between the different investigators. In fact while the results of Dehn et al. (2000) show consistently lower tensile strengths (only around 7% of the compressive strength approximately), results of Holschemacher (2002) were fluctuating widely (between 7% and 13%) and his own investigations suggest them to be around 8%. Liu's experimental investigations resulted in an acceptably conservative best fit equation:

$$f_c = 8.3(f_s)^{1.26}$$

where
  f_c is the compressive strength
  f_s is the split tensile strength

In a broad sense, the various investigations on SCCs generally indicate that the currently available provisions of CEB-FIP can be safely adopted.

The shear strength characteristics have not yet been adequately addressed, in spite of the fact that it is this aspect that probably is very much in doubt as the significantly lowered coarse aggregate content will have an adverse effect on the aggregate interlock shear transfer capability. Though his investigations suggest that the improved microstructure of an SCC will adequately compensate for the aggregate interlock shear transfer effect, it is prudent to neglect the term representing this contribution in the shear calculations of structural engineering practice. Among all the critical failure factors that influence the ultimate strength of members and even collapse of structures, shear is probably one of the most significant and needs specific attention particularly because of its nature of sudden failure. This aspect is not given importance enough in SCCs and needs to be looked at. The failure pattern and failure planes showing large powder content explained in compression are certainly a matter for concern. It is suggested that the shear reinforcement be augmented sufficiently to address this aspect. Also as reported the bond strength development appears to be dependent on the confinement as discussed later, which is a sign of shear strength inadequacy. From this consideration also augmented shear reinforcement with closer spacing may be essential.

9.6 Applicability of Conventional Concrete Relations to SCCs

The specific aspect that is being discussed here is in fact not the relationships between the different strength and modulus parameters but actually the measurement aspects of the compressive and tensile strengths in general. One major difference between the European and American methods of assessing the compressive strength of concrete is that the European assessments are based on the 150 mm cube compressive strength while the American relationships are all based on 150 × 300 mm cylinders. There are a number of suggested conversion practices but for the purpose of this chapter the conversion table that was proposed by the Euro norms is followed. This aspect of the conversion and equivalence between the water–cement ratio to strength relationships of the European (for the C42.5 cement) and American standards for high-strength concretes was earlier discussed appropriately. The second aspect that might be of interest is the tensile strength test. Unlike the compressive strength test where the failure appears to be dominated by a lot of powder material at the failure planes (the broken surfaces) even in very-high-strength concretes or in normally vibrated concretes, the cleavage and separation do not show any such powder material, and the failure in split tension test is indeed clean and is very similar to that in the normally vibrated concrete.

9.7 Modulus of Elasticity

The modulus of elasticity of an essentially binary composite as defined by the moduli of the individual constituents at their respective proportions is well known. Naturally, with significantly lower aggregate content SCCs could exhibit a slightly lower modulus compared to their counterparts of the same strength. Investigations by Ambroise and Pera (2002), on SCCs ranging from 25 to 40 MPa containing fly ash or LSP, have shown a general reduction of about 10% in the elastic modulus compared to normal concretes, which was attributed by them to the high amount of paste and fine particles. Walraven (2003) also reported that the modulus of elasticity of the SCCs investigated by him was slightly lower than that of conventional concretes. However, investigations of Bonen and Shah (2004) on SCCs appear to confirm that the ACI 318-89 equation for normal-weight concretes given here could adequately predict the modulus of elasticity of SCCs almost up to the strength of about 80 MPa.

$$E = 4.73(f_c)^{0.5}$$

Their investigations also show that if the strengthening of the concretes is extended to about 150 MPa, the CEB equation below appears to be more appropriate.

$$E = 9.5(f_c + 8)^{0.33}$$

where
  E is the elastic modulus
  f_c is the compressive strength

9.8 Bond with Reinforcement

Studies on characteristics in terms of the pullout tests and also beam tests as discussed earlier clearly show that there is no serious influence due to the changes in the matrix characteristics of SCCs in most cases. Investigations by Domone (2007) have shown that the bond strength of steel to SCCs is certainly no lower than that of equivalent traditionally vibrated concretes and in some cases may be even higher. However, studies by Pandurangan (2010) on the bond characteristics of lap splices in tension reinforcement in concrete beams with specifically incorporated crack initiators clearly show that the beams with the confinement index less than 2.0 exhibited poor ductility and increasing the confinement index to 3.0 resulted in improved ductility, indicating the importance of confinement to adequately address the problems of bond strength.

Bond characteristics of prestressing wires particularly the anchorage zones of pretensioned components are always a matter of concern even in normally vibrated concretes. The fact that concretes of strengths around 40 MPa with fairly limited cement content in normally vibrated concretes appear to have adequate capacity in terms of transmission lengths indicates that more coherent and denser SCCs may not pose a serious problem. Comparative studies by Khayat (2003) indicate that there is no significant difference between the bond characteristics of conventionally vibrated concretes and SCCs investigated.

9.9 Creep and Relaxation

The phenomenon of creep, the plastic deformation under sustained loading, is a parameter that is of interest particularly in terms of prestressed concrete applications, both pretensioned and posttensioned. Relaxation is actually a corollary that occurs because of the creep deformation and a constant stain. Invariably performance of SCCs is expected to be certainly better than that of normally vibrated concretes, as the internal microstructure refinements due to the additional powder content (particularly so if they are of a pozzolanic nature) will indeed present significantly better in compressible mass. It is to be remembered that this comes with the rider that the concrete does not contain excessive quantities of powder materials and also the associated higher water content, which will invariably result in a highly porous mass that proves to be contrary to the previous expectations. However, there has not been any serious investigations reported in the area, though concretes of very high strengths or self-consolidation capabilities have been used in several highly loaded structures, superhigh-rise buildings, support masts of large-span cable-stayed bridges, and bridge piers and pylons to name a few. Another fact that is probably more relevant is also that these concretes of high strength tend to be significantly more elastic in compression, with a pronounced linear strain behavior till the peak stress, which will have much lower creep or relaxation problems.

9.10 Prestressing and Anchorages

Bond characteristics of prestressing wires particularly the anchorage zones of pretensioned components are always a matter of concern even in normally vibrated concretes. The fact that concretes of strengths around 40 MPa with fairly limited cement content in normally vibrated concretes appear to have adequate capacity in terms of the transmission lengths indicates that more coherent and denser SCCs may not pose a serious problem. Comparative studies by Khayat (2003) indicate that there is no significant difference between the bond characteristics of conventionally vibrated concretes and SCCs investigated.

9.11 Applicability of NDT

At the very outset, it may be stated that the discussions here pertain to only NDT characteristics relevant to the mechanical aspects being discussed, maybe more specifically to the information regarding the rebound hammer and UPVs alone. The higher paste content associated with SCCs in some cases may pose some problems in terms of the rebound numbers associated with low-strength concretes. However, experiments in the laboratory on several different types of self-compacting compositions have shown that there is no appreciable difference between the results related to these and normally vibrated concretes. Several investigations on the UPVs reported as discussed earlier have not shown any major variation between the results from the two types of concretes. However, it may be appropriate to look at some of the assessment criteria for NDT tests in its proper perspective. The rebound numbers associated with concretes in general range from about 15 to 45, but the values below 20 indicate serious defects while the values above 45 indicate a high integrity in the composite. It is almost impossible to delineate good compact concretes of strengths higher than even 40–50 MPa that exhibit rebound numbers beyond 35–40. In a nutshell, this is only a broad indicator of the surface hardness and integrity of the surface layer with the inner core to a large extent and no more. It helps in understanding the delamination problems associated with cover concrete and gives a broad idea of the concrete's homogeneity in general. Table 9.1 presents the assessment criteria as advocated by the CEB norms.

It is also important to note that to have a better insight into the integrity and presence of possible defects in a block of concrete the UPV scan at a few pertinent locations along with the rebound numbers mentioned here can be a useful tool in an experienced hand. In fact it is advised that the rebound number and the UPV are measured together to have a better control of the strength predictions. There are even some such test machines

TABLE 9.1

Comprehensive Assessment Criteria for Strength Evaluation

Concrete Quality	Rebound Number	Ultrasonic Pulse Velocity (km/s)
Very good	>40	>4
Good	30–40	3.5–4
Average	25–30	3–3.5
Poor	20–25	>0.35
Very poor	<20	Hallow sound
After reference	CEB, 1989	CEB, 1989

Source: After CEB, Diagnosis and Assessment of Concrete Structures—State-of-Art Report, Bulletin No 192, 1989, p. 120.

available though by and large the results from such universal assessments are highly questionable as the calibrations are not effective for all concrete types. Obviously, the limitations as observed for rebound numbers are also true for UPV values below 3.0 and above maybe 4.5. It can be clearly seen that the range of these values is far too small to be able to present a realistic picture of the concrete strength directly but can be a guide to the homogeneity and to a lesser extent porosity of the concrete mass. Table 9.1 presents the assessment criteria as advocated by the CEB for the UPVs in concrete.

Investigations on fly ash–based SCCs by Liu (2010) have clearly shown that there is almost negligible difference between normally vibrated concretes of the laboratory and the SCCs investigated. His investigations resulted in a relation between the UPV and compressive strength:

$$f_c = 0.003\, e^{2.5\,(\text{UPV})}$$

which was almost the same as that of normally vibrated concretes studied in their laboratory. Apart from this, the compressive strength and dynamic models of elasticity relation, which is generally obtained from the longitudinal and transverse UPVs, was given as

$$f_c = 0.0002\,(E_d)^{3.4}$$

In both these equations, f_c is the compressive strength and E_d is a dynamic modulus of the SCCs investigated. However, his comparison of the results of Khatib (2008), the relationship between the UPV and compressive strength, appears to show lower pulse velocities, which was attributed by him to be possible effects of the different materials, test instruments, and methods.

References

Ahmadi, M.A., Alidoust, O., Sadrinejad, I., and Nayeri, M., Development of mechanical properties of self compacting concrete contain rice husk ash, *World Academy of Science, Engineering and Technology*, 34, 2007, 168–171.

Ahmed, F.M., Nuruddin, F.M., and Shafiq, N., Compressive strength and workability characteristics of low-calcium fly ash-based self-compacting geopolymer concrete, *Proceedings of World Academy of Science, Engineering and Technology*, 5(2), 2011, 64–70.

Akcay, B. and Tasdemir, M.A., Mechanical behaviour and fibre dispersion of hybrid steel fibre reinforced self-compacting concrete, *Construction and Building Materials*, 28, 2012, 287–293.

Al Salami, S.T., Mechanical properties of conventional and self-compacting concrete as related to the mechanical properties of their binding mortar, MSc thesis, University of Mustansiriya, Baghdad, Iraq, April 2008, 119pp.

Altin, M., Saritas, I., Cogurcu, M.T., Tasdemir, S., Kamanli, M., and Kaltakci, M.Y., Determination of the resistance characteristics of self-compacting concrete samples by Artificial Neural Network, in *Proceedings of the Ninth International Conference on Computer Systems and Technologies and Workshop for PhD Students in Computing, CompSysTech'08*, Technical University-Gabrovo, Bulgaria, 2008.

Ambroise, J. and Pera, J., Design of self-leveling concrete, in Shah, S.P., Daczko, J.A., and Lingscheit, J.N. (Eds.), *First North American Conference on the Design and Use of Self-Consolidating Concrete*, ACBM, North Western University, Chicago, 2002, pp. 89–94.

Appa Rao, C.V., Behaviour of concretes with metakaoline, MS thesis, submitted to Indian Institute of Technology Madras, Chennai, India, 2001.

Askari, A., Sohrabi, M.R., and Rahmani, Y., An investigation into mechanical properties of self-compacting concrete incorporating fly ash and silica fume at different ages of curing, *Advanced Materials Research*, 261–263, 2011, 3–7.

Atan, M.N. and Awang, H., The compressive and flexural strengths of self-compacting concrete using raw rice husk ash, *Journal of Engineering Science and Technology*, 6(6), 2011, 720–732.

Bonen, D. and Shah, S.P., Effects of formulations on the properties of self-consolidating concrete, in *Concrete Science and Engineering: A Tribute to Arnon Bentur, International RILEM Symposium*, Kovler, K., Marchand, J., Mindess, S., and Weiss, J. (Eds.), in *Third RILEM Symposium on Self-Compacting Concrete*, Reykjavik, Iceland, 2004, pp. 43–56.

Bosiljkov, V.B., SCC mixes with poorly graded aggregate and high volume of limestone filler, *Cement and Concrete Research*, 2003, 33, 1279–1286.

Boukendakdji, O., Kadri, E.-H., and Kenai, S., Effects of granulated blast furnace slag and superplasticizer type on the fresh properties and compressive strength of self-compacting concrete, *Cement and Concrete Composites*, 34(4), 2012, 583–590.

Cassagnabere, F., Mouret, M., Escadeillas, G., Broilliard, P., and Bertrand, A., Metakaolin, a solution for the precast industry to limit the clinker content in concrete: Mechanical aspects, *Construction and Building Materials*, 24, 2010, 1109–1118.

Caverzan, A., Cadoni, E., and Di Prisco, M., Dynamic tensile behaviour of self-compacting steel fibre reinforced concrete, *Applied Mechanics and Materials*, 82, 2011, 220–225.

CEB, Diagnosis and Assessment of Concrete Structures—State-of-Art Report, Bulletin No 192, 1989, p. 120.

Dehn, F., Holschemacher, K., and Weibe, D., Self-compacting concrete (SCC) time development of the material properties and the bond behavior, *LACER*, 5, 2000, 115–124.

Dehwah, H.A.F., Mechanical properties of self-compacting concrete incorporating quarry dust powder, silica fume or fly ash, *Construction and Building Materials*, 26, 2012, 547–551.

Desnerck, P., De Schutter, G., and Taerwe, L., Bond behaviour of reinforcing bars in self-compacting concrete: Experimental determination by using beam tests, *Materials and Structures*, 43, 2010, 53–62.

Domone, P.L., A review of the hardened mechanical properties of self-compacting concrete, *Cement and Concrete Composites*, 29(1), 2007, 1–12.

Druta, C., Tensile strength and bonding characteristics of self-compacting concrete, MSc thesis, Louisiana State University, Baton Rouge, LA, August 2003, 108pp.

El-Dieb, A.S., Mechanical, durability and microstructural characteristics of ultra-high-strength self-compacting concrete incorporating steel fibers, *Materials and Design*, 30, 2009, 4286–4292.

Esping, O., Effect of limestone filler BET(H$_2$O)-area on the fresh and hardened properties of self-compacting concrete, *Cement and Concrete Research*, 38, 2008, 938–944.

Felekoglu, B., Turkel, S., and Baradan, B., Effect of water/cement ratio on the fresh and hardened properties of self-compacting concrete, *Building and Environment*, 42, 2007, 1795–1802.

Filho, F.M.A., Barragan, B.E., Casas, J.R., and El Debs, A.L.H.C., Hardened properties of self-compacting concrete—A statistical approach, *Construction and Building Materials*, 24, 2010, 1608–1615.

Ganesh Babu, K. and Siva NageswaraRao, G., Efficiency of fly ash in concrete with age, *Cement and Concrete Research Journal*, 26(3), 1996, 465–474.

Gencel, O., Ozel, C., Brostow, W., and Martinez-Barrera, G., Mechanical properties of self-compacting concrete reinforced with polypropylene fibres, *Materials Research Innovations*, 15(3), 2011, 216–225.

Gesoğlu, M., Güneyisi, E., and Özbay, E., Properties of self-compacting concretes made with binary, ternary, and quaternary cementitious blends of fly ash, blast furnace slag, and silica fume, *Construction and Building Materials*, 23, 2009, 1847–1854.

Gesoğlu, M. and Özbay, E., Effects of mineral admixtures on fresh and hardened properties of self-compacting concretes: Binary, ternary and quaternary systems, *Materials and Structures/Materiaux et Constructions*, 40, 2007, 923–937.

Gheorghe, M., Saca, N., Ghecef, C., Pintoi, R., and Radu, L., SELF compacted concrete with fly ash addition [Beton Autocompactant Cu Cenusa Zburatoare], *Revista Romana de Materiale/Romanian Journal of Materials*, 41(3), 2011, 201–210.

Glasser, F.P., Chemistry of cement—Solidified waste forms, in Spence, R.D. (Ed.), *Chemistry and Microstructure of Solidified Waste Forms*, Lewis Publishers, Boca Raton, FL, 1993, pp. 1–40.

Güneyisi, E. and Gesoğlu, M., Properties of self-compacting mortars with binary and ternary cementitious blends of fly ash and metakaolin, *Materials and Structures*, 41, 2008, 1519–1531.

Güneyisi, E., Gesoğlu, M., and Özbay, E., Strength and drying shrinkage properties of self-compacting concretes incorporating multi-system blended mineral admixtures, *Construction and Building Materials*, 24(10), 2010, 1878–1887.

Holschemacher, K. and Klug, Y., A database for the evaluation of hardened properties of SCC, *LACER*, 7, 2002, 124–134.

Hunger, M., Entrop, A.G., Mandilaras, I., Brouwers, H.J.H., and Founti, M., The behavior of self-compacting concrete containing micro-encapsulated Phase Change Materials, *Cement and Concrete Composites*, 31, 2009, 731–743.

Jahren, P., Use of silica fume in concrete, ACI SP-79, *First CANMET/ACI International Conference*, SP-79, Malhotra, V.M. (Ed.), Vol. II, 1983, pp. 625–642.

Jalal, M., Mansouri, E., Sharifipour, M., and Pouladkhan, A.R., Mechanical, rheological, durability and microstructural properties of high performance self-compacting concrete containing SiO$_2$ micro and nanoparticles, *Materials and Design*, 34, 2012, 389–400.

Khatib, J.M., Performance of self-compacting concrete containing fly ash, *Construction and building materials*, 22(9), 2008, 1963–1971.

Khayat, K.H., Petrov, N., Attiogbe, E.K., and See, H.T., Uniformity of bonds strength of prestressing strands in conventional flowable and self-consolidating concrete mixtures, in *Third RILEM Symposium on Self-Compacting Concrete*, Reykjavik, Iceland, 2003, pp. 703–712.

Kheder, G.F. and Al Jadiri, R.A., New method for proportioning self-consolidating concrete based on compressive strength requirements, *ACI Materials Journal*, 2010, 490–497.

Kou, S.C. and Poon, C.S., Properties of self-compacting concrete prepared with recycled glass aggregate, *Cement and Concrete Composites*, 31, 2009, 107–113.

Leemann, A. and Hoffmann, C., Properties of self-compacting and conventional concrete-differences and similarities, *Magazine of Concrete Research*, 57(6), 2005, 315–319.

Li, B., Guan, A., and Zhou, M., Preparation and performances of self-compacting concrete used in the joint section between steel and concrete box girders of Edong Yangtze River Highway Bridge, *Advanced Materials Research*, 168–170, 2011a, 334–340.

Li, J., Configuration of low-strength self-compacting concrete adapted to local raw materials, *Advanced Materials Research*, 335–336, 2011, 1159–1162.

Li, J., Qu, X., Chen, H., Li, J., and Jiang, L., Experimental research on mechanical performance of self-compacting reinforced concrete beam with recycled coarse aggregates, *Advanced Materials Research*, 2012, 374–377, pp. 1887–1890.

Li, J., Qu, X., Wang, L., Zhu, C., and Li, J., Experimental research on compressive strength of self-compacting concrete with recycled coarse aggregates, *Advanced Materials Research*, 306–307, 2011b, 1084–1087.

Liu, M., Self-compacting concrete with different levels of pulverized fuel ash, *Construction and Building Materials*, 24, 2010, 1245–1252.

Maghsoudi, A.A., Mohamadpour, S., and Maghsoudi, M., Mix design and mechanical properties of self compacting light weight concrete, *International Journal of Civil Engineering*, 9(3), 2011, 230–236.

Mehta, P.K., Condensed silica fume, Concrete technology and design, in Swamy, R.N., *Cement Replacement Materials*, Vol. 3, Surrey University Press, Guildford, U.K., 1986, pp. 134–170.

Mohamed, H.A., Effect of fly ash and silica fume on compressive strength of self-compacting concrete under different curing conditions, *Ain Shams Engineering Journal*, 2, 2011, 79–86.

Mohamed, M.A.S., Ghorbel, E., and Wardeh, G., Valorization of micro-cellulose fibers in self-compacting concrete, *Construction and Building Materials*, 24, 2010, 2473–2480.

Mohammadhassani, M., Jumaat, M.Z., Ashour, A., and Jameel, M., Failure modes and serviceability of high strength self compacting concrete deep beams, *Engineering Failure Analysis*, 18, 2011, 2272–2281.

Najim, K.B. and Hall, M.R., Mechanical and dynamic properties of self-compacting crumb rubber modified concrete, *Construction and Building Materials*, 27, 2012, 521–530.

Narasimhulu, K., Natural and calcined zeolites in concrete, PhD thesis, submitted to Indian Institute of Technology Madras, Chennai, India, 2007.

Neville, A.M., *Properties of Concrete*, 4th ed., Longman, London, U.K., 1995.

Nepomuceno, M., Oliveira, L., and Lopes, S.M.R., Methodology for mix design of the mortar phase of self-compacting concrete using different mineral additions in binary blends of powders, *Construction and Building Materials*, 26, 2012, 317–326.

Pandurangan, K., Kothandaraman, S., and Sreedaran, D., A study on the bond strength of tension lap splices in self compacting concrete, *Materials and Structures/Materiaux et Constructions*, 43, 2010, 1113–1121.

Parra, C., Valcuende, M., and Gomez, F., Splitting tensile strength and modulus of elasticity of self-compacting concrete, *Construction and Building Materials*, 25, 2011, 201–207.

Prakash, P.V.S., Development and behavioural characteristics of silica fume concretes for aggressive environment, PhD thesis, submitted to IIT Madras, Chennai, India, May 1996.

Rahman, A., Nondestructive tests of self-compacting concrete with compressive strength (20–80 MPa), MSc thesis, University of Mustansiriya, Baghdad, Iraq, May 2008, 108pp.

Rao, G.S.N., Effective utilization of fly ash in concretes for aggressive environment, PhD thesis, submitted to IIT Madras, Chennai, India, May 1996.

Roy, D.M., Silsbee, M.R., Sabol, S., and Scheetz, B.E., Superior microstructure of high-performance concrete for long-term durability, *Transportation Research Record*, 1478, 1995, 11–19.

Sahmaran, M., Christianto, H.A., and Yaman, I.O., The effect of chemical admixtures and mineral additives on the properties of self-compacting mortars, *Cement and Concrete Composites*, 28, 2006, 432–440.

Scheetz, B.E., Silsbee, M.R., and Roy, D.M., Fundamental aspects of the durability of cementitious waste forms, in *Proceedings of the Indo-US Workshop on Durability of Concrete Structures*, Indian Concrete Institute, Bangalore, India, 1993.

Sebaibi, N. et al., Composition of self-compacting concrete (SCC) using the compressible packing model, the Chinese method and the European standard, *Construction Building Materials*, 43, 2013, 382–388.

Sellevold, E.J. and Radjy, F.F., Condensed silica fume (Micro silica) in concrete: Water demand and strength development, First CANMET/ACI International Conference, SP-79, Malhotra, V.M. (Ed.), ACI SP 79, Vol. II, 1983, pp. 677–694.

Smith, I.A., The design of fly ash concretes, in *Proceedings of the Institution of Civil Engineers*, London, U.K., Vol. 36, 1967, pp. 769–790.

Sree Rama Kumar, V., Behaviour of GGBS in concrete Composites, MS thesis, submitted to Indian Institute of Technology Madras, Chennai, India, 1999.

Su, N., Hsu, K.C., and Chai, H.W., A simple mix design method for self compacting concrete, *Cement and Concrete Research*, 31, 2001, 1799–1807.

Sua-Iam, G. and Makul, N., The use of residual rice husk ash from thermal power plant as cement replacement material in producing self-compacting concrete, *Advanced Materials Research*, 415–417, 2012, 1490–1495.

Sukumar, B., Nagamani, K., and Raghavan, R.S., Evaluation of strength at early ages of self-compacting concrete with high volume fly ash, *Construction and Building Materials*, 22, 2008, 1394–1401.

Surekha, S., Performance of rice husk ash concretes, MS thesis, submitted to Indian Institute of Technology Madras, Chennai, India, 2005.

Tayeb, B., Abdelbaki, B., Madani, B., and Mohamed, L., Effect of marble powder on the properties of self-compacting sand concrete, *The Open Construction and Building Technology Journal*, 5, 2011, 25–29.

Tittarelli, F. and Moriconi, G., Use of GRP industrial by-products in cement based composites, *Cement and Concrete Composites*, 32, 2010, 219–225.

Turk, K., Caliskan, S., and Yazicioglu, S., Capillary water absorption of self-compacting concrete under different curing conditions, *Indian Journal of Engineering and Materials Sciences*, 14, 2007, 365–372.

Turk, K., Turgut, P., Karatas, M., and Benli, A., Mechanical properties of self-compacting concrete with silica fume/fly ash, in *Ninth International Congress on Advances in Civil Engineering*, 27–30, 2010, 1–7.

Valcuende, M. and Parra, C., Bond behaviour of reinforcement in self-compacting concretes, *Construction and Building Materials*, 23, 2009, 162–170.

Vejmelkova, E., Keppert, M., Grzeszczyk, S., Skalinski, B., and Cerny, R., Properties of self-compacting concrete mixtures containing metakaolin and blast furnace slag, *Construction and Building Materials*, 25, 2011, 1325–1331.

Vilanova, A., Fernandez-Gomez, J., and Landsberger, G.A., Evaluation of the mechanical properties of self compacting concrete using current estimating models: Estimating the modulus of elasticity, tensile strength, and modulus of rupture of self compacting concrete, *Construction and Building Materials*, 25, 2011, 3417–3426.

Walraven, J., Self-compacting concrete: Development and applications, in *Proceedings of Fib (CEB-FIP) Workshop*, Chennai, India, November 2003.

Zhao, H., Sun, W., Wu, X., and Gao, B., Effect of initial water-curing period and curing condition on the properties of self-compacting concrete, *Materials and Design*, 35, 2012, 194–200.

10

Performance and Service Life of Self-Compacting Concrete

10.1 Introduction

Understanding the performance and service-life behavior of constructed facilities is a matter of serious concern, particularly given the fact that many infrastructural facilities of several cities in the most advanced nations have also been found to be affected by aging. The upkeep, not to mention the renovation, rehabilitation, and replacement requirements, of even a fraction of these facilities that are urgently needed runs into astronomical figures that are unsustainable. The comprehensive assessment of United States of America's infrastructure as reported in the 2017 infrastructure report card (ASCE) summarizes the cumulative infrastructure's GPA to be just D⁺, which is not very encouraging. An estimate of the needs of the system, based on current trends extended to 2025, is at $4590 billion. This is one of the better managed system scenarios in the developed world, and it is not possible even to guess the status of others, particularly with increasing urbanization and the unplanned and totally disorganized, if not wild, growth of several of our megacities. Not to be an alarmist, but yet being pragmatic, it is also important to point out that there are estimates of corrosion losses particularly related to the industrial infrastructure that probably also need to be examined to understand the sheer size of the problem. It is only imperative that any material developed and promoted should go through specific checks and balances to ensure its performance over the expected life span of the structure. These are essentially the factors that are proposed to be discussed briefly in the following paragraphs to ensure that one is at least aware of the need for some of these investigations and assessments.

The performance of structures and materials invariably depends upon their environment, and this aspect has been discussed already along with the international recommendations. The present effort is to examine some of the parameters more closely while understanding that it is almost impossible to discuss them in a comprehensive manner. As far as the behavior of self-compacting composites is concerned, this understanding of the performance could be broadly classified into two: deterioration of the material composite due to environmental forces and the effect of environmental forces on the corrosion of the embedded reinforcement, which is invariably the larger of the two problems being studied. In both cases it is the permeation and ingress of either moisture or chemicals from the environment that are responsible for durability and service life. Thus, it is important to understand the physical and chemical dynamics of the permeable pore space in terms of both how to modify it and the effects of the environmental permeation that is taking place.

10.2 Durability of Concrete

Concrete has been accepted for a long time as a material of construction, even for structures situated in the very severe environment of the oceans. Concrete structures are generally designed for ensuring a maintenance-free (though not totally accurately) operating life span of over 50 years. However, experience in the recent past shows that there have been a number of cases of severe damage or failure, some of them even catastrophic, very much earlier. There have been a large number of investigations on the problems of the deterioration of concrete and the consequent corrosion of steel in concrete. A comprehensive review of all these works and the related ones is almost too much even to attempt. However, without a proper understanding of at least the significant parameters, a specific understanding of the problems and the remedial measures will be difficult.

The durability of concrete has been an important consideration in aggressive environments in spite of its other advantages like moldability, low cost, and so on. The deterioration, in general, may be caused by physical, chemical, or mechanical actions. Particularly, the constituents of the cement paste are susceptible to the sulfate ion in aggressive environments such as seawater. On the one hand, the sulfates chemically react with the hydrated C_3A compounds present in the concrete forming ettringite (C_3A 3CS 32H). The ettringite formed in the pores may cause considerable expansion and even cracking. On the other hand, the calcium hydroxide liberated during the hydration of cement reacts with the sulfate solution to form gypsum (CSH_2). This reaction is due to the acidic nature of sulfate solutions, which, contrary to the ettringite expansion cracking phenomenon, results in surface damage of the softening-spalling type (Mehta and Gjorv, 1974).

It is an established fact that the corrosion of metal in aqueous media is governed by an electrochemical mechanism. The current flows between the local anodes and cathodes, wherever a potential difference between the two exists. Steel in concrete is generally known to be highly resistant to corrosion.

One view is that steel is protected from corrosion by ferric oxide (Fe_2O_3) a surface film (approx. 10,000 Å thick) passivating the steel against corrosion (Shalon and Raphael, 1959). For corrosion to begin, this film must be broken or depassivated. Depassivation could happen if the alkalinity (basicity) of the pore water (electrolyte) in the concrete decreases. This can be caused by carbonation, especially near cracks, or dilution, which accompanies cracking. However, the depassivation due to a basicity decrease is considered unlikely because the reserve basicity of concrete from the crystalline $Ca(OH)_2$ present is so high that the pH is essentially constant even when a great amount of chloride ions penetrate the concrete (Gjorv, 1976). Thus, the depassivation must be caused directly by corrosive anions, which is the most important. Depassivation is thus induced directly by reaching a threshold concentration of Cl^- in concrete in the immediate vicinity of the steel surface. Another view is that no protective oxide film might exist (in the anodic area) and corrosion might be initiated by the formation of a chloride-ion film at the steel surface. Since the walls of pores in cement paste adsorb great amounts of Cl^-, the Cl^- film can form at the steel surface only if the Cl^- concentration exceeds a certain threshold value. However, the reactions within the interface layer are quite complicated and are not yet fully understood. The black rust may also form at oxygen-limited areas through a direct reaction with hydroxyl ion. It is the red rust that is responsible for the cracking of concrete because its volume is four times as large as that of steel, while the black rust volume is only twice as large (Stratfull, 1973).

A large amount of information is available on the rebar corrosion behavior under marine environmental conditions, as a result of the major research programs carried out by several investigators. A critical review of this literature shows that the rebar corrosion is influenced by a host of parameters, namely, environmental parameters, characteristics of the concrete constituent materials, and properties of the reinforcements. A brief review of the major parameters that influence the rebar corrosion behavior is discussed here, in order to emphasize their importance in understanding the corrosion behavior of steel in concrete.

10.2.1 Environmental Parameters

The exposure classes for concrete in relation to the environmental conditions and the corresponding limitations are all best presented in the latest CEB-FIP Model Code (1994). The recommendations available here are the best for achieving a satisfactory performance of the structure. It is clear from this as well as from other reports available that permeability of concrete is the most important parameter for overall durability. The corrosion performance of the materials in marine environment is commonly discussed according to specific environmental zones involved, namely, atmospheric, splash, tidal, shallow ocean, deep ocean, and mud zones. A classification of a typical marine environment and the probable behavior of steel in concrete

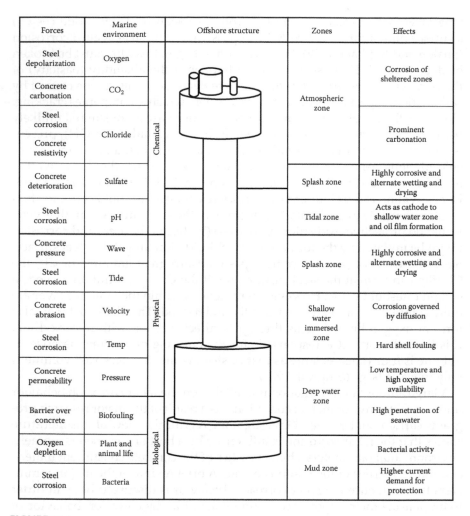

Forces	Marine environment	Offshore structure	Zones	Effects
Steel depolarization	Oxygen			Corrosion of sheltered zones
Concrete carbonation	CO_2		Atmospheric zone	
Steel corrosion	Chloride			Prominent carbonation
Concrete resistivity				
Concrete deterioration	Sulfate		Splash zone	Highly corrosive and alternate wetting and drying
Steel corrosion	pH		Tidal zone	Acts as cathode to shallow water zone and oil film formation
Concrete pressure	Wave		Splash zone	Highly corrosive and alternate wetting and drying
Steel corrosion	Tide			
Concrete abrasion	Velocity		Shallow water immersed zone	Corrosion governed by diffusion
Steel corrosion	Temp			Hard shell fouling
Concrete permeability	Pressure		Deep water zone	Low temperature and high oxygen availability
Barrier over concrete	Biofouling			High penetration of seawater
Oxygen depletion	Plant and animal life		Mud zone	Bacterial activity
Steel corrosion	Bacteria			Higher current demand for protection

FIGURE 10.1
Marine environment and its effects.

are presented in Figure 10.1. The properties of the seawater, which affect the corrosion behavior, may be subdivided into chemical, physical, and biological parameters. An overall summary of these environmental factors and their probable relationship with rebar corrosion is presented in Figure 10.1 (Ganesh Babu et al., 1990). Most of the research programs carried out on steel corrosion behavior in concrete composites are limited to the shallow water zone, atmospheric zone, and splash zone. It is reported that among the factors presented in Figure 10.1, the influence of dissolved oxygen and chlorides present in seawater is considered as the prime factor that influences rebar corrosion.

10.2.2 Concrete Parameters

Cement content, type of cement, water–cement ratio, aggregate type, size and distribution, nature of the water, and water content, apart from the mix proportions, will all affect the corrosion behavior of a rebar (Figure 10.2). Their individual effects on corrosion have been studied by several researchers. An important observation, at this point, is that many studies are done for cement paste and mortar, and there is limited information regarding concrete and more so on well-designed concrete mixes. Here, only such properties of concrete that are directly governing the time to depassivation or initiation of corrosion are discussed. These are alkalinity of concrete, resistivity of concrete, and the effect of cracking in concrete.

10.2.2.1 Alkalinity of Concrete

The alkalinity of concrete is one of the primary factors controlling depassivation of steel. Normally, concrete has a pH in the range of 12.5–13.2, due to the high calcium hydroxide content of the concrete (Gjorv, 1976). In this high alkaline environment, a thin protective coating (Fe_2O_3) is formed on the steel surface and remains intact up to pH 12.5. The protection of the film continues so long as the alkaline environment adjacent to the steel does not decrease to values of pH below 12.5. However, in the course of time, the alkaline environment surrounding the steel may be lowered due to the penetration of seawater (at a pH of about 8) and diffusion of salts through the porous concrete.

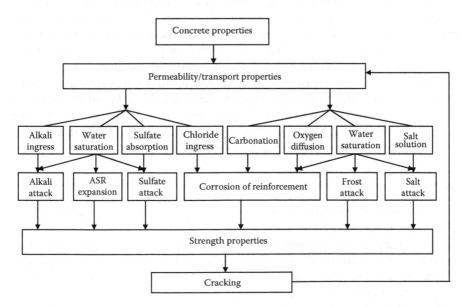

FIGURE 10.2
Review of parameters influencing degradation of concrete.

Also among the various salts present in the ocean waters, magnesium sulfate has a pronounced effect on the alkalinity of calcium hydroxide.

10.2.2.2 Resistivity of Concrete

The resistivity of concrete may influence the rate of corrosion by governing the flow of current through concrete from the anodic area of the steel to its cathodic area. Various factors governing the electrical resistivity of concrete were studied by many researchers, and it was observed that the water–cement ratio has a significant influence on the resistivity. Gjorv et al. (1977) have studied the resistivity of concrete in synthetic seawater considering a number of variables such as water–cement ratio, chloride concentration, moisture content, and hydration of concrete. Water–cement ratio and chloride concentration are found to have a significant influence on concrete resistivity.

10.2.3 Reinforcement Parameters

In most of the investigations on rebar corrosion to date, the influences of steel properties have seldom been considered. Some of the primary factors affecting rebar corrosion include the type of steel reinforcement (mild steel; high-strength deformed bars, both cold twisted and hot rolled; alloy steels; and prestressing steels), metallurgical composition, and surface condition. It is probably essential to also consider the effects of mechanical working on the steels during reinforcement cage fabrication leading to additional stresses at specific locations apart from welding. It is important to be able to quantify the status of deterioration of a reinforced concrete during its lifetime, to assess the need for repair, the performance of protection mechanisms in existence, and the need for application of protection methods. The study of mechanistic phenomena, or of proposed ameliorative measures, in the laboratory likewise requires measurement techniques. An array of methods are available, from mundane to the highly sophisticated, to study the corrosion behavior of reinforcement, and they can be broadly classified under tests related to concrete and reinforcements (Figure 10.3).

10.2.3.1 Tests Related to Concrete

The most important tests on concrete relating to the corrosion of reinforcements are alkalinity (pH), resistivity, diffusion to chlorides, sulfates, and oxygen permeability. An important factor to be kept in mind is that the factors affecting the durability of concrete will also directly influence the corrosion of steel in concrete, and many of them are interrelated. The alkalinity or pH of concrete, a parameter influencing the passivation film on steel, is primarily dependent upon the cement content and its interaction with the environment. Penetration of salt solutions in the case of marine environment and

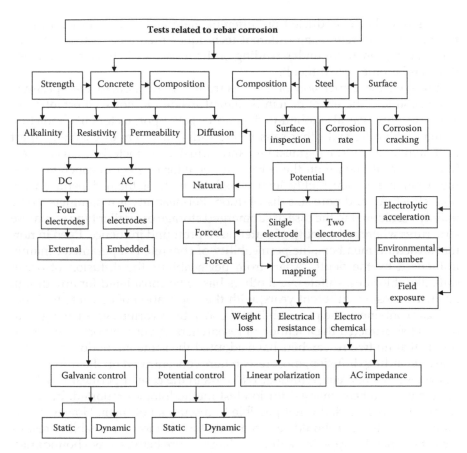

FIGURE 10.3
Overview of the methods of testing rebar corrosion.

carbonation in industrial areas are two of the important factors to be considered. Information on this parameter is scanty. The basic test methodology is to prepare an extract of the concrete and measure the pH directly.

Resistivity, an important parameter influencing the conductivity of the medium, which in turn allows the galvanic currents to be set up in the structure, is a parameter that has attracted a lot of attention in recent years. A comprehensive survey of the various methodologies used by different researchers showed basically two different systems using DC and AC circuits. Long arguments have been put forth for and against these two approaches. Apparently, it was observed that the DC systems are more suitable to establish the resistivity parameter relevant to the material parameters of concrete as an insulator material. Also in both the approaches, researchers have used both four-probe and two-probe techniques for measurements. In the two-probe technique, methodologies with surface mounted as well as embedded electrodes have been used. The reported literature shows that studies have been conducted on different materials such as cement paste, mortar, and concrete using different techniques mentioned earlier. At this stage a comprehensive understanding of the resistivity behavior of concrete thus has become very difficult.

However, the dimension of the electrodes in relation to the specimen dimension and the effect of reinforcement inside are not well established. The resistivity was also found to depend largely upon factors like moisture content, chloride, and other salts in the different layers of concrete, apart from temperature. The permeability and diffusion characteristics of concrete have a great influence on the corrosion behavior of steel in concrete. Both these parameters are highly interrelated, as both of them depend on the pore structures and pore size distribution. Permeability is a parameter to be mainly considered in the case of submerged storage or similar type of marine structures wherein the depth of water has a profound influence. Even in normal structural members, the permeability of concrete is a parameter mainly influenced by the depth. Apart from permeability, the diffusion of ocean salts like chlorides, sulfates, and others has to be considered for any design against corrosion. In recent years, with the application of concrete to liquefied gas containers, the permeability of concrete to various gases along with liquids has attracted the attention of researchers. A very important observation is that no two researchers have adopted the same methodology for the tests. Regarding diffusion, most researchers concentrated on the diffusion of chlorides into concrete (Whitting, 1983). As in the case of permeability tests, there have also been many diffusion test methodologies reported. In such a concise report as this, it is not possible to discuss all of them. However, one important factor that should be noted is that while most permeability results are established by applying a differential pressure between two boundaries, diffusion tests can be either through natural diffusion or by forced diffusion by applying a potential difference between two boundaries.

The depth of the carbonation layer in a concrete component can be assessed by several different methods. X-ray or chemical analysis of suitable samples, collected from different depths, to determine their content of calcium hydroxide and calcium carbonate, may be done in the laboratory. However, for the purpose of assessing the potential hazards of corrosion of embedded metal, adequate indications of the extent of carbonation are generally provided by the simple procedure of making the reduction in alkalinity visible by spraying appropriate and freshly broken surfaces of the concrete sample with solutions of suitable acid–base indicator compounds possessing different colors according to the alkalinity or pH value. It must be recognized, however, that tests with these indicator compounds only distinguish between the zones where the alkalinity of the concrete has been maintained at a high level and the zone where the alkalinity has been reduced to a pH value in the region of 8 or lower by carbonation or possibly even by leaching of the lime and alkali hydroxide with water. It is possible, therefore, that some carbonation has occurred in the zones that are shown to be highly alkaline and that complete carbonation has not been accomplished in the zones showing reduced alkalinity (Schiessl, 1988).

Because of the clear role of chloride ion in initiating corrosion of the steel, using the techniques to determine chlorides has become a common practice both in the laboratory and in the field. Most of the techniques involved entail sampling of the concrete above and around the reinforcing steel generally by coring, slicing the core as necessary, grinding, dissolving either in nitric acid or in water (depending on whether a water-soluble or a total chloride determination is required), and then conducting either an electrometric or a conventional titration against silver nitrate solution to determine the chloride level.

10.2.3.2 Tests Related to Steel Reinforcement

There have been many tests reported in the literature for the assessment of the corrosion of steel in concrete. Most of the earlier tests, primarily through field exposure, could provide only qualitative information, while in the recent years researchers have been using some techniques such as electrochemical tests in metallurgy to establish precise and quantitative results. A broad classification of these techniques is reported here. The first visible sign of reinforcing steel distress in a conventional reinforced concrete structure is frequently the appearance of rust staining on the surface. Obviously, this occurs only where the surface can be visually examined. At a later stage, cracks that may have developed at the subsurface reach the surface and spalling of the concrete ultimately ensues. At the earlier stage, it is possible to assess damage due to internal cracking by using a "sounding" technique involving chain drag or striking of the surface with a hammer, or alternatively by using an ultrasonic reflective method to determine the presence of subsurface delaminations. This technique is obviously useful on the upper surfaces of bridge decks, but it also has utility on vertical surfaces, and the ultrasonic technique may be modified to be of value even on submerged surfaces.

The relatively low conductivity of concrete, even when it is contaminated with chloride, and the fact that macrocell action is involved in the corrosion process have allowed the use of potential measurement on the top surface of the concrete to delineate anodic and cathodic areas of the reinforcing steel. This technique pioneered by Stratfull (1973) has been widely used by others since its inception, and it has been applied to structures such as buildings, piers, offshore structures, and others, besides the bridge decks for which it was originally used. Some fundamental problems also exist with the potential measurement technique. First, just because a number is obtained from the high impedance voltmeter used to take the readings, it should not be considered that this number represents anything regarding a rate measurement. It is exceedingly difficult, if not impossible, to assign any rate of corrosion to a given potential, except in a very qualitative sense. Second, the development of the potential is still not completely understood. By this it is meant that little critical laboratory work has been done to confirm the effects of coupling active and passive reinforcing steel in concrete under different conditions of oxygen, chloride, and moisture levels of the concrete. Different types of reference electrodes are in use, the suitability of which to specific environments has not yet been well established. Notwithstanding these difficulties of use and interpretation, the potential measurement technique has proved useful and reliable as a means of delineating the corrosion activity on a reinforced concrete, particularly when used with other nondestructive or semi-destructive techniques. These include coring for cover depth and chloride level determination and ultrasonic delamination detection.

10.2.4 Methods of Corrosion Control

The nature and severity of the problem of chloride induced deterioration of reinforcing steel in concrete have often been studied. It is not surprising that considerable research and development have been undertaken into ameliorative methods. These methods should either prevent the problem of corrosion of steel during initial design and installation or, conversely, allow efficient repair or rehabilitation procedures, which will extend the life of a structure that is deteriorating due to corrosion of reinforcing steel. The three basic types of protection techniques are—changing the environment around the reinforcing steel, either by decreasing the amount of chloride reaching the reinforcing steel to a value low enough to retard or prevent severe corrosion or by using admixtures in the concrete to essentially increase the concentration of chloride necessary to initiate corrosion; changing the nature of the rebar surface so as to be resistant to corrosion, either by surface treatment or by bulk alloying; and changing the electrochemical nature of the surface or the rebar by impressed current (cathodic protection).

In the last few years, new specifications and codes of practice have been introduced for the design and construction of concrete structures, and the latest among these, which needs special mention, is the CEB-FIP code (1994). This code specifies the limitations on the various concrete constituents, like cement content, water–cement ratio, and others applicable for the various environmental exposures. Some of the other codes are specifically meant for offshore concrete structures. Det norske veritas (DnV), the Norwegian Petroleum Directorate (NPDO), the British Standards Institution (BSI), the American Concrete Institute (ACI), and the Bureau of Indian Standards (BIS) have all introduced specific recommendations. For long-term durability in marine environment, areas of special concern have been mix proportions, cover to reinforcement, chloride content, and crack widths. Table 10.1 presents the maximum and minimum limits of these, recommended by various specifications. From the table it can be seen that these limits vary widely. They observed that with the normally accepted limit of 0.45 for the water–cement ratio, target strength of 47.5 MPa can be obtained and suggested that these higher strengths should be considered at the design stage.

TABLE 10.1

Appropriate Specifications for Offshore and Marine Concrete

Specifications	CP110 (1972)	DnV (1977)	NPD (1983)	ACI (1978)	BS (1985)
Mix Details	Cement Content kg/m³ and (w/c) Ratio				
Splash zone (critical)	360	400	400	—	400
	(0.50)	(0.45)	(0.45)	(0.45)	(0.40)
Clear Cover (c)	in mm				
Splash zone (critical)	60	50	50	65	60
Chloride Content	Values in % Cl⁻ by Weight of Cement				
Reinforcement	0.35	0.19	—	0.10	0.40
Prestressing steel	0.06	0.19	—	0.06	0.10
Crack Control					
Max crack width (mm)	0.3 or 0.004 × c	As agreed	—	—	0.30
Max steel stress (MPa)	—	160	—	120	

Source: After Browne, R.D. and Baker, A.P., The performance of structural concrete in marine environment, in *Developments in Concrete Technology*, Lydon, F.D. (ed.), Applied Science Publishers, London, U.K., 1979, pp. 111–149.

Coating in most cases serves as a means of isolating the embedded steel from the surrounding environment. Thus, an intact coating shields the steel from the various adverse conditions occurring at the concrete steel interface, which can cause corrosion of the steel and subsequent failure of the structure. In general, the coatings can be said to shield the steel from corrosive solutions and water and greatly reduce the area of exposed steel and consequently lower the probability that an area of exposed steel coincides with an area of protection failure by the concrete; and to increase the resistance of the steel to stray current corrosion, stress corrosion cracking, and hydrogen embrittlement. More specifically, coating of reinforcing steel is generally used to eliminate the effect of some anticipated factor that could promote corrosion. Coatings or impregnations must be classified with respect to their effect; they may reduce or impede Cl penetration, gas penetration (to decrease carbonation), or water penetration. To stop ongoing corrosion, the access of oxygen to the reinforcement must be either totally impeded or brought to a standstill. In the first case, the possibility of macrocell (local separation of anodes and cathodes) must be considered, which means that the whole structure is to be coated to guarantee absolute efficiency. In the case of reduction of water content within the structure, the influence of chlorides on the electrolytic conductivity must be considered. In any case, the durability of the protective measure itself must be checked.

In the previous paragraphs, we discussed the various parameters that influence the corrosion behavior of steel in concrete. At this stage it is also relevant to note that a complete and comprehensive investigation considering all the above parameters (environmental, concrete and steel materials, admixtures, and coatings) even for a small set of concretes is almost impossible because of the substantial requirements in time and budgetary provisions. However, it is to be recognized that there have been a few major investigations that cover a majority of these parameters, through a few national and international efforts, as already reported. One important yet general observation is that most of the investigations on the corrosion of steel in concrete did not lay enough emphasis on the characteristics of concrete in which the steel is embedded, but only concentrated on the corrosion aspects in mostly nominal strength or lower-grade concretes. There are two primary reasons for this. First, the researchers looking into corrosion were not aware of or were not interested in the significance of the parameters related to concrete. Second, and more important, corrosion is a long-term process by itself and will not lead to any significant distress of the cover concrete in a reasonable time frame. This time frame will be higher for larger covers and better concretes of even marginally higher strengths, restricting the ingress of the environment into the concrete.

This problem was sought to be remedied by accelerating the corrosion process, either by deliberately lowering the quality of concrete or by forcing the environmental ingress (wetting and drying, etc.) or by impressing a DC current to anodically corrode the steel in concrete. While the above methods have been useful to a degree in arriving at the comparative estimates between specific systems, the actual process of corrosion in different environments is impossible to duplicate by such accelerated methods. Also, in many cases even the basic characteristics of the concrete were not studied properly. It is thus important to note that studies on the corrosion of steel in concrete should be based on well-designed concretes over a wide strength range. As discussed earlier, many investigations reported to date were limited to only one or two specific characteristics, which make it difficult to understand the overall relationships between the different parameters. In brief, while many of these investigations help in understanding a specific parameter in relation to a few others, a comprehensive view would require a detailed investigation of at least the major parameters related to both steel and concrete.

10.2.5 Durability Investigations

Having broadly understood the complex interactions of the various forces that influence the performance of concrete structures, it is obvious that there is no one single method or rule for establishing the service life of constructed facilities. However, a broad guideline for an effective assessment may consist of the following. First, the performance requirements of the structure need an enumeration. The environment in which it is supposed to perform needs to be defined. The loading and deformations and the effects on the structure should be considered. Then the choice of materials and their combinations as well as the construction and maintenance procedure have to be ascertained. Even with all the care about these factors, one will have different levels of distress, and these have to be analyzed and understood. At this stage the causes of the signs of distress need to be analyzed. Some of the deterioration effects—such as modes of concrete deterioration, corrosion of reinforcing steel, mechanism of corrosion of steel in concrete, factors affecting corrosion of steel in concrete, effect of reinforcement corrosion on structural behavior, carbonation, freeze–thaw damage, alkali–aggregate reactions, sulfate attack, attack by seawater, acid attack, chloride penetration, influence of cracks on corrosion, chloride binding, and so on—have to be studied.

The previous paragraphs present a brief introduction to the intricacies of the deterioration of cementitious composites and their possible modifications through pozzolanic materials and superplasticizers. It is now possible to appreciate some of the individual efforts in the study of various facets of the deterioration behavior of SCCs appropriately. As was the case of the green state and hardened state characteristics, each of these investigations have to be first looked at through the fundamental parameters of materials, compositions, and interactions. This helps to place the reported literature in perspective for the later discussions. At this point the discussions are on specific investigations with the background of what was already discussed. Naturally, the studies that were discussed earlier under the green and hardened state characteristics do contain information on durability characteristics as well, and they will be discussed along with the present information appropriately.

A comparison between the effect of the filler in relation to a viscosity modifying agent on two self-compacting concretes (SCCs) of 40 and 60 MPa strengths, in terms of permeability, absorption characteristics, and chloride diffusivity, was presented by Zhu (2003). The fillers used were limestone powder and pulverized fuel ash (PFA) at different percentages. It is not clear with such a wide variation in filler characteristics, the free water contents, and even superplasticizer content if the results should be interpreted more in terms of the strength characteristics of the concretes as most of the parameters that are discussed are directly related to permeation characteristics. Craeye (2011) studied the effect of initial hydration and characteristics of SCCs with different mineral admixtures—limestone powder, quartzite, and fly ash. The study indicates that there is a certain initiation of hydration due to the mineral admixtures that was varying with the admixture. The study included mercury intrusion porosimetry (MIP) and also variation of ultrasonic pulse velocity (UPV) with time during hydration to establish these characteristics appropriately.

The characteristics of different SCCs with different combinations of limestone powder were studied by Persson (2003) through the salt frost scaling tests, immersed in 3% NaCl solution with a daily temperature cycling of −20°C to 20°C for 28 and 90 day concretes. The study indicated that at 6%, air content resisted the salt scaling effects better than 8%. Siad (2010) studied twelve concretes containing limestone powder, natural pozzolanic, and fly ash at three different strengths of 30, 50, and 70 MPa for the fresh and hardened state characteristics. In specific the mass loss characteristics in 5% H_2SO_4 and HCl solutions up to a period of 90 days were also reported. The study shows linearity in the mass loss characteristics.

A study essentially to characterize Helix fiber mortars at a very low water–cement ratio of about 0.27 with silica fume content of approximately 15%–17.5% was attempted by El-Dieb (2009). He achieved a strength of about 100–130 MPa at 90 days. The total cementitious materials content ranged from 775 to 900 kg. He studied the effect of steel fiber content on the electrical resistivity apart from the sorptivity and alkali attack on these materials. The study included chloride permeability and bulk chloride diffusion apart from the rapid chloride permeability test (RCPT) values with different fiber concentrations of 0%–0.52%, resulting in the charge passing levels of 100–500 coulombs. Boulekbache (2010) studied the compressive strength and direct shear behavior of 30 normally vibrated concrete (NVC), 60 (SCC), and 80 (HSC) MPa concretes at 0%, 0.5%, and 1% fiber volume fractions. The study generally indicates that fiber orientations and the fiber volume fraction significantly influence the yield stress and direct shear characteristics. One important observation is that concretes with a lower workability resulted in inadequate distribution and orientation of fibers.

The effect of aggregate grading on limestone powder–based SCCs was reported by Brouwers (2005). He also compared the effect of Fuller, Anderson and Anderson, and Su grading on the actual distribution of the aggregates with various particle sizes. The study included both the green and hardened state characteristics of concretes made with these different aggregate grading. One factor that could be learned from his study appears to be the usefulness of the 2–4 mm particles required in the sand grading as is obvious from the DIN characteristics that are suggested in this particular study. Ioani (2009) reported a broad outline of the quality assurance and quality control characteristics required to implement the durability and strength characteristics of SCCs in the precast industry in Romania. An important observation is that the recommended range of aggregate gradation for SCCs with 16 mm maximum size aggregates was seen to be a small band near the DIN-B grading, and his preferred range was between DIN-A and DIN-B. This corroborates admirably with the present suggestion of the average between the two of these two gradings as the best for achieving SCCs of the highest performance. His suggestion of exposure classification as the maximum water–cement ratios, minimum strength class, and minimum cement content appears to be in line with what was already suggested in the requirements of high-performance characteristics in the present publication. In a study using three different cements - ordinary Portland cement (OPC), an ordinary Portland cement with limestone powder (~14 mass%) and fly ash (~18 mass%), and an ordinary Portland cement combined with slag (~70 mass%) - with varying strengths, Leemann (2011) studied the influence of paste volume on E-modulus, flexural and compressive strengths, drying shrinkage, creep, and stress development under restrained conditions. The concretes contain different cements at a water–cement ratio of 0.40. The normal concretes contained cement content of 320 kg/m³ while the SCCs contained 520 kg/m³. The aggregate was continuously graded at 32% of 0–4 mm, at 16% of 4–8 mm, at 17% of 8–16 mm, and at 35%

of 16–32 mm, which corresponds to approximately the DIN–AB grading. SCC reaches lower values for E-modulus and compressive and flexural strengths but higher values for shrinkage, creep, and strain under restrained conditions. The comparisons in this particular investigation are not to be taken directly as a representation of the characteristics of SCCs because the cement content in normal concretes was 320 kg while that in the SCCs was 520 kg, which will automatically lead to a much higher shrinkage and creep levels. However, the comparisons between the normal concretes of 70 MPa with the SCCs at 50 MPa are also not really appropriate.

Barrita (2003) attempted to look at the characteristics of a specific SCC at different temperatures of curing including steam curing. She observed that curing at an elevated temperature of 50°C produced a durable concrete compared to moist curing through studies on unidirectional absorption characteristics of cylindrical specimens (Mnahoncakova 2008). Apart from this the vapor diffusion, thermal conductivity, and freeze–thaw resistance of these concretes were also studied. The study is essentially an addition to the database on SCC behavior as claimed by the authors. Anagnostopoulos (2009) studied the effect of temperature (300°C and 600°C) on SCCs with different fillers like slag, limestone powder, and glass filler in comparison with normal concrete. However, the limited cement content in the case of normal concrete of 330 and 375 kg may not be directly comparable with concretes containing an additional filler, particularly slag. The fillers, however, were added about 30%–40% in these variations with the water content at 190 kg/m^3. Normal concretes were obviously not self-compacting, while concretes with powders were seen to be self-compacting. The study reports the residual strengths and ultrasonic pulse velocities after exposure to the temperatures studied. The variation of the stress–strain characteristics after exposure to these temperature regimes was also reported. Studies on SCCs containing limestone filler and one normal concrete with 32.5 and 52.5 cements were reported by Fares (2010). The study included a comparison of differential thermal analysis and SCM images of these concretes apart from the room temperature. In a way these could give the effect of hydration and the pozzolanic effects apart from the high-temperature effects on these concretes. Uysal (2011a) studied the characteristics of SCCs containing 10%, 20%, and 30% replacement of limestone, marble, and basalt powders in a concrete containing a total cementitious material content of 550 kg at a water content of 182 kg resulting in a water–powder ratio of 0.33. The weight loss and ultrasonic pulse velocity characteristics of these concretes at different temperatures of 200°C, 400°C, 600°C, and 800°C were reported with and without polypropylene fibers.

Roziere (2007) studied different concretes with paste volume, water content, and limestone powder content variations through shrinkage and fracture characteristics of these SCCs. The restrained shrinkage ring test was used for characterizing the shrinkage effects on the fracture effects of measured-through beam specimens of 100 × 150 × 700 mm. The fracture behavior is characterized through fracture toughness and crack tip opening displacement

(CTOD) values. Acoustic emission maps of the fracture specimens were also attempted. It was seen that water content has a more significant effect than pace content on the strength and elastic modulus of these concretes. Li (2011) studied concrete of the joint section between steel and concrete of the box girders of the hybrid cable-stayed bridge across the Yangtze River. The high-strength SCC was made with 55 grade cement containing both steel and polypropylene fibers. The concretes achieved a strength of about 70–80 MPa. The split tensile and elastic moduli of these concretes were also studied. The fracture toughness characteristics of I_5, I_{10}, and I_{30} for these concretes along with the plastic shrinkage cracking and creep were measured. The chloride-ion permeability values of the 28-day concrete were observed to be less than 4.0×10^{-12} m^2/s. The freeze–thaw resistance showed that the relative dynamic modulus of elasticity was more than 60% after 200 cycles.

The characteristics of SCCs containing different powder extenders like fly ash, ground granulated blast furnace slag (GGBS), zeolite, limestone powder, marble powder, and basalt powder at different percentages were reported by Uysal (2011b). The cementitious materials content was constant at 550 and water content of 182 resulting in a water–powder ratio of 0.33. The effect of temperature was studied at temperature regimes of 200°C, 400°C, 600°C, and 800°C apart from room temperature. The green state characteristics were also studied. The hardened state characteristics in terms of the compressive strengths at room temperature appear to indicate strengths in the range of 60–80 MPa, which reduced to the range of around 10 MPa after exposure to about 800°C. The author used an artificial neural network methodology to understand the results of these concretes at the various temperatures, which could be a good predictive method for concretes of similar types exposed to fire and high-temperature effects in practice.

As the parameters that influence the durability of concrete are many, researchers have chosen different routes for investigating the several different combinations of SCCs. Some of the broader parameters and specifically their influence are discussed in the following paragraphs to draw a clear picture of their influence along with the possible relations to ensure higher durability and service life. In each one of these, a brief summary of some of the attempts reported in the literature is presented, and at the end some basic comments based on the author's experimental investigations in the laboratory have also been appended.

10.3 Strength and Porosity

In concrete, there is a continuous conflict between the need for lower water content to ensure a concrete of the highest strength with minimum cement content and thus cost, to the requirement for a higher amount of water to

ensure an appropriate compaction and thus avoid honeycombs and significant air entrapment with minimum energy inputs. The quantity of entrapped
air also depends on other parameters like the size and shape of the member,
reinforcement congestion, and ultimately the method of competition. In fact
the entire engineering of concrete technology actually revolves around this
one single aspect. While the science and technology can modulate the durability in the performance of cementitious composites, it may only be appropriate to realize that concrete is finally an inherently porous material, and an
understanding of this porosity will go a long way to have a handle on the
durability and performance of concrete.

The porosity of concrete is indeed well known, but what is not known
is the fact that there is such a large quantity of it with such great variation
that it cannot be measured easily by any simple system or for that matter
even by one that is more advanced. This porosity with the larger ones like
the entrained air bubbles ranging from about a few millimeters to the space
between the interlayers of the C-S-H gel that is only a few nanometers is the
one that dictates the strength as well as the durability and performance of the
cementitious composites. A very broad outline of the types of void spaces,
their causes and characteristics, and how they influence the moisture and
chemical transport phenomena along with the possible remedial avenues is
presented in Table 10.2. This table is only a broad indicator of the various systems of voids in concrete and show what could be the complexity in trying
to understand, assimilate, and address the various factors to ensure a better
performance of the cementitious composites.

10.4 Transport Characteristics

The transport characteristics of porous media generally refer to either the
permeation of fluids or the diffusion of ions in the system. It is only the permeation of fluids that is discussed here as the diffusion of ions in the system
actually influences the corrosion characteristics more often, which will be
discussed later. The permeation of fluids can also be broadly subdivided into
permeability due to a pressure differential or due to capillary absorption. The
test methodologies that evolved characterize a material like concrete through
several of these aspects. In the first place the sorptivity in terms of absorption
as well as desorption of water is easy to assess because concrete undergoes
no major physical and chemical changes during the process. In the laboratory a concrete sample is dried (generally at a temperature of 105°C) in a hot
air oven while the moisture loss is measured with time till two subsequent
readings have an insignificant difference between them (mostly predefined
depending upon the accuracy required). The absorption characteristics of a
fully dried concrete sample can be easily obtained by immersing the same

TABLE 10.2

Schematic Outline of the Complex Void System in Concrete

Void Type	Size	Approx. Vol. (%)	Measurement	Cause	Transport Processes	Technology Remedies
Production						
Surface air bubbles	0.5–2.0 mm	—	Visual, OM	Viscosity	Absorption permeation	Mix design
Entrapped air bubbles	0.2–2.0 mm	8	Visual, OM	Paucity of compaction	Absorption permeation	Mixer compaction
Entrained air	0.1–1.0 mm	Variable	Visual, OM	Admixtures	Absorption	Design
Intrinsic						
Macrocapillaries	50 nm–10 μm	6	OM, SEM	Bleed channels	Absorption permeation	w/c ratio pozzolans
Microcapillaries	5–50 nm	4	SEM, NA	Bleed channels	Diffusion	w/c ratio pozzolans
Macrogel pores	2–10 nm	20	NA, EIS	Hydration	Diffusion	Pozzolans
Microgel pores	>2 nm	10	NA, EIS	Hydration	Diffusion	—
Interfacial transition zone	10–50 μm	—	SEM	Hydration	Diffusion	Pozzolans
Bleed water lenses	10–100 μm	—	SEM	w/c ratio	Diffusion	w/c ratio
External causes						
Microcracks	>50 μm	—	OM, SEM	Shrinkage thermal	Permeation	Curing
Macrocracks	<50 μm	—	OM	Stress	Permeation	Loading
Honeycombs	mm/cm	Large	Highly visual	Improper practices	Direct filling	QA/QC

Note: OM, optical microscopy; SEM, scanning electron microscopy; NA, nitrogen adsorption; EIS, electrochemical impedance spectroscopy.

in water and measuring the rate of absorption with time. One can also use a similarly dried concrete specimen to evaluate capillary absorption (weight of water absorbed with time) by exposing one side and the water surface (with the sides appropriately sealed and the top exposed if required). Table 10.3 presents the typical assessment criteria based on a 30-minute absorption to understand the quality of concrete. The fact that in 30 minutes primarily major permeable porosity is affected by absorption is the principle behind such a specification. In fact the comparison of this as a percentage of the total absorption either over a longer period of time like 3–5 days or through the boiling water method suggested by ASTM will be able to present the relative proportion between the macro- and microporosities in a concrete composite.

If one needs a more accurate understanding of these results, the square root of time to weight gain/loss relationships could be very useful in all these evaluations. This simple study of the absorption, desorption, and moisture migration through capillary porosity will be able to give almost all that is there to be understood in terms of porosity and pore size distribution (at least for an engineering assessment) as this represents permeable porosity, which influences both permeability and diffusion characteristics to a large extent. Understandably the chemical attack characteristics of cementitious composites are also largely influenced by this particular permeable porosity, of course with the exception that the deterioration at the exposed surface that is directly in contact with the chemical solutions is dictated by reaction kinetics of cementitious materials and the chemicals involved in the system primarily.

However, for either a more detailed understanding or addressing more specific modulations of the material and its interaction with specific environmental aspects, it is possible to assess the porosity and pore size distribution more accurately by using MIP. Further the level of understanding pore structure at and below the nanolevels possible with MIP is to use nitrogen (gas) adsorption or impedance (electrochemical) spectroscopy. Carbonation driven by carbon dioxide in the environment interacting with moist concrete surface could enable understanding the porosity at such microlevels. Attempts have also been made to directly assess carbon dioxide permeability. A further extension of this is the assessment of the oxygen permeability that influences corrosion characteristics. Though there have been attempts to understand O_2 permeability from the earlier CO_2, the fact that molecular sizes are different is considered relevant by a few researchers.

Sorptivity measurements can also effectively assess changes due to microcracking occurring from temperature effects, thermal cycling, and even drying shrinkage characteristics if only the specimens in the measurements are examined carefully. Instead of the moisture migration test described earlier, the surface water absorption characteristics of concrete can be measured through the ingress of water into the concrete by attaching a small reservoir

TABLE 10.3

Comprehensive Assessment Criteria for Evaluation of Durability of Concrete

Concrete Quality	30 mts Absorption (%)	30 mts Surface Absorption (mL/m²/s)	Permeability (ms)	Oxygen Diffusion (m/-/s)	Resistivity (Ω-cm)	Potential w.r.t. SCE (Converted) (mV)	Cl⁻ Penetrated (%)	Chloride Permeability (Coulombs)
Very good	—	—	—	—	—	—	—	>4000
Good	<3.0	<0.17	$<10^{-12}$	$<5 \times 10^{-8}$	>20,000	>−120	<0.4	2000–4000
Average	3.0 to 5.0	0.17 to 0.35	10^{-12} to 10^{-10}	5×10^{-8} to 5×10^{-7}	10,000–20,000	−120 to 270	0.4–1.0	1000–2000
Poor	>5.0	>0.35	$>10^{-10}$	$>5 \times 10^{-7}$	5,000–10,000	<−270	>1.0	100–1000
Very poor	—	—	—	—	<5,000	—	—	<100

Sources: After CEB, Diagnosis and Assessment of Concrete Structures—State-of-Art Report, Bulletin No. 192, 1989, CEB, p. 120, 1989.

or cup to the surface by means of a capillary tube. The rate of water absorption calculated per unit area in unit time presents the quality of the concrete (Table 10.3). A similar test for air permeability was also proposed where the drop in air pressure in the reservoir attached to the surface is measured by using a U-tube arrangement.

One can alternatively use a more direct method for permeability by measuring the rate of the flow of water through concrete. There are several test devices available for this, which measure the flow rate through a sealed specimen by applying water pressure on one face and measuring the flow rate through the thickness. Apart from some of these, the Germann water permeation (GWT) test apparatus was also used in the laboratory to measure the permeability of both SCCs and normally vibrated concretes. It can be said with confidence that the permeability characteristics of both SCCs and normal concretes are nearly similar if one compares concretes containing the same mineral admixture at the same percentage. The permeability criteria for the assessment of concrete quality are given in Table 10.3.

There are several other methods and systems that were used by various researchers for measuring the permeability of concrete, which are probably not very different from the ones described earlier in either theory or practice. However the permeability of gas, be it carbon dioxide or oxygen, is also measured in a way similar to the surface water absorption tests than those described earlier. In this case the specimen sealed in a steel ring with the cut surfaces exposed is applied with the gas at a known temperature, pressure, and flow rate on one side. A helium gas stream passing on the other side at the same temperature, pressure, and flow rate is analyzed through a gas chromatograph, and the diffusion coefficient is determined after an equilibrium state is reached. The assessment criteria for the test are presented in Table 10.3.

After looking at some of the criteria for assessing the porosity and permeability of cementitious composites in general, one can see that investigators have followed a wide variety of these and combinations thereof to assess the characteristics of SCCs they studied. The transport of either moisture or ions into the body of a concrete member is governed by the porosity and particularly the permeable pores that are generally larger fractions. Therefore, reducing the permeability of concrete significantly improves the durability as well as corrosion performance. SCCs with their augmented fine particles and superplasticizers present a relatively dense structure that will reduce the porosity and permeability. One of the simplest measures of permeable porosity is its sorptivity characteristics as already stated.

The porosity/permeability of concrete are probably the most important parameters influencing both the strength and durability characteristics in general. The main cause for corrosion of steel in concrete is the penetration of CO_2, Cl^-, O_2, and other aggressive ions to the level of steel. Concretes of lower permeability physically resist the permeation of these aggressive ions. Understandably, this part of the ion permeability or diffusion is more relevant in addressing corrosion characteristics, and thus studies relating to these aspects are discussed later in the appropriate section. MIP is an excellent tool not just to understand porosity but also pore size distribution at all levels of permeable porosity. Ludwig's (2001) investigations on the porosity of SCC and normally vibrated concretes made with fly ash with similar water to binder ratios through MIP observed increased porosity under the coarse aggregate particles in vibrated concrete, while SCC showed a dense microstructure in the transition area. In a study on the effect of limestone filler and fly ash in relation to a viscosity modifying agent with two SCCs of 40 and 60 MPa strength, Zhu (2003) reported oxygen permeability and capillary water absorption. The results indicate that the SCC mix using no additional powder but a viscosity agent exhibits higher permeability than the other SCC mixes that used limestone powder or PFA at both strength grades. De Schutter (2003) reported that the water permeability and gas permeability of SCCs were lower than those of normal concrete at the same water–cement ratio. The water absorption characteristics of SCCs containing 30% fly ash and 10% silica fume cured for periods of 3, 7, 14, and 28 days under different curing conditions studied by Turk (2007) showed a good correlation with compressive strength regardless of the curing regimes. Rougeau (1999) studies through MIP also show a better performance of SCCs. Even investigations by Assie (2003) show a different microstructure in SCCs compared to that of vibrated concretes.

Khatib (2008) studied a concrete with fly ash replacement levels of 20%, 40%, 60%, and 80% at a constant cementitious materials content of 500 kg/m^3. The study showed water absorption increases with the increasing level of fly ash as was also obtained in some other studies of the author earlier. A study by Atan (2011) on six SCCs incorporating rice husk ash (RHA) along with other fine materials like limestone powder, fly ash, and silica fume showed no major difference in water absorption characteristics. Studies on slag and metakaolin concretes were reported by Vejmelková (2011). It was seen that the SCC with metakaolin also exhibited a higher durability, water absorption, and water penetration apart from lower capillary porosity. Studies on fresh, mechanical, as well as durability characteristics of SCCs containing micro- and nanosilica 10% and 2% individually as well as together by Jalal (2012) showed that water absorption and capillary absorption significantly decreased. Barrita (2003) attempted to look at the characteristics of a specific SCC at different temperatures of curing including steam curing. She observed that water uptake experiments showed higher penetration at the waterfront, higher water diffusivity, and also higher sorptivity for the specimens cured at higher temperatures for a short period of time.

There have been several studies in the laboratory on both SCCs as well as conventionally vibrated concretes over the years. The studies effectively suggest that there is almost a 6%–12% permeable porosity depending on the strength of the concrete. However, the CEB-FIP recommendation for broadly assessing the quality of performance is based on the 30-minute absorption value (of a 75 mm diameter and 75 mm long core, oven dried at 105°C, immersed in water) as suggested in Table 10.3. One fact that should be noted is that concretes containing fly ash as the powder extender in SCCs or as supplementary cementitious material in the normally vibrated concretes have always shown a much higher water absorption characteristics compared to their counterparts without fly ash in the laboratory. The fact that this characteristic is not observed with the other pozzolanic materials like silica fume should also be noted. Invariably compared with the concretes containing the same amount of the specific pozzolanic mineral admixture, be it fly ash, GGBS, silica fume, and so on, the corresponding water absorption characteristics of SCCs are always similar to those of normally vibrated concretes with the same admixture at the same percentage. Not knowing this fact has resulted in some researchers interpreting that SCCs could show a higher absorption. Incidentally, in ternary mixtures containing a small quantity of high-end pozzolan silica fume or metakaolin along with fly ash, the initial water absorption characteristics in 30 minutes were not affected though these concretes contain a fair amount of fly ash. This may probably be due to the fact that the internal pore structure is significantly broken down through the pozzolanic action of the high-end pozzolan even at such low percentages. In a way this is one aspect that is taken advantage of in very high self-compacting strength concretes used in some of the high-rise constructions all over the world.

10.5 Environmental Degradation

Structures are always exposed to the vagaries of nature, be it rain or shine. The effects of higher temperature differentials during the day and night and between the seasons in the tropics as well as the freezing and thawing effects in the colder regions exert specific forces on both green and hardened concretes in various ways. The expansion of water in the pores of concrete due to freezing can set off forces that are significant due to the confinement effects, and in fact the freeze–thaw durability test (ASTM C666) itself is one that evolved from it. The thermal expansion and contraction in concretes exposed to thermal cycling is particularly relevant to the most critical structures of the nuclear power industry.

The effect of simple wetting and drying or heating and cooling has also been used for understanding the changes in the microstructural characteristics and microcracking of structural concretes exposed to temperature differentials continuously, as in nuclear structures. Experience in the laboratory shows that simple water absorption can also reflect to a certain extent some of these changes, though sometimes hampered by the autogenous healing of cracks. A more appropriate understanding could be through ultrasonic pulse velocity measurements. The ultrasonic pulse velocities of dry concrete samples are always subject to considerable variation, but these assessments have to be made only in the dry state with a suitable coupling agent. A much more accurate method is to assess the dynamic modulus through sonic measurements. It is also possible to arrive at the dynamic modulus by measuring the longitudinal as well as transverse ultrasonic pulse velocities and calculating the same through theoretical equations.

Investigations on freezing and thawing of SCCs containing fly ash by Ludwig (2001) showed that though SCCs achieved higher strengths, they were also exhibiting higher weathering losses, which were attributed to their lower porosity. SCCs containing limestone powder and fly ash studied for their freeze–thaw resistance by Mnahoncakova (2008) showed no major differences between their behaviors. Studies by Persson (2003) indicated that the frost resistance of SCCs was better than the corresponding conventional concrete at similar air content and water–cement ratios, though the salt scaling effects were nearly the same. Studies on two concretes containing cement and 40% metakaolin and a blast furnace slag cement with 56% slag by Vejmelková (2011) show that the metakaolin concretes performed better in terms of the freeze–thaw resistance because of the finer pore structure compared to the slag concrete, with a mass loss of 0.396 kg/m² after 56 cycles.

Several investigations were undertaken in the laboratory on the thermal cycling effects of concretes in the dry state as well as through alternate wetting and drying in water and in seawater. Studies were also done by exposing concrete samples to such alternate wetting and drying in sodium animation chloride solutions apart from the corresponding sulfates to understand their effect on concrete as part of the environmental exposure studies. Though the numbers of studies on SCCs are fairly limited, these investigations specifically show that the difference between SCCs and normally vibrated concretes is almost negligible so long as both these concretes have the same pozzolanic admixtures at the same percentage. This information is extremely helpful as the amount of information available on normally vibrated concretes with different pozzolanic admixtures at various percentages is enormous and will at the least form the basic background for studies on the SCCs or even for their understanding and adoption. In fact unless there is a specific aspect that is known to alter the behavior of an SCC in comparison to the normally vibrated concrete with essentially the same constituent composition, chemical and physical characteristics will not be significantly different from the facts and the chemical interactions are primarily dependent on the chemical kinetics of the system. Also, the physical characteristics are in fact primarily dependent upon strength, though there could be some minor variations in terms of the porosity and pore size distribution.

10.6 Chemical Degradation

Degradation or deterioration of concretes can be due either to the naturally available chemicals in soils and water (like sulfates in seawater) or maybe from any specific industrial environment. The two primary effects of chemicals on concrete are generally due to acid attack or sulfate attack. Studies on these forms of chemicals in particular are attempted through exposure to their weak solutions, and some of them are used as specific tests for investigations on the durability of concrete composites designed.

10.6.1 Acid Attack

The modern world's increasingly urbanized and industrialized environments have shown a propensity toward the acid rain phenomenon. The effect of acidity in the waters will also have a neutralizing effect on the predominantly basic concrete pore solutions and destroy the chemical equilibria that are responsible for the strength of concrete. The softening and dissolution of exposed concrete surfaces in acidic industrial environments is well known. Though not a direct representation of the specific phenomenon, studies on the effects of weak acids (5%–15% HCl, H_2SO_4, or HNO_3) have all been used as a representation of the chemical resistance of concretes in general. The parameters used for this assessment are normally the weight loss and the strength loss from the change in solid diagonals of the cube specimens used in the tests sometimes. Persson (2003) studied the characteristics of different SCCs with different combinations of limestone powder through the salt frost scaling tests, immersed in 3% NaCl solution with a daily temperature cycling of −20°C to 20°C for 28- and 90-day concretes. The study indicated that a 6% air content resisted the salt scaling effects better than 8%. SCCs made with fly ash and silica fume were exposed to a sulfuric acid solution for 35 days by Stark (2002). The mass loss in SCCs was lower at all ages during the exposure period of 90 days compared to the control concretes.

10.6.2 Sulfate Attack

The effects of sulfate has been the primary topic of concern particularly in relation to the behavior of concretes in marine environments and also in sulfate laden soils. Sulfate concentrations (water-soluble sulfates expressed as SO_4) ranging from 150 to 10,000 ppm are categorized by ACI 318 into mild, moderate, severe, and very severe exposure classes, while the EN 206 classifies sulfates from 600 to 6000 ppm into the categories moderately aggressive and highly aggressive to indicate the significance of durability concerns with using them. Most structural design codes prescribe specifications in terms of cements, maximum water–cement ratios, and minimum cement contents to counter the effects of sulfates on the structure. It is not possible or even appropriate to deal with all this at length, and what is presented here is to show the importance of the subject in terms of durability. In principle the deterioration in sulfates chemically reacts with the hydrated C_3A compounds present in the concrete forming ettringite (C_3A 3CS 32H) in the pores causing considerable expansion and cracking or resulting in softening–spalling. Stark (2002) investigated SCC containing fly ash and silica fume for the effect of sulfate resistance. Cores exposed to 5% sodium sulfate solution for 84 days and 168 days in a 5% sodium sulfate solution showed that the SCCs had higher sulfate resistance than the reference concrete in terms of the weight loss and tensile strengths studied. SCCs containing limestone powder exhibited a significant decrease in weights when exposed to sodium sulfate solution (Persson, 2003). It was felt that this could be due to the large quantity of limestone powder in the mixes. Nehdi (2004) studied the durability of SCC with high-volume replacements of fly ash, slag, and silica fume or rice husk ash combinations. The sulfate expansion of SCCs immersed in 5% Na_2SO_4 for 9 months showed that the sulfate expansion decreased substantially with fly ash addition. SCCs with ternary and quaternary mixtures have shown even lower expansions probably because of the effects of silica fume or RHA.

10.6.3 Carbonation

The reduction in alkalinity of concrete over a period of time due to the environmental interaction of the atmospheric carbon dioxide interacting with the calcium hydroxide in moist concrete is all but well known. This reduction in alkalinity due to carbonation would cause corrosion of steel in concrete when it reaches the level of steel after the cover concrete is neutralized. The phenomenon is essentially the same in SCCs too and is of interest in different perspectives. Firstly the improved material compaction of the SCCs leading to the void space being minimized will certainly ensure a better performance. On the other hand the porosity of the concrete of a particular strength generally remains the same and thus will not make a major difference from the well-compacting concretes. Even so one should realize that the use of mineral admixtures of SCCs can invariably reduce the available alkalinity by a very small margin, which is still a matter of concern. Notwithstanding this, the microstructure from these pozzolanic admixtures in terms of the discontinuous pores that they generate might still be a positive effect. Chi (2002) found that the corrosion rates of carbonated concretes through AC impedance studies are higher than those of non-carbonated concretes and SCCs have shown even lower corrosion rates than ordinary Portland cement concretes. The normal concrete specimens show higher carbonation depths than those of SCCs. They attributed that the slag in SCC due to pozzolanic reaction and filling effects results in minimizing the pore size and volumes, thus reducing the carbonation rate. Ludwig et al. (2001) investigated the carbonation depths of SCC and vibrated concretes made with fly ash designed for similar water-to-binder ratios. They observed that SCC concrete had about 40% better resistance in terms of carbonation depths, and the values obtained for SCC and vibrated concrete were 3.7 and 6.3 mm. Rougeau et al. (1999) compared the carbonation performance of SCCs with high-performance concrete, and they observed that the carbonation depths obtained for SCCs are slightly higher than those of the high-performance concrete.

10.7 Alkali–Aggregate Reactivity

The alkali–aggregate reactivity, which is also sometimes known as alkali-silica reactivity, is very specific to certain types of aggregates and predominantly affects water-retaining structures or structures exposed to water or moisture. It is essential to have a handle on these aspects particularly for applications of that nature. Though there are not many studies in this direction, Kou (2009) studied mortars and concretes containing glass as both fine aggregate and coarse aggregate replacement in SCCs reaching a strength of 75 MPa. The drying shrinkage as well as the alkali–silica reaction (ASR) expansions were not significantly affected in any way, probably because the glass was used in the inert crystalline form as aggregate and not as a powder. The complexity of the entire process is far too complicated to be simply explained in this short report. However, the use of supplementary cementitious materials like fly ash, GGBS, silica fume, and so on has always proved to be an excellent way to address alkali–aggregate reactivity in general and probably is still one of the best solutions in the case of self-consolidating concrete too.

10.8 Thermal Degradation

Thermal effects can be broadly subdivided into two categories, one that is really ameliorative in promoting the strength gain rate of concretes to dehydration and the second that promotes thermal cracking due to differential expansion and even spalling due to the loss of even the chemically bound water in cementitious composites. The effect of the first one is used in establishing the strength of concrete through accelerated test methods such as warm water method, boiling water method, autogenous curing method, steam curing, and autoclaving to name a few. It is also to be recognized that the marginally increased temperature is highly beneficial and helps in improving the strength gain rate of the low-end pozzolanic admixtures like fly ash in particular. The advantages of such marginally higher temperatures (preferably 60°C–70°C) will substantially improve the same gain rates of concretes containing higher proportions (40%–60%) of fly ash and GGBS. Not many studies of this particular nature have been reported so far on SCCs, but the fact that chemical reaction kinetics remains the same in both normally vibrated concretes and SCC can always be taken advantage of. The second aspect of high-temperature degradation of concrete includes those that represent fire effects studied in several ways, be it in hot air ovens or muffle furnaces, as required by the temperatures considered in the investigations. A few of the attempts reported in the literature are discussed here. SCCs containing limestone filler and a normal concrete with C32.5 and C52.5 cements were studied by Fares (2010) through differential thermal analysis and SCM images after exposure to 150°C, 300°C, 450°C, and 600°C. The microstructural changes were found to be similar in terms of the high-temperature effects on these concretes. Fire resistance studies on SCCs and normal concrete with the same water–cement ratios by Boström (2003) show that SCCs had higher damage.

There have been laboratory investigations on the thermal effects in concretes particularly when they are subjected to very high temperatures. In one of the test methods, concretes had to be exposed to temperatures ranging from 200°C to 700°C for varying periods of a few hours, then allowed to be cured for specified periods, and the strength variations during the process had to be recorded. Some of these concretes formed the massive foundations of blast furnaces that encounter such thermal stresses. The cement in this case had to be a blast furnace slag cement with a minimum 50% slag, and the aggregates were also specified to be of higher strength and density. The concretes were all made to be nearly self-compacting for these requirements. Most studies of thermal cycling were basically for concretes designed for nuclear power applications. These concretes have been studied for various combinations of temperature differentials, number of cycles, durations, and rest periods, and the effects were investigated through a variety of tests both destructive and nondestructive. Some of the concretes designed had to be necessarily created using heavyweight hematite in aggregates of all sizes including fine aggregates and had to be effectively nearly self-compacting to be placed in highly confined steel shells that will form shields. In particular, reliably establishing the deterioration due to thermal cycling was always a major requirement.

10.9 Corrosion Characteristics

In the following paragraphs, a brief review of the studies that were reported on a few self-consolidating composites is presented. One important observation is that most of these investigations and the conclusions drawn from them are very specific to the materials and their combinations in the study, and many comparisons made with normally vibrated concretes are in general those that have not been made with the same material in the same proportions. This makes it very difficult to conclude that they are comparable. Comparing them with normal cement concretes not even of the same strength makes it even more difficult.

Concrete with a huge reservoir of calcium hydroxide arising out of the hydration of cement is considered to be an ideal material for the reinforcement to be situated in, because of the protection offered by the passivating layer formed on the steel due to such high alkalinity of over a pH of 13.0. Under most circumstances there is no significant depletion of this alkalinity level below a pH of 11.0 or 10.0 to be a cause for serious concern by the destruction of this passivating layer even over the entire lifetime of the structure under normal circumstances. Also, if the water–cement ratio of concrete composites is adjusted to be below 0.45, it is recognized that the porosity becomes discontinuous and will not allow the ingress of harmful chemicals and ions to disturb the equilibrium at the surface of the passivating layer on steel. This effect significantly improves the electrical resistivity of the concrete hindering the passage of the electrochemical current in the corrosion circuit. This brings to the fore the importance of electrical resistivity of concrete in inhibiting the corrosion of steel in concrete. One of the specific test methods that evolved out of such a recognition is the resistivity of concrete. As already discussed, there are several different methods proposed for its evaluation that are not discussed at this particular stage. However, the assessment of the likelihood of corrosion is assessed through the criteria given in Table 10.3. Chi (2002) studied the effect of carbonation time on the resistivity behavior of SCCs. It was seen that the resistivity increases with an increase in carbonation time. The resistivity of SCCs was observed to be much higher than that of normal concretes due to the pozzolanic reaction of the slag present.

The second aspect of the corrosion assessment is to understand what is known as the dissolution potential of the reinforcing steel at the concrete interface. Primarily, it is known that this potential is very low for noble metals and is always measured in terms of them as the reference. In principle a standard reference electrode (like saturated calomel electrode, CSE, or a copper/copper sulfate electrode) for which the reference potentials are already known (to facilitate conversion between the different electrodes) is used to measure the half-cell potential of the embedded reinforcement to understand the likelihood of corrosion. Such potentials taken at different locations in a grid on a flat surface like a slab or a bridge deck will result in equipotential lines, which are known as potential mapping. The assessment criteria for likelihood of corrosion through the observed potentials are given in Table 10.3.

However, Chi (2002) studied the half-cell potential measurements of SCCs and normal concretes of both carbonated and non-carbonated specimens by immersing them in 3.5% NaCl solution with respect to saturated calomel electrode (SCE). They found that, as time increased, all non-carbonated

SCCs and normal concretes have more than 90% probability to corrode. In addition, all carbonated concretes have over 90% probability to corrode after carbonation. They concluded that this may be the reduction of the pH value of the pore water causing chloride ions to diffuse or fill the concrete and corrosion may follow. In the same investigation, Chi (2002) studied the corrosion rates of the same SCCs and normal concretes with respect to AC impedance. They found that the corrosion rates of carbonated concretes are higher than those of non-carbonated concretes, and SCCs showed lower corrosion rates than normal OPC concretes. They also stated that carbonation increases the rate of corrosion of reinforcement. No significant difference was found in open circuit potential for SCC and OPC. However, AC impedance test results have illustrated different corrosion rates among all mixtures.

Apart from this the chloride content at the level of steel could also be a cause for the corrosion of steel in concrete. It is understood that while the passivation layer protects the reinforcement inside the concrete from corrosion, chlorides if present over a threshold limit can freely pass through the passivation layer allowing the electrochemical currents to pass through. There have been several discussions regarding these threshold limits. Table 10.3 presents the assessment criteria given by the CEB (1989). The ingress of chloride into the concrete is generally through a diffusion process. The characteristics of this diffusion process and the assessment of service life through the same are discussed later.

Ludwig (2001) studied the chloride diffusion of SCC containing fly ash and observed that SCC concrete with fly ash had better resistance in terms of chloride diffusion coefficients than similar vibrated concretes. The effects of the filler and a viscosity modifying agent on two SCCs were studied by Zhu (2003). He reported the permeability, absorption, and chloride diffusivities. With the fillers, limestone powder, and pulverized fuel ash at different percentages apart from variations in water and superplasticizer contents, the results are difficult to interpret. However, the normal concrete and the SCCs using pulverized fuel ash showed much lower values of chloride migration coefficient than other mixes. SCCs containing ternary and quaternary cements were seen to have lower chloride-ion penetrability compared to that of a reference SCC made with 100% OPC (Nehdi, 2004). A study by Tragardh (2003) on cores from tunnel linings, bridges, and retaining walls indicated that SCC had a higher resistance against chloride permeability than control concretes at similar water–cement ratios, maybe due to the dense interfacial transition zone. The effect of initial water curing for different periods on compressive strength, carbonation, and chloride diffusion characteristics of concretes containing 20% fly ash was studied by Zhao (2012). The NT-BUILD 355 test was used to establish the chloride migration coefficient. Gesoglu (2009) presented results on ultrasonic pulse velocity and resistivity characteristics apart from sorptivity, chloride permeability, and shrinkage on 22 concrete mixtures having fly ash, GGBS, and silica fume. His conclusions indicate that the ternary mixtures containing slag and silica fume performed best in terms of the durability characteristics. Jalal (2012) studied fresh, mechanical, and durability characteristics of SCCs containing micro- and nano-silica 10% and 2% individually as well as together. For these concretes of compressive strengths ranging between 50 and 90 MPa, the chloride-ion concentrations at different depths were measured apart from resistivity to represent the durability characteristics.

10.10 Service-Life Prediction or Residual Life Evaluation Methods

For studying concrete structures for service life, several different avenues have been detailed in the literature so far. A complete overview of formulations and understanding the deterioration processes and evaluations are essential for a general understanding of service-life prediction methodologies. In general some simple laboratory techniques that establish basic characteristics of factors like chloride diffusivity of carbonation particularly with some time parameters involved have been found to be possible methods for extending them for establishing some sort of a life estimate. Some of these are very simple laboratory investigations on simulation and the others are more elaborate, while some others looked into the assessment of the actual structures in the field for a more realistic assessment. Several techniques including nondestructive methods have been used in these efforts.

A study on service-life prediction or the evaluation of the remaining life of an existing structure can be done in several ways probably based on the type of the degradation and failure expected and the previous experience with such structures. In case a more realistic assessment of the effects of these forces as reflected in the structure is not possible through both sampling and non-destructive testing (NDT) measures, or in cases where it is explicitly directed to have an elaborate investigation on such sources, a combined investigation strategy can be adopted. After studying the concrete mix designs used in the construction of the structure, a detailed experimental program could be formulated to understand the deterioration processes and to evaluate the service life of the structure based on a comprehensive set of parameters relating to concrete as well as steel. These experimental investigations can be broadly categorized into the following:

- Laboratory investigations on representative concrete specimens
- Primarily sampling and nondestructive evaluations on the structure

The program consists of the various experiments conducted on the representative concrete specimens. While determining the mechanical properties, these specimens were also subjected to various NDT tests such as UPV and rebound hammer. Thereafter, the representative concrete specimens were tested to determine the various durability characteristics by conducting experiments like absorption/desorption studies, moisture migration, and acid attack that could provide information relevant to the behavior of such concretes. Also, the resistivities of concretes could be measured at different ages to establish the likelihood of the corrosion to take place. This might not be a direct representation of the concretes existing in the structure, which could be determined only by conducting similar tests on the structure or on samples procured from the existing structure.

One of the first things that come to mind is obviously the most talked about and the most intensely investigated parameter, the chloride diffusion into concrete, closely followed by another similar diffusion parameter, which is carbonation. Most of these can be appropriately addressed through mathematical procedures that probably relate to the diffusion of the service life of the structure. In the initial days the service life was also known as corrosion, which was discussed as a two-part phenomenon, the time to initiation (the time that it takes for the chloride front to reach the reinforcement level through the concrete and build up to a concentration that will initiate corrosion of steel) and the second part that depicts the time taken to result in an unacceptable cracking and spalling that will expose the reinforcement directly to the environment. Later it was realized that the second part will not really contribute much to the service life and the first part itself could be taken as the service life so that one can at least attempt repair and rehabilitation of the structure. This in fact is the service life in terms of the corrosion of steel in concrete. The carbonation or the degradation of the alkalinity in concrete is also affected by a similar diffusion process, and the visuals are formulated appropriately to address this; one can always arrive at the service life in terms of carbonation.

10.10.1 Chloride Diffusivity

Laboratory studies based on exposure of the chlorides (sometimes even concentrations higher than those normal in nature) or specific accelerated methods like salt spray, wetting, and drying, or even through forced migration or diffusion studies using an applied potential differential of an electrical field have all been used extensively. In a typical experimental investigation in the laboratory, a set of painted specimens (one face of the specimen was kept open for the chloride ingress to take place) were kept in various curing conditions for simulating the unidirectional ingress of chloride into concrete. An attempt was made to establish the relation between the amount of chloride ingress using a traditional long-term procedure of direct immersion and a procedure involving an alternate wetting and drying cycle. In the latter procedure, wetting was done for 4 days and drying for 3 days at a temperature of 105°C. The various exposure conditions related to the wetting are as follows.

- Wetting and drying in seawater
- Wetting and drying in 15% NaCl Solution
- Wetting and drying in a salt mist chamber

As a standard procedure, the specimens were drilled to a depth of 65 mm (with a 5 mm interval), and the powder samples obtained at the corresponding depths were to be used to determine the chloride diffusion profiles in the concrete. But at the very start, we always attempt to establish the definite depth of chloride ingress through a silver nitrate test so that the drilling process could be controlled accordingly. Unfortunately for the natural diffusion process to be of significance even in the lower strength concretes, the concretes have to be exposed for significantly longer periods than the one and a half years that was done presently. In the specimens immersed in saltwater only for a period of 1.5 years, the ingress of chloride into the concrete was found to be too low for determining the diffusion profiles. However, the results corresponding to the accelerated techniques here have shown sufficient insight into the possible effects of environment on concretes of such type. It may be noted that these accelerated studies were chosen and designed to represent a possible deterioration of 1 year through one cycle (based on earlier reports) so that we get some idea of the total depth of ingress and its effect on such concrete. Also, the chloride penetration in the accelerated technique, Rapid Chloride Penetration test, has been used along with these results for a comparison.

10.10.2 Macrocell Corrosion Test

Also, certain reinforced concrete specimens with embedded reinforcement at various cover depths were subjected to the earlier mentioned exposure conditions to determine the variation of potentials in steel and to replicate the behavior of reinforced concrete specimens subjected to severe environments. Apart from this tests were undertaken to study the macrocell corrosion test proposed by the ASTM standards. The specimen in this test procedure were modified from that used as per ASTM standards, for obtaining more results within the same amount of time. This is also a proven test that has been investigated earlier by others in the laboratory to test the corrosion effects. Concrete specimens were cast with reinforcements; a 150 mm concrete cube was reinforced with 5 steel bars, out of which 4 bars of 8 mm diameter are kept at a cover of 10, 20, 30, and 40 mm from the 4 corners of the cube. And the fifth bar of 25 mm diameter was kept at the center of the cube, such that it remains equidistant from all the steel rods. Then, each individual 8 mm diameter was connected to the central rod via a 100 Ω resistor. All these specimens were placed in 3% NaCl solution, and the solution was kept at a height of 120 mm. The four outer rods act as anodes and the central rod acts as a cathode, and the current passing in between each individual outer rod and central rod was measured by measuring the voltage difference between the two ends of the resistor connected.

10.10.3 Thermal Cycling of Concrete

Another important aspect investigated was the effect of thermal cycling on these representative concrete specimens. In this the representative concrete specimens were subjected to thermal cycling. The concrete specimens were subjected to a daily thermal cycle consisting of 8 hours of heating in a hot air oven at 105°C followed by immersion in water for 15 hours after the specimens have cooled reasonably for almost about an hour. The effects of thermal cycling were studied through the parameters that are most relevant to porosity and microcracking as discussed earlier at the end of 0, 10, 20, 30, and 50 cycles. One important fact to be noted is that due to an inadvertent failure of the circulating fan of the hot air oven, the specimens were kept immersed in water after the 25th cycle for a period of about 1 month before further cycling, allowing the specimens to recuperate and recover from the effects of the earlier thermal cycling.

The actual parameters studied include the variations in the wet and dry weights and absorption and desorption characteristics (to understand the overall pore volume) in these specimens due to the temperature cycling. Apart from these the electrical resistivity and the longitudinal and transverse pulse velocities in both in the wet and dry states after the specific number of cycles as given were also measured. The ultrasonic pulse velocities were measured by an Ultrasonic Tester (Marui, Japan), having both the longitudinal and transverse 50 kHz transducers. Finally, the degradation was assessed through the changes in strength at the end of the different number of cycles. These ultrasonic pulse velocities were used for calculating the variations in the Poisson's ratio, dynamic shear modulus, and the dynamic Young's modulus.

These discussions prove that there could be several modifications and modulations of the existing methodologies in terms of specimen type, size, and reinforcement locations; test procedure for accelerating the desired effect; and the assessment methodology in adopting standard laboratory techniques in the investigation to ensure an appropriate result that is comprehensive. In fact it is a firm belief of many research workers that a single test or a couple of tests on a few samples will not be able to present the true picture of any of the environmental and loading effects on the service life of structural facilities. Some other methods of evaluation of the experimental results and their appraisal through relevant coding provisions are discussed later. At this stage it is very important to understand the diffusion characteristics particularly chlorides in the concrete (the representative parameter) to have a critical understanding of the theory and the interpretation of the service-life criteria and service-life assessment.

10.10.4 Service-Life Determination Using Chloride Diffusivity

Reinforced concrete is a versatile, economical, and widely used construction material. It can be molded to a variety of shapes and finishes. Usually, it is durable and strong, performing well throughout its service life. However, it does not perform adequately as a result of poor design, poor construction, and inadequate materials selection, exposed to a more severe environment than anticipated, or a combination of these factors. There could be various possibilities like ingress of harmful liquids and gases in concrete, freezing and thawing, alkali–aggregate reactions, leaching, and several other factors that could alter the originally intended concrete formulations to act differently as a structure undergoes changes due to the deterioration and corrosion. As such the service-life assessment by itself is not to be based on any specific aspect or aspects but should be a much more comprehensive assessment of the structure, the intended use, the environment in which they are supposed to perform, and maybe even a large number of unforeseen yet possible scenarios that could affect the life of the structure. Such an assessment is too complicated, and it is normally relegated to aspects like risk assessment, which would still be almost as complicated by having to reflect on the cost of replacing the structure and, at that point of time, the risk involved to human life and the lives of other living creatures, the possible aftereffects in terms of contamination, trauma, and social rebuilding, to name just a few. This aspect of the effect of service life is normally understood through what is generally planned in terms of the proactive disaster mitigation strategies in various organizations. In the present case, it was only expected to examine a few more specific aspects of well-established service-life problems, particularly in terms of the durability and corrosion of the steel in concrete and the methods of the number and the service life more accurately in the mathematical assessment of the phenomena involved.

A majority of concrete structures in the marine environment shows signs of degradation due to the corrosion of the reinforcement in the presence of chlorides. In some cases, the degradation is visible within a few years of construction completion. Despite the fact that the marine environment is particularly severe, other factors affect the premature degradation due to corrosion, such as poor construction quality as a result of poor workmanship, inadequate standards based on prescriptive measures, and poor design as a result of insufficient information with regard to the parameters that influence the degradation process.

The service life of a structure is broadly depicted by the statement—A structure shall be designed and executed in such a way that it will perform during its intended life with an appropriate degree of reliability and in an economic way, remain fit for use for which it is required, and sustain all actions and influences likely to occur during execution and use (Euro code). Specific reference is made to the inclusion of durability in these requirements. It continues to say—The above requirements shall be met by the choice of suitable materials, by appropriate design and detailing, and by specifying control procedures for design, production, execution, and use relevant to the particular project. Service life can also be defined as the time for which a structure is expected to be able to fulfill its requirements with sufficient reliability with or without periodic inspection and maintenance and without unexpected high costs for maintenance and repair. Associated with the definition of service life is the concept of limit state based on performance criteria. The performance of a structure is referred to as its ability to fulfill demands and requirements set by the owner, authorities, and users. Concrete structures are usually expected to have a long service life since they often require large investments. This means that it is necessary to have an understanding of the interrelations between design, material, deterioration, and future maintenance. Furthermore, it is necessary to have an understanding of how aging of concrete changes the design parameters. Depending on the type of structure different definitions of the service life are used.

There are different methods to predict the service life of the concrete structure. Two of the well-known models are described here. But the precise validity of the predictions produced by these mathematical models is not confirmed. If it is assumed that the structure is close to the sea then it would be mainly affected by the chloride ingress into the concrete. Service life is defined as the time from the construction until the chloride content at the depth of reinforcement is high enough (threshold value) to initiate corrosion.

Two different methods for assessing the service life in terms of chloride diffusion are explained hereunder. One of the models for the calculation of service life can be summarized in the following five steps.

1. To increase the accuracy of service-life prediction, it is recommended to test two sets of concrete specimens at two different ages. This could be at a time t_1, a few years after the structure has been exposed to chloride. The concrete cores are drilled from the structure and used to measure the chloride content at different depths to obtain the values of $C(x, t_1)$ distribution. The second sample is taken after a few more years of exposure at time t_2, and the data of $C(x, t_2)$ is established. The time interval between these two could be varied depending upon the severity of the environment and the quality of concrete, but should be enough to represent the changes in diffusion.

2. To theoretically calculate the chloride diffusion coefficients that varied with both depth x and time t for the experimental results at time

t_1 and time t_2, Tumidajski (1995) developed a Boltzmann–Matano method that proposed the equation for $D(c^x)$ given here:

$$D\left(c^x\right) = -\frac{1}{2t}\left(\frac{dx}{dC}\right)_{c^*} \int_{c^i}^{c^*} x \cdot dC$$

where c^* is an arbitrary reference value.

3. Calculate the age parameter (β), a variable that depends on several variables such as the type of cementitious materials, the mix proportions, environmental changes, and so on, and which represents the rate of decrease of the chloride diffusion coefficient with time through

$$\beta = \frac{\ln\left(\dfrac{D_2}{D_1}\right)}{\ln\left(\dfrac{t_1}{t_2}\right)}$$

4. Calculate the reduction parameter (δ), based on the assumption that the diffusion coefficient D depends on time t, depth x, and chloride content C; the reduction parameter, δ, in the service-life prediction model is calculated by using

$$\delta = \frac{x_{cr}}{\sqrt{D_{LT} t_{LT}}} = \text{erfc}^{-1} \frac{\displaystyle\int_{C_{sm}}^{C_r} D(C)dC - \int_{C_{sm}}^{C_i} D(C)dC}{\displaystyle\int_{C_{sm}}^{C_s} D(C)dC - \int_{C_{sm}}^{C_i} D(C)dC}$$

C_r is the threshold chloride concentration

C_s is the surface chloride content

x_{cr} is the cover depth

C_{sm} is the nominal chloride surface concentration

The corrosion of reinforcement is assumed to occur when the chloride concentration has reached the threshold chloride content level defined.

5. Calculation of service life t_{LT} using the equation

$$t_{LT} = t_1 \left(\frac{x_{cr}}{\left(\delta\sqrt{D_1 t_1}\right)}\right)^{2/1-\beta}$$

The second method is known as the LIGHTCON model and the evaluation procedure to determine the service life of an existing structure using it as given here (Maage, 1995).

1. A few years after the exposure to the chloride environment, at time t_c the concrete cover c is established.
2. Then the cores are drilled from the structure and are tested according to the test method APM 302 or equivalent, arriving at the potential chloride diffusion coefficient D_{pc} at time t_c.
3. Also the surface region of the drilled cores are tested according to test method APM 207 or equivalent to arrive at c_s in the achieved chloride diffusion coefficient D_{ac} at time t_c, chloride concentration at the surface of the concrete c_s, and the initial chloride concentration c_i.

$$c(x,t) = c_i + (c_s + c_i)\,\mathrm{erfc}\left(\frac{x}{\sqrt{4tD}}\right)$$

4. The age parameter α is calculated according to

$$\alpha = \frac{\ln\left(\dfrac{D_{a,avg}}{D_{p0,avg}}\right)}{\ln\left(\dfrac{t_0}{t_c}\right)}$$

5. Calculate the reduction parameter (ξ)

$$\xi = 2\,\mathrm{erfc}^{-1}\left(\frac{C_{Cr} - C_i}{C_s - C_i}\right)$$

   where
      C_{cr} is the threshold chloride concentration
      C_s is the surface concentration

6. Calculation of service life

$$t_{LT} = t_0 \left\{ \left(\frac{x_{cr}}{\xi\sqrt{t_0} + D_{p0,avg}}\right)^{\left(\frac{2}{1-\alpha}\right)} \right\}$$

This is only a brief summary of two models available in the literature to evaluate the service life based on the diffusion of chloride in concrete.

10.10.5 Chloride Diffusion Studies

Permeation of chloride ions through the concrete occurs either through diffusion when immersed in, say, seawater or by capillary suction, which occurs because of wetting and drying like in the splash zone. The porosity and permeability of concrete are the primary factors that influence the intrusion of chloride ion into concrete. One of the tests proposed by Whiting (1984), which was later approved by both AASHTO T277 (1983) and ASTM (1994), is known as the chloride-ion permeability test and is also known as RCPT. The test setup utilizes a 100 ø × 50 mm long specimen fitted with two chambers on either side, one filled with 3% sodium chloride solution and the other with 0.3M sodium hydroxide solution. A 60 V DC power supply was used to drive the chloride ions through concrete. The total charge passing in 6 hours was a measure of the chloride-ion permittivity of the concrete. The concrete quality can be assessed based on the limits given by ASTM C1202 (Table 10.3).

Raghavan (2002) investigated the RCPT values of SCCs with that of a normal concrete at 28 days. They observed that the RCPT values of 1100–1500 coulombs for SCCs compared to about 4000 coulombs for normal concretes may be due to the high filler. In a study that was essentially the characterization of Helix fiber mortars at very high strength with silica fume (100–130 MPa), El-Dieb (2009) investigated the effect of steel fiber content on electrical resistivity, chloride permeability, and bulk chloride diffusion apart from the RCPT values for fiber concentrations of 0%–0.52% resulting in the charge passing levels of 100–500 coulombs.

10.10.6 Corrosion Rate Studies

Corrosion rate is studied through the electrochemical potentiodynamic polarization technique. The standard test specimen and the appropriate parameters that are suitable for the corrosion rate study of reinforcing steel in concrete through potentiodynamic polarization have all been studied in depth to arrive at the set of specifications most appropriate for the concrete compositions. For concrete strengths that are very dry or for specimens with highly different open circuit potentials (OCPs), the drive currents and scan rate can be suitably modulated to arrive at the appropriate Tafel plot. The test was conducted in the laboratory on an 8 mm diameter, 100 mm long reinforcing bar embedded centrally in concrete cylinders of 100 mm diameter and 200 mm height. The sample immersed in 3% NaCl solution was connected to a scanning potentiostat (EG & G Princeton Applied Research). The potentiostat has a capability of IR drop compensation up to 10 kohms. The OCP for each specimen was measured prior to application of the potentiodynamic scan. The initial (cathode) and final (anode) potentials in the potentiostat were adjusted to about 250 mV on either side of the OCP. In all the tests, a scan rate of 1 mV/s was selected. The corrosion rates were obtained from the potentiodynamic records, first, by using the Tafel plot technique (through I_{corr}). The record is a semi-log graph that was used to obtain the corrosion rates. Tangents were drawn at 50 and 100 mV from the OCP on anodic and cathodic regions to cut the horizontal line at the OCP at the same point. The drop to the current axis from this intercept gives the corrosion current (I_{corr}). Thus, the corrosion current can be obtained at 50 and 100 mV on either side of the OCP. The corrosion rate can be determined electrochemically by a number of methods that result in a corrosion density that is converted to corrosion rate through Faraday's law (Fontana, 1987) as follows:

$$\text{Corrosion rate} (\text{in mpy}) = 0.1288\, i e / d$$

where
   i is the current density (I_{corr}/area of bar) = $(I_{corr})/(\pi \times 0.8 \times 10)$ ($\mu A/cm^2$)
   e is the equivalent weight of metal (g) [27.95 (g)]
   d is the specimen density (g/cm^3) [7.86 (g/cm^3)] for the steel specimen of
      8 mm diameter and 100 mm length

The corrosion rates obtained for the self-compacting fly ash concretes and normally vibrated fly ash concretes of different grades show that as the strength of the concrete increases, the corrosion rate decreases. It was seen from this investigation in the laboratory that both the self-compacting fly ash concretes and the corresponding normally vibrated fly ash concretes were showing almost similar results, and in fact the normally vibrated fly ash concretes were showing slightly better performance in terms of corrosion rates.

10.10.7 Electrolytic Accelerated Corrosion Studies

The electrolytic accelerated corrosion test is a comprehensive test developed and perfected in the laboratory over the past three decades. The test emulates the RCPT test proposed by Whiting (1984) that was discussed earlier, which is actually a test on a 50 mm thick cylinder specimen through which the current passing in 6 hours under a potential difference of 60 V, driving the chloride ions in a 3% NaCl solution from one side to the other, is established, representing the chloride-ion permeation through the specimen. The test is also accepted as a standard test, ASTM C1202. The electrolytic accelerated corrosion test consists of a very similar arrangement in which the specimen is actually similar to the lollipop specimen as described earlier for establishing the corrosion rate tests. In this investigation the electrolytic accelerated corrosion test was conducted by using the applied voltage technique. The test is performed for a specific period or continued till the specimen cracks (Figure 10.4).

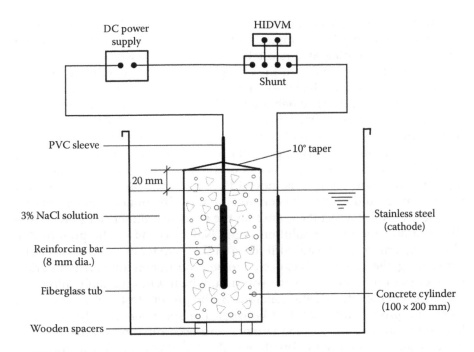

FIGURE 10.4
Accelerated corrosion cracking test setup for service life.

The accelerated corrosion setup consists of a cylinder of 100 mm diameter and 200 mm height with an 8 mm diameter and 100 mm length cold twisted high yield strength deformed bar (the reinforcement under investigation) centrally placed. The specimens are cured for the required period as desired. In principle the corrosion rates are established as discussed earlier before the test. A suitable rest period of at least 3 days under proper conditions is required before proceeding with the potentiodynamic polarization test as discussed hereunder. The test specimen with the embedded reinforcement is placed in a 3% NaCl electrolyte bath along with a stainless steel counter electrode. At this stage this setup can be used to measure a variety of different parameters that actually establish the characteristics of both the concrete in the cylinder and the embedded reinforcing bar. Before connecting the circuit of the electrolytic test, the OCP of the reinforcing bar before the test is established through a saturated calomel electrode. This OCP can be used to study the changes that occur in the potentials during the several rest periods in the test, which can provide a wealth of information for a discerning research investigator.

At this stage a 60 V DC current is connected to drive the chlorides from a 3% NaCl electrolyte between the embedded steel reinforcement to be tested and a stainless steel counter electrode (cathode). The current and voltage (resistance) of the cell are measured every half hour up to 6 hours using a high-impedance digital voltmeter (HIDVM). These can also be monitored continuously and integrated to establish the charge passing 6 hours, which incidentally is very similar to the charge passing in the RCPT. It may be noted that the annular thickness that the charge passes through is nearly 50 mm, the only difference being an additional polarization resistance on the surface of the steel, which is probably not very significant at such high driving potentials. Incidentally a very similar test was recommended with a significantly lower driving potential of 6 V, in a test proposed by the Florida DOT (1978), in which instead of the total current passing, the average daily resistance (which is indeed the same as measuring the current) is taken as an indicator of the concrete quality. The minor differences that probably may not affect these are that the specimen dimensions are a little different and the testing solution is 5% NaCl instead of the present 3%. However, a more significant difference in this is that after 28 days of moisture curing, the specimen is expected to be conditioned for a further period of 28 days in the same 5% NaCl solution, which will invariably allow the diffusion of chloride into the concrete. In the present test that is proposed, to ensure proper initial contact between the specimen and the solution, before the start of the test the laitance and accumulation on the surfaces were cleaned with a nylon brush and the specimen was left in the solution for a period of almost a day, primarily to facilitate the potential and resistance measurements as was discussed earlier. Also the initial current at the start of the test can directly present the resistivity of the concrete that is studied. The resistivity (R_2) can be calculated from this data as follows (Mc Carter, 1984).

$$R_2 = (2\pi R_1 L) / \left[\log_e (r_1 / r_2) \right]$$

where
   R_2 is the resistivity in kohm (cm)
   R_1 is the resistance at 60 V (V/I) (kΩ)
   I is the initial current in (A)
   L is the length of the steel bar (cm)
   r_1, r_2 are the radii of the cylinder and the steel bar, respectively, in cm

The potentials with reference to SCE were also noted before and after the test for 12 hours each day generally. After a 12-hour test in a day, the sample was kept at rest for the remaining 12 hours of the day in the 3% NaCl solution to attain a stable potential as before the start of the test next day. OCPs are measured both immediately after and before the start of the test next day, which once again can be used to understand the corrosion status of the reinforcement inside.

The test was restarted by applying the same current at 60 V (DC) each day till the specimen cracked, and the charge to cracking was calculated. The current, voltage, and resistances were noted every 6 hours during this period. If the sample cracked, the test was stopped; otherwise it continued for a 15-day (180 hours) period. Some of the very high-strength concrete samples do not show any crack even at this stage with the current passing, or the crack was very small. The test can be discontinued at this particular stage if required and the assessments proposed could be extrapolated.

A methodology of arriving at the service-life period of the concrete through the values of the total charge to cracking (Q_c) and corrosion current (I_{corr}) from the potentiodynamic polarization studies was reported earlier (Raju, 1990). The assumption that the corrosion rate is not going to change over the entire period of the structure is probably not totally accurate, but yet with the increased strength gain in concrete (augmented by a proper maintenance and repair strategy), it may not be totally inappropriate to make such an assessment.

$$\text{Life} = \frac{Q_c}{I_{corr}}$$

There have been several investigations in the laboratory on a variety of concretes ranging from normal strengths of about 20 MPa to strengths as high as 120 MPa, containing a variety of chemical and mineral admixtures (particularly fly ash, GGBS, silica fume, RHA, metakaolin, and zeolite). The studies also included a range of SCCs containing fly ash and also more recently ternary mixtures based on the above pozzolanic mineral admixtures, which are not discussed in detail here.

10.10.8 Service-Life Determination Using Carbonation

As already stated a further aspect of the assessment of the likelihood of corrosion of steel in concrete is what is understood as passivation due to the effects of the carbon dioxide present in the environment. Though carbon dioxide may not directly react with the calcium hydroxide in concrete in the dry state, with the help of a little moisture the dissolved carbon dioxide will neutralize the calcium hydroxide, resulting in the lowering of the pH, particularly in the cover concrete. This lowering of the pH will automatically result in the depassivation of the protective layer on the reinforcement, resulting in corrosion. A simple method of assessing this lowering of alkalinity in the concrete is to spray a phenolphthalein solution on a broken surface in the cover zone. It can be readily seen that at locations where alkalinity is not affected through carbonation, the concrete surface turns bright pink. This test is equally useful to understand the depth of carbonation both in the field and in the laboratory. In the laboratory accelerated carbonation studies can be done in specially controlled chambers that contain at a specified level.

While there are not many tests reported on the carbonation characteristics of SCCs, it is safe to say that these will be no different from those of normally vibrated concretes of the same cementitious composition. Zhao (2012) studied the effect of initial water curing for 3, 7, and 14 days on compressive strength, carbonation, and chloride diffusion characteristics in a concretes containing 20% fly ash with a total cementitious materials content of 460 kg/m³ resulting in a strength of about 60 MPa. Accelerated carbonation studies indicated carbonation depths ranging from 6.5 to 3 mm in concretes of strength ranging from 58 to 63 MPa. A very similar approach to the diffusion process of chloride into the concrete is also possible with the carbonation process as was explained by a few researchers. A highly comprehensive overview of these particular processes has been presented by Schiessl (1988).

A brief review on the durability and service life of self-consolidating composites was presented in the previous paragraphs. One important observation is that most of these investigations and the conclusions drawn from them are very specific to the materials and their combinations of the study, and many comparisons made with the normally vibrated concretes are in general those that have not been made with the same material in the same proportions, which makes it very difficult to consider them as comparable. Comparing them with normal cement concretes sometimes not even of the same strength makes the comparisons even more difficult.

10.10.9 Service-Life Management of Constructed Facilities

Recent years have seen the adoption of third-party inspection, assurance, and approval protocol as an important part of the activities in all major constructions projects. This is to ensure that the contractor is not encumbered with this task, though it is his primary responsibility according to the contractual obligations. This provides for an independent evaluation of the entire set of operations in the construction activity. A look at the CEB-FIP design guide for durable concrete success (CEB-FIP, 1992) categorically defines that such an independent third-party authority has to be constituted even before the planning stage itself, primarily to ensure that there is a continuity in the processes—planning, design, construction, maintenance, and service quality assurance activities, particularly for major constructed facilities. While this body can coopt persons for specific activities if unforeseen, it should have the competence to be able to interact and advice at various stages with an accepted plan of action. It is obvious that this body is directly and independently responsible to the local governmental administrative authority through the owner of the constructed facility. The various facets of the different entities and their subsidiary activities are all generically listed in Table 10.4. It is felt that no further explanation on these activities is essential.

TABLE 10.4

Service-Life Management of Constructed Facilities

Task	Operations
Planning	Structure
	Form and shape—(frame, tube, tube in tube)
	Material—(steel, concrete, composite)
	Construction
	Sequence—(erection, transient)
	Incremental—(height, overhang, delayed)
	Approval—(third party)
Design	Structure
	Members—(size, shape, connectivity)
	Loading
	Load exceedance in service—(vehicular, superimposed)
	Catastrophic—(earthquake, flood, aircraft impact)
	Approval and inspection—(third party)
Materials	Concrete
	Mix design—(materials selection, compatibility, mixture)
	Loading
	Environment—(tropical, cold, water, sea, chemical)
	Tests
	Green—(consistency, temperature, shrinkage)
	Hardened—(strength, temperature, shrinkage)
	Durability—(porosity, permeability, chemical attack)
	Corrosion—(pH, Cl^- diffusion, carbonation)
	Quality assurance and control—(third party)
Service	Construction
	Schedule—(sequence, delay, modifications)
	Maintenance
	Importance—(periodicity, inspection, level)
	Rehabilitation
	Cause—(remedial measures, acceptance criteria)
	Overall record—(third party/owner/public authority)

References

AASHTO T277, *Standard Method of Test for Rapid Determination of the Chloride Permeability of Concrete (AASHTO T277)*, American Association of State Highway and Transportation Officials, Washington, DC, 1983.

Anagnostopoulos, N., Sideris, K.K., and Georgiadis, A., Mechanical characteristics of self-compacting concretes with different filler materials, exposed to elevated temperatures, *Materials and Structures*, 42, 2009, 1393–1405.

Assie, S., Escadeillas, G., and Marchese, G., Durability of self compacting concrete, in *Third International, RILEM Symposium*, Reykjavik, Iceland, 2003, pp. 655–662.

ASTM C1202, *Standard Test Method for Electrical Indication of Concrete's Ability to Resist Chloride Ion Penetration*, ASTM Standards, West Conshohocken, PA, 1994.

Atan, M.N. and Awang, H., The mechanical properties of self-compacting concrete incorporating raw rice husk ash, *European Journal of Scientific Research*, 60(1), 2011, 166–176.

Barrita, C.F.D.J., Bremner, T.W., and Balcom, B.J., Effects of curing temperature on moisture distribution, drying and water absorption in self-compacting concrete, *Magazine of Concrete Research*, 55, 2003, 517–524.

Boström, L., Self-compacting concrete exposed to fire, in *Third RILEM Symposium on Self-Compacting Concrete*, Reykjavik, Iceland, 2003, pp. 863–869.

Boulekbache, B., Hamrat, M., Chemrouk, M., and Amziane, S., Flowability of fibre-reinforced concrete and its effect on the mechanical properties of the material, *Construction and Building Materials*, 24, 2010, 1664–1671.

Brouwers, H.J.H. and Radix, H.J., Self-compacting concrete: Theoretical and experimental study, *Cement and Concrete Research*, 35, 2005, 2116–2136.

Browne, R.D. and Baker, A.P., The performance of structural concrete in marine environment, in *Developments in Concrete Technology*, Lydon, F.D. (ed.), Applied Science Publishers, London, U.K., 1979, pp. 111–149.

CEB, Diagnosis and assessment of concrete structures—State-of-art report, Bulletin No. 192, Comité Euro-International du Béton (CEB), Lausanne, Switzerland, 1989, p. 120.

CEB-FIP, *Durable Concrete Structures: Design Guide*, Thomas Thelford, London, U.K., 1992, p. 128.

CEB-FIP Model Code, *Committee Euro-International du Beton*, Thomas Telford, London, U.K., 1994.

Chi, J.M., Huang, R., and Yang, C.C., Effects of carbonation on mechanical properties and durability of concrete using accelerated testing method, *Journal of Marine Science and Technology*, 10, 2002, pp. 14–20.

Craeye, B., De Schutter, G., Wacquier, W., Van Humbeeck, H., Van Cotthem, A., and Areias, L., Closure of the concrete supercontainer in hot cell under thermal load, *Nuclear Engineering and Design*, 241, 2011, 1352–1359.

De Schutter, G., Audenaert, K., Boel, V., Vandewalle, L., Dupont, D., Heirman, G., Vantomme, J., and Hemricourt, D., Transport properties in self compacting concrete and relation with durability: Over view of a Belgium Research Project, in *Self-Compacting Concrete, Third International, RILEM Symposium*, Reykjavik, Iceland, 2003, pp. 799–807.

El-Dieb, A.S., Mechanical, durability and microstructural characteristics of ultra-high-strength self-compacting concrete incorporating steel fibers, *Materials and Design*, 30, 2009, 4286–4292.

Fares, H., Remond, S., Noumowe, A., and Cousture, A., High temperature behaviour of self-consolidating concrete, Microstructure and physicochemical properties, *Cement and Concrete Research*, 40, 2010, 488–496.

Florida DOT, An accelerated laboratory method for corrosion testing of reinforced concrete using impressed current, Research report 206, Florida Department of Transportation, Tallahassee, FL, 1978, p. 10.

Fontana, M.G., *Corrosion Engineering*, 3rd edn., McGraw-Hill Book Company, New York, 1987.

Ganesh Babu, K., Raju, P.V.S.N., and Ranga Raju, U., Evaluation of reinforcement corrosion in concrete under marine environment, in *Proceedings of the International Conference on Offshore Mechanics and Arctic Engineering*, Houston, TX, 1990.

Gesoglu, M., Guneyisi, E., and Ozbay, E., Properties of self-compacting concretes made with binary, ternary, and quaternary cementitious blends of fly ash, blast furnace slag, and silica fume, *Construction and Building Materials*, 23, 2009, 1847–1854.

Gjorv, O.E. and Vennesland, O., Sea salts and alkalinity of concrete, *American Concrete Institute Journal*, 73(9), 1976, 512–516.

Gjorv, O.E., Vennesland, O., and EI-Budsaidy, A.H.S., Electrical resistivity of concrete in oceans, in *Proceedings of the Offshore Technology Conference*, OTC paper No. 2803, Houston, TX, 1977, pp. 581–588.

Ioani, A., Domsa, J., Mircea, C., and Szilagyi, H., Durability requirements in self-compacting concrete mix design, in *Concrete Repair, Rehabilitation and Retrofitting II— Proceedings of the Second International Conference on Concrete Repair, Rehabilitation and Retrofitting*, Cape Town, South Africa, 2009.

Jalal, M., Mansouri, E., Sharifipour, M., and Pouladkhan, A.R., Mechanical, rheological, durability and microstructural properties of high performance self-compacting concrete containing SiO_2 micro and nanoparticles, *Materials and Design*, 34, 2012, 389–400.

Khatib, J.M., Performance of self-compacting concrete containing fly ash, *Construction and Building Materials*, 22, 2008, 1963–1971.

Kou, S.C. and Poon, C.S., Properties of self-compacting concrete prepared with recycled glass aggregate, *Cement and Concrete Composites*, 31, 2009, 107–113.

Leemann, A., Lura, P., and Loser, R., Shrinkage and creep of SCC—The influence of paste volume and binder composition, *Construction and Building Materials*, 25, 2011, 2283–2289.

Li, B., Guan, A., and Zhou, M., Preparation and performances of self-compacting concrete used in the joint section between steel and concrete box girders of Edong Yangtze River Highway Bridge, *Advanced Materials Research*, 168–170, 2011, 334–340.

Ludwig, H.-M., Ehrlich, N., Hemrich, W., and Weise, F., Selbstverdichtender beton— Grundlagen und Praxis (Self-compacting concrete—Principles and practice), *Betonwerk und Fertigteil-Technik/Concrete Precasting Plant and Technology*, 67(6), 2001, 58–67; 76–80.

Maage, M., Poulsen, E., Vennesland, Ø., and Carlsen, J.E., Service life model for concrete structures exposed to marine environment initiation period, LIGHTCON Report No. 2.4, STF70 A94082 SINTEF, Trondheim, Norway, 1995.

Mc Carter, W.J. and Curran, P.N., The electrical response characteristics of setting cement paste, *Magazine of Concrete Research*, 36, 1984, 42–49.

Mehta, P.K. and Gjorv, O.E., A new test for sulphate resistance of concrete, *Journal of Testing and Evaluation (JTEVA)*, 2(6), 1974, 510–514.

Mnahoncakova, E., Pavlikova, M., Grzeszczyk, S., Rovnanikova, P., and Cerny, R., Hydric, thermal and mechanical properties of self-compacting concrete containing different fillers, *Construction and Building Materials*, 22(7), 2008, 1594–1600.

Nehdi, M., Pardhan, M., and Koshowski, S., Durability of self-consolidating concrete incorporating high-volume replacement composite cements, *Cement and Concrete Research*, 34, 2004, 2103–2112.

Persson, B., Internal frost resistance and salt frost scaling of self-compacting concrete, *Cement and Concrete Research*, 33(3), 2003, 373–379.

Raju, P.V.S.N., Corrosion behaviour of steel in concrete, PhD thesis, Indian Institute of Technology Madras, Chennai, India, 1990, p. 331.

Raghavan, K.P., Sivarama Sarma, B., and Chattopadhyay, D., Creep, shrinkage and chloride permeability properties of self-consolidating concrete, in *First North American Conference on the Design and Use of Self-Consolidating Concrete, ACBM*, North Western University, Chicago, IL, 2002, pp. 341–348.

Rougeau, P., Maillard, J.L., and Marry-Dippe, C., Comparative study on properties of self-compacting concrete and high performance concrete used in precast construction, in *First International RILEM Symposium on Self-Compacting Concrete*, Skarendahl, A. and Petersson, Ö. (eds.), Stockholm, Sweden, 1999, pp. 251–262.

Roziere, E., Granger, S., Turcry, Ph., and Loukili, A., Influence of paste volume on shrinkage cracking and fracture properties of self-compacting concrete, *Cement and Concrete Composites*, 29(8), 2007, 626–636.

Schiessl, P., Corrosion of steel in concrete, Report of the Technical Committee 60-CSC RILEM, Chapman and Hall, London, 1988.

Shalon, R. and Raphael, M., Influence of sea water on corrosion of reinforcement, *ACI Journal*, 55, June 1959, 1251–1268.

Siad, H., Mesbah, H.A., Khelafi, H., Kamali-Bernard, S., and Mouli, M., Effect of mineral admixture on resistance to sulphuric and hydrochloric acid attacks in self-compacting concrete, *Canadian Journal of Civil Engineering*, 37(3), 2010, 441–449.

Stark, D., *Performance of Concrete in Sulfate Environments*, Portland Cement Association, Skokie, IL, 2002.

Stratfull, R.E., Half-cell potentials and corrosion of steel in concrete, Highway Research Record, No. 433, 1973, p. 12.

Tragardh, J. and Kalinowski, M., Investigation of the conditions for a thaumasite form of sulfate attack in SCC with limestone filler, in *Third RILEM Symposium on Self-Compacting Concrete*, Reykjavik, Iceland, 2003, pp. 844–854.

Tumidajski, P.J., Chan, G.W., Feldman, R.F., and Strathdee, G., A Boltzmann-Matano analysis of chloride diffusion, *Cement and Concrete Research*, 25(7), 1995, 1556–1566.

Turk, K., Caliskan, S., and Yazicioglu, S., Capillary water absorption of self-compacting concrete under different curing conditions, *Indian Journal of Engineering and Materials Sciences*, 14, 2007, 365–372.

Uysal, M. and Sumer, M., Performance of self-compacting concrete containing different mineral admixtures, *Construction and Building Materials*, 25(11), 2011a, 4112–4120.

Uysal, M. and Tanyildizi, H., Predicting the core compressive strength of self-compacting concrete (SCC) mixtures with mineral additives using artificial neural network, *Construction and Building Materials*, 25(11), 2011b, 4105–4111.

Vejmelkova, E., Keppert, M., Grzeszczyk, S., Skalinski, B., and Cerny, R., Properties of self-compacting concrete mixtures containing metakaolin and blast furnace slag, *Construction and Building Materials*, 25, 2011, 1325–1331.

Whitting, D., Insitu measurement of the permeability of concrete to chloride ions, Special Publication of ACI, SP-82, American Concrete Institute, Detroit, MI, 1983, pp. 501–524.

Whiting, D., In situ measurements of the permeability of concrete to chloride ions, in *ACI SP-82, American Concrete Institute*, Farmington Hills, MI, 1984, pp. 501–524.

Zhao, H., Sun, W., Wu, X., and Gao, B., Effect of initial water-curing period and curing condition on the properties of self-compacting concrete, *Materials and Design*, 35, 2012, 194–200.

Zhu, W. and Bartos, P.J.M., Permeation properties of self-compacting concrete, *Cement and Concrete Research*, 33(6), 2003, 921–926.

11

Frontiers and Research Needs

11.1 Introduction

There have been several modifications to the science of engineering and technology of construction materials over the millennia, since the times of the earliest civilizations known to humankind. It is obvious that in spite of the progress that has been made over this long a period, it is only during the last 50 years that things have changed significantly, even though it all started with the first so-called invention of hydraulic cements by Joseph Aspdin in the early 1800s. In fact for over 100 years, the then cementitious materials have not significantly changed though a lot of the chemistry behind them has been comprehensively researched and placed on record. But in the last 50 years concrete has undergone a sea change, both in terms of the expected strengths of the cementitious composites and in terms of performance levels expected; it has become the choicest material for superhigh-rise buildings, long-span bridges, shoreline, and offshore constructions including sea-links and channel tunnels, and so on, to name a few. These structures naturally needed much higher strengths and higher performance to be able to withstand both the imposed structural loadings as well as the environmental impact from the increasingly urbanized society. In more recent years with the advancements in computational facilities on the one side and the understanding of the possible modifications through superplasticizers, high-end pozzolanic admixtures, and fibrous reinforcement on the other, a variety of new and radically different concepts of cementitious composites have all come into existence, which have started redefining the concept of structural concrete. Self-consolidating concrete (SCC) composites are probably the simplest of the modifications possible as envisaged at present in trying to make the conventional low- to medium-strength concretes behave slightly differently, facilitating the concreting in operations without the help of an external energy input for compaction. It is neither possible nor is the idea of this book to go into these different aspects in depth. However, it may not be out of place to bring together a few concepts and a few possible modifications that also need an understanding of the significantly improved consistency concepts in their evolution and performance.

One of the first among these modifications is the possibility of making very-high-strength concretes, or even superhigh-strength concretes, which has already been reported by a few researchers. Chapter 7 exclusively deals with the concepts, development, and the direct design methods for such a group of cementitious composites. In laboratory investigations, it was clearly seen that concretes of strengths in the range of 90–120 MPa, which necessarily have to be very low water–cementitious materials ratios, lose a lot of strength in not being compacted appropriately. To be able to comfortably manufacture these concretes for field applications, it is imperative that they are of a totally flowing or self-consolidating character. It is also felt that some of the information that is available with the industry will not really percolate into the public domain understandably for the commercial value that is attached, though there are always broad outlines and sporadic information nuggets that will be available in publicity literature and in the concept publications of a few international conferences. The next few paragraphs have been devoted to a general overview of such information as available on the various topics relating to self-consolidating cementitious composites that are probably being introduced or being explored at present.

11.2 Applications and Prospects

The very first application that probably needs a mention is the massive end anchorages of the Akashi Kaikyo Bridge by Okamura (Okamura et al., 2005; El-Dieb, 2009). The massive cable end anchorages of the Akashi Kaikyo Bridge, a three-span suspension bridge of 3910 m length (with its central span of 1990 m), were probably the very first application of a significant magnitude of SCC. It was to be the world's longest suspension bridge when completed in 1998. The concrete to be placed (including the foundations) in each one of the spans was 520,000 m^3 in the first and 250,000 m^3 in the next, each of them housing the cable anchor frames weighing 4000 ton in their main bodies. An estimated 1900 m^3 of concrete was expected to be placed each day in these anchorages. The concrete mix proportion used was low-heat cement, limestone powder (LSP), water, superplasticizer, sand, and gravel at 260, 150, 145, 7.8/6.4, 609/769, and 1121/965 kg/m^3 for the concrete in the anchorages at either end. The strength of the concrete was not clearly specified for the two variations but with a water–cement ratio of about 0.56, may be reducing to an effective 0.50 with the contribution from LSP, it could have been around 45–50 MPa at 180 days. The designers could easily take advantage of the increased strength with that age for the low-heat cement with almost 35% LSP in the total cementitious materials.

SCC of 90 MPa strength grade was used in the construction of the International Commerce Centre, a building 485 m high with 118 stories, in Hong Kong. The structure had a central core tube of 30 × 30 m with a thickness of 2 m, along with 8 mega columns of size 2.85 × 3.50 m and 3.00 × 3.25 m with four outriggers. The self-compacting concrete designed was of 90 MPa strength with high modulus, which was required to limit the deflections due to the high wind forces of about 60 m/s. The concrete had a slump flow filling ability of 650–800 mm, T_{50} of 2–5 seconds, and L-box passing ability of 0.8–1.0 with a GTM screen stability of less than 15% as acceptance criteria. The elastic modulus ranged from 40 to 50 GPa when tested.

The 163-storied, 828 m tall Burj Khalifa tower utilized several different concretes for the piles, foundation mat, and the super structure. The tower, founded on a 3.7 m thick solid reinforced concrete raft was supported by 194 bored cast-in-situ piles of 1.5 m diameter and 43 m long, with a capacity of 3000 ton each. SCC of 60 MPa grade tremie concrete was used in the piles. The mat of 12,500 m^3 consisted of a 50 MPa SCC of 0.34 water–cement ratio in four separate pours. Considering the limitations on the peak differential temperature, the raft mix adopted a 40% fly ash content with ice. The tower itself was a reinforced concrete structure, with the buttressed core system extending through to the occupied space Level 156. Above this, a structural steel braced frame supports a 230 m tall spire. High-modulus concretes (about 44 GPa at 90 days) containing fly ash with strengths ranging from 60 to 80 MPa were used for the tower's columns and walls. However, the final concrete delivered had a strength of 100 MPa with an elastic modulus of 48 GPa.

It is reported that for the 165 m high Blackfriars residential tower, one of a string of towers planned (due for completion in 2018) in London, an SCC of C80/95 was specified. It is probably appropriate at this particular stage to at least look at the characteristics of the SCCs in a few megastructures that have been built in the recent past. In the construction of the Petronas Towers in Malaysia, it was reported that the concrete strengths specified were on the order of 100 MPa, achieved through a combination of silica fume and fly ash. In more recent times in the design of Trump Tower in Chicago it was reported that SCC of 110 MPa strength, achieved through 12 mm maximum size limestone coarse aggregate and a cementitious composite containing slag, fly ash, and silica fume, was adopted for certain sections. Once again it is not envisaged to give a complete account of these aspects even in these specific structures to reinforce the cause for the adoption of SCC in any way. The primary idea is to look at the material itself and the structural configuration and concept where it was felt appropriate to adopt such very-high-strength concretes. Such a brief overview is only to suggest an outline for a future application possibility and for the need to investigate further to arrive at specific tailored compositions for some of these requirements.

11.3 SCCs in Repair and Rehabilitation Practice

While summarizing the development of SCC over the past few years, Okamura (2005) reported that the shortened structural life of reinforced concrete was primarily due to insufficient workmanship and compaction. Explaining this as the background for the first move toward SCC, he stated that the first term for this compaction-free concrete was high-performance concrete (HPC). As there was an opinion in the academia that HPC referred to mainly high-strength concrete, to elucidate the original concept of its development, the term SCC was introduced focusing on the high quality of concrete that can be achieved. Against this background, one can now see that the original idea of SCC was to achieve a high-strength concrete, through the use of superplasticizers, which has not been appreciated fully. The fact that with only a marginally increased powder content (about 400–500 kg/m^3) particularly through fly ash at about 30%, and restricting the water content of about 160 kg/m^3 while using superplasticizers, it is possible to achieve strengths of around 50–60 MPa without much difficulty. It is this aspect that concretes of medium and high strength can easily be achieved even with the relatively low-end pozzolanic admixtures like fly ash that has not been realized by not only the construction industry but even in several of the research efforts that are under way.

At this juncture, it is only appropriate to mention that several of the existing concrete structures in the industrial sector (chemical industries in particular) and also in transport sector (bridges, waterfront structures, marine structures) invariably suffer from a large amount of spalling of the cover concrete. The unfortunate problem in the case of industries is that the operations cannot be suspended even for a day as the process equilibrium gets disturbed and restarting operations requires substantial amount of investments in terms of time and product wastage, forcing the industry to postpone or avoid as far as possible these structural defects till they become unavoidable. Simple plaster repairs with even more modern chemical plaster compounds have not been successful in many cases both from the point of view of compatibility and also differential shrinkage problems. SCCs offer an extremely viable alternative in addressing such narrow and confined locations through their ability to flow around and past obstructions, filling all the spaces needed to be covered. SSCs containing reasonably small-sized aggregates can be made with different combinations of cement and pozzolanic admixtures to suit the needs of the industry. In the case of waterfront and marine structures use of ground granulated blast furnace slag (GGBS) as a pozzolanic admixture for supplementing the cement will help not only in repair but GGBS in the system has the property of binding chloride that will help in arresting the corrosion as well. The advantages of higher performance as suggested above by Okamura can be reasonably utilized in many such repair and rehabilitation problems through an appropriate choice of cementitious materials and their application procedure.

11.4 Re-Alkalization of Concrete

The possibility of carbonation-led corrosion cracking is one of the major problems associated with some specific structures. One of the locations where the major problem was identified to be carbonation-led corrosion was obviously in the cover concrete of the massive abutments and piers of bridge structures, including those crossing highways and railways in particular. Two specific aspects were invariably noticed as the primary cause. In the first place, these being massive in nature, for reasons of economy, were invariably made with generally much lower slump concrete than the structural beams or the pre-stressed girders that are actually assumed to be the only members that matter for carrying the loads. The second aspect is that the joints between the spans invariably allow water to drain through these gaps; even in spite of the drain holes provided, which are rarely maintained. Invariably with the availability of moisture and also to a certain extent the carbon dioxide from vehicular exhaust, carbonation of the cover concrete is certainly a normal prospect. If this is associated with the deicing salts that are used in colder regions, the prospect of cover failure is only to be expected. If the problem is essentially a carbonation-generated failure of the cover, the simplest and probably the most effective remedy is removing the cover concrete as required, and even augmenting the concrete cover thickness to redress the situation. While the additional cover concrete will obviously provide the necessary barrier, one can look for a re-alkalization of the old internal concrete that is still carbonated to a certain extent so that the reinforcement is appropriately protected from the diffusion of the calcium hydroxide into the parent concrete. This exercise needs a specific understanding of the chemistry involved, the amount of re-alkalization possible, by using a slightly richer mix than the parent concrete. The use of an abnormally rich mix of this application can also result in excessive shrinkage of the cover layer leading to delamination and failure within a very short period. A similar problem can also exist in structures associated with thermal power plants and cement manufacturing units where both carbonation and temperature differential between the cover and internal concretes could be a cause for failure, and attempts to re-alkalize these structures after removing the cover to moderately severe concrete composites have been generally successful.

11.5 Chloride Binding and Extraction

The ingress of chloride and the diffusion process into concrete have been explained in the earlier chapters. There have been various tests and methodologies to predict and to design SCCs with greater confidence with an understanding of the investigations reported. While this is so, the fact that the use of GGBS as the pozzolanic powder extender in these SCCs could help the chloride binding capacity of the concretes if used for marine environment (as already reported in normally vibrated concretes by several researchers) is not well appreciated. It is only apt to suggest that the slight changes in the microstructure of the macrostructure of SCCs may not show a significant difference if adopted in marine concrete applications. The second part is that as already discussed the use of self-compacting concrete compositions for repairing the concrete cover particularly in marine structures will certainly help not only in reducing the chloride ingress but also, if it can be modulated as a reserve of alkalinity, in reversing the diffusion process and the interface of the old and new concretes, reducing the level of chlorides in the steel reinforcement. This in a way is the reversing of the diffusion process and the interface that extracts already existing chlorides back into the concrete, helping to lower the levels of chlorides near the reinforcement. The actual thickness of concrete that is required and the effects of the interfacial bonding coat for integrating the old and new concretes have to be properly addressed and studied before actual theoretical calculations can be attempted.

11.6 Tunnel Lining and Grouting Applications

The advantage of using normally vibrated concretes particularly with collapse slump in prefabricated construction is well acknowledged. However, for the curved segmental casting required for tunnel linings, it is only apt to suggest self-compacting concretes to ensure larger and more dimensionally stable segments. This being the first part in the construction will help significantly in ensuring appropriate fitting in intricate parts of the tunnel lining. The next operation after the linings are in position is to ensure an effective grouting of the space between the lining and the rock face of the tunnel as it is invariably highly uneven and the concrete has to reach all corners and spaces of varying thickness and tortuosity. One cannot even think in terms of an alternative that is appropriately proportioned like self-compacting concretes for this purpose. It not only helps in occupying all the available space but can also be an effective barrier to the permeating moisture and water that need to be addressed.

11.7 Underwater Concrete Applications and Repair

The need to impress upon the effectiveness of self-compacting concretes for underwater applications and repairs is superfluous to say the least. In fact, it can even be said that self-compacting concrete concepts essentially originated from the earlier well-known underwater concretes. The essential prerequisite for underwater concrete is to be cohesive and thixotropic to have the ability of not being washed away (even the surface layers) not just underwater but in moving and oscillatory water bodies as well. Broadly such cohesiveness is brought about by the use of a specific binding agent, which was basically a cellulose-based material. The viscosity modifiers of today are also similar materials to ensure a very similar property in the case of self-compacting concretes as well. This aptly summarizes the fact that SCC composites are obviously some of the best candidate materials and applications underwater and for underwater repairs. Even so it is only advisable to assess the acceptability and performance integrity in flowing and oscillatory water bodies before adoption. It is possible to appropriately modify the cohesion by addressing the type or percentage of the viscosity modifier to suit the requirements.

11.8 Applications in Marine Environment

The advantages of utilizing self-compacting concretes in marine environment particularly through the use of GGBS as powder extenders has already been explained in earlier paragraphs, more in terms of the repair and rehabilitation strategy. The idea of repeating this more specifically is that while most applications are used to date even in such marine and estuarine and channel link structures, it is seen invariably that the concretes used are always fly ash based. One can only assume that this is primarily due to the fact that fly ash is more easily available and the knowledge and confidence in terms of its use as important structures are always a factor that gravitates the users more towards fly ash rather than the well-acknowledged alternative of GGBS. Another and probably more important consideration was the nonavailability of GGBS for such applications and even so the advantages offered by GGBS as pozzolanic admixtures in concretes for marine environment far outweigh both considerations of costs as well as availability in the interests of the overall performance of the structure over its entire life period.

11.9 Ultra-High-Strength Grouts and Composites

Ultra-high-strength grouts were originally started for ensuring to fill the gap between reciprocating or vibrating machinery after levelling and its foundation to ensure an appropriate transmission of loads so that the combined mass of the total system will participate in the vibration damping mechanisms as per design. Initially, some of these have been simple cementitious compounds and later on more complex polymeric systems have also come into existence. In more recent times, a large number of alternatives are available that contain essentially high-strength cementitious mortars sometimes in combination with microfiber additions. Special cementitious composites like microdefect-free cements and reactive powder compositions in more recent times could all be used for making this particular type of cementitious compositions. Incidentally, for some of the cementitious chemical admixtures, manufacturers have gone in for spatiality grouts, which contain formulations that are essentially made with combinations of specific superplasticizers and silica fume almost invariably. Some of these were tested for their characteristics in terms of mechanical and durability performance in the laboratory. Some of these manufacturers also advocate the addition of an appropriate quantity of very small-sized graded aggregates (generally below 10 mm) for applications requiring larger filling thicknesses or for filling the foundation bolt holes of heavier steel structures and machinery. It is also advised that such a composition with graded aggregates could also be used for the repair and rehabilitation of the cover concrete in several cases where the spalling occurred due to corrosion. In such cases, the integration of the old concrete and the new concrete becomes a matter of concern, and the problem is effectively addressed through either the use of styrene butadiene rubber latex, either as an integral part of the superimposed cementitious cover or maybe in a slurry that forms the interfacial bonding coat between the old and new concretes. The reason why a discussion on this particular aspect has been taken at this stage is to show that with minor modifications or use of a bonding coat as suggested the available self-compacting concretes are also excellent, if not the most suitable candidate materials for both the repair applications as well as to perform as the foundation integrating levelling grouts that are presently in existence. Some of these forms of applications have been successfully applied in several integration and repair and rehabilitation jobs both for normal structures, in industrial and chemical plants, and also more importantly in ocean and marine structures. It is to be noted that while other such repair applications have not performed adequately, these modified flowing cementitious composites, with admixtures like fly ash, GGBS, and silica fume as the case may be depending on the environment in which they have to be used, have been very successful.

11.10 Reinforced Fibrous Composites

The use of fibrous materials in structural composites is a practice that was used by humankind in several ways over the millennia. The addition of straw in the making of clay bricks was used by Egyptians long before concrete ever came into existence. Even the use of horse hair to reinforce lime mortars is known from early civilization. In fact, the initial birth of reinforced concrete started with well-distributed reinforcement in the form of a mesh, which came to be known as Ferro cement, whose later modification was indeed fiber-reinforced concrete. The basic advantages of discrete fiber-reinforced cementitious composites are predominantly twofold. First, they help in distributing the shrinkage and thermal microcracking during setting and hardening of concretes. Apart from this, once the crack occurs it bridges the two faces and does not allow the crack progression during the service loads and imparts ductility to the matrix. This induced ductility has been the primary characteristic that is a much sought after in addressing cyclic and fatigue loadings that are natural during the service life of constructed facilities. Several different fibrous materials have come into existence over the years—steel, polymer, carbon, and ceramic in their several forms and modifications. The percentage of fiber addition in concrete composites is generally limited to 2%, particularly in the case of inflexible steel fibers that have diameters ranging from 0.3 to 0.4 mm with an aspect ratio of about 80–100 (resulting in a length of approximately 30–40 mm). Beyond this, the effects of the fiber balling, due to the interaction between fibers, becomes problematic which is further enhanced in the case of hooked end or crimped fibers in particular. In an attempt to increase the fiber content in a significant proportion, effort was made to prepack the fibers and later allowing fine cement slurry to percolate onto the void spaces resulting in what is known as SIFCON, a material that has an excellent ductility.

In more recent years according to a few, the most comprehensive and utilitarian development has been made in arriving at what is presently termed as ultra-high-performance concrete (UHPC). UHPC is a new class of advanced cementitious composite that exhibits mechanical and durability properties far superior to the hitherto known conventional and even fiber-reinforced concretes. UHPC is defined generally as a cementitious composite containing microfibers and only fine aggregates to arrive at a highly homogeneous mix that exhibits compressive strengths above 150 MPa along with the significant pre- and post-cracking tensile strengths of over 5 MPa. Richard (1995) has shown that it is reasonably one of the best developments in HPCs, capable of achieving compressive strengths ranging up to 800 MPa, depending upon the production (including high-pressure compaction) and thermal curing (like high-temperature curing including steam or autoclaving). Information reported by several investigators (El-Dieb, 2009; Park, 2012; Rossi, 2013) shows that due to the absence of coarse aggregate, to make it extremely homogeneous with the steel microfiber reinforcement, the design mixes contain higher cementitious materials contents in the range of 900–1100 kg/m^3, at very low water–cement ratios (in the range of 0.2). It is important to note that the cementitious materials contain invariably silica fume at about 15%–20% and also a microfine filler or generally quartz powder of another 15%–20%.

11.11 Research and Developmental Requirements

The structural rigidity that is offered by a concrete section can never be appropriately replicated through steel owing to the thin sections that are prone to buckling and will have to be appropriately stiffened. A second advantage of concrete constructions is the fact that concretes are relatively more resilient to the effects of fire while the steel counterparts have to be protected against fire hazards. Also though the steel construction itself could be several times faster because of the possibility of fabrication in parts and off site, the advancements in concretes in prefabrication have ensured that even for high-rise structures concrete alternatives would be relatively more economical. It is obvious from these observations that properly integrated concrete performs as a superior cover to even fabricated steel structural members both as an additional strengthening component and also as a protection layer on environmental forces. This aspect in particular can be used both in repair and rehabilitation of structures that are already in existence, or concepts can be further enhanced to innovate structural compositions that are at the interface of the steel and concrete applications known hitherto.

Another important fact that we still remember at this particular stage is that it is not the idea to recount several examples and possibilities of the use of high-strength self-compacting concretes in recent years, but to point out that there is a great opportunity that is not really seized by the research community in general and the construction community in particular. There are several advantages in terms of not just ease of construction, but much more beyond that, in terms of slender structures, more elegant designs, and newer forms, and the list is unending, apart from a considerable improvement in performance and service life. The few paragraphs on the various avenues for such innovations and interventions are essentially steps toward the goal of making a difference in terms of superhigh-strength high-performance cementitious composites through a judicious use of the superplasticizers of mineral admixtures and the concepts of highly flexible self-consolidating cementitious composites.

11.12 Concluding Remarks

The idea of this compilation is not to repeat and restate the available information on self-compacting or SCCs into certain accessibly delineated segments, but to look at the entire material and its utility from an outsider's perspective to be useful to the construction practice. Also, it is envisaged to look at the data available through the more traditional concrete technology principles and to project the same with appropriate reasoning. The fact that self-contracting concrete in particular is yet another facet of concrete in all its various modifications and modulations has been reported by many, and this is an attempt to look at that particular statement in its own perspective. In this attempt, most often the earlier concepts and philosophy of ensuring high performance in cementitious composites have been maintained to the extent possible while extending some of that information and axioms to appropriately depict and represent the various research findings in more recent times.

References

El-Dieb, A.S., Mechanical, durability and microstructural characteristics of ultra-high-strength self-compacting concrete incorporating steel fibers, *Materials & Design*, 30, 2009, 4286–4292.

Okamura, H., Maekawa, K., and Mishima, T., Performance based design for self-compacting structural high-strength concrete, in *Seventh International Symposium on the Utilization of High-Strength/High-Performance Concrete*, Washington, DC, ACI SP 228, 2005, pp. 13–34.

Park, S.H., Kim, D.J., Ryu, G.S., and Koh, K.T., Tensile behaviour of ultra-high performance hybrid fibre reinforced concrete, *Cement and Concrete Composites*, 34, 2012, 172–184.

Richard, P. and Cheyrezy, M., Composition of reactive powder concretes, *Cement and Concrete Research*, 25(7), 1995, 1501–1511.

Rossi, P., Influence of fiber geometry and matrix maturity on the mechanical performance of ultra-high-performance cement-based composites, *Cement and Concrete Composites*, 37, 2013, 246–248.

Index

A

Abrams cone, 302, 309
Absolute volume method, 111, 135
Accelerated curing methods, 319–321
ACI, *see* American Concrete Institute
Additives, 18
Admixtures, 14
    chemical admixtures, 26–27, 110
    mixing water, 27
    pozzolans, *see* Pozzolanic admixtures
    superplasticizers, 26–27
    VMAs, 27
Aggregates
    CEB-FIP model code, 28–29
    coarse and fines aggregates,
        characteristics of, 28–34
    reinforcing fibers, 34–35
Air-entraining admixture (AEA)
        content, 293
Akashi Kaikyo Bridge, 76, 414
Alkali–aggregate reactivity, 29, 391
American Concrete Institute (ACI)
    nomogram, SCCs, 193–194, 204–205,
        216–217
    recommendations, 65–66
Anderson and Anderson model, 280, 312
Anti-washout underwater concrete, 81

B

Binder–space ratio, 346
Bingham model, 274–275
Blackfriars residential tower, 415
Blast furnace slag cement (BFSC), 15
Blocking ring, *see* J-ring test
Boiling water method, 321
Bond strength, 356–357
Burj Khalifa tower, 151, 415

C

Calcined clays, 18, 247–248
Calcium carbonate-based mineral
        additives, 18

Carbonation-led corrosion cracking, 417
CEB-FIP model code, 367, 374
    aggregate grading, 28–29
    efficiency value, fly ash, 120
    environmental exposure, durability
        recommendations, 111–113
Cementing efficiency and strength
        relations, fly ash, 118–119
    lean concretes, 120
    maximum percentage replacement,
        131–132
    replacement percentages, 119–121
    single efficiency value, 120–121
    standards, 120
    total powder content, 147–148
    water–cementitious materials ratio,
        145–146
        compressive strength variation,
            124–129
        Δw concept, 121–124
        water–cement ratio, 133–135
Cements, 109–110; *see also* Portland
        cements
    Euro standard classification of, 15–16
    fineness, 17
    low-heat cement, 15
    maximum water–cement ratio, 111
    minimum and maximum cement
        contents, 111
    rapid-hardening cement, 15
    strength grades, 14–15
    sulfate-resisting cement, 15
Chemical admixtures, 26–27,
        327–328, 339
Chloride diffusivity
    chloride-ion permeability test, 403
    diffusion profiles, 397
    laboratory studies, 396
    marine environment, 399
    performance criteria, 400
    premature degradation, 399
    proactive disaster mitigation
        strategies, 399
    rapid chloride penetration test, 397

9 780367 572112